Introduction to PSpice®
Using OrCAD®
for Circuits and Electronics

Third Edition

Muhammad H. Rashid
Electrical and Computer Engineering
University of West Florida

PEARSON
Prentice
Hall

Upper Saddle River, New Jersey 07458

Library of Congress Cataloging-in-Publication Data

Rashid, M. H.
  Introduction to PSpice Using OrCAD for Circuits and Electronics/Muhammad H. Rashid. --3rd ed.
    p. cm.
  Rev. ed, of: SPICE for Circuits and Electronics Using PSPICE. 1995.
  Includes bibliographical references and index.
  ISBN 0-13-101988-0
  1. SPICE (Computer file) 2. PSpice. 3. Electric circuit analysis—Data processing.
I. Title.

TK454.R385 2003   2003048208
621.319'2'0113--dc21   CIP

Vice President and Editorial Director, ECS: *Marcia J. Horton*
Publisher: *Tom Robbins*
Associate Editor: *Alice Dworkin*
Editorial Assistant: *Andrea Messineo*
Vice President and Director of Production and Manufacturing, ESM: *David W. Riccardi*
Executive Managing Editor: *Vince O'Brien*
Managing Editor: *David A. George*
Production Editor: *Craig Little*
Director of Creative Services: *Paul Belfanti*
Creative Director: *Carole Anson*
Art Director: *Jayne Conte*
Cover Designer: *Bruce Kenselaar*
Art Editor: *Greg Dulles*
Manufacturing Manager: *Trudy Pisciotti*
Manufacturing Buyer: *Lisa McDowell*
Marketing Manager: *Holly Stark*

The author and publisher of this book have used their best efforts in preparing this book. These efforts include the development, research, and testing of the theories and programs to determine their effectiveness. The author and publisher make no warranty of any kind, expressed or implied, with regard to these programs or the documentation contained in this book. The author and publisher shall not be liable in any event for incidental or consequential damages in connection with, or arising out of, the furnishing, performance, or use of these programs.

This is a revised edition of the book *SPICE for Circuits and Electronics Using PSPICE*, Second Edition (0-13-124652-6).

The Author gratefully acknowledges Cadence Design Systems, Inc. in making OrCAD and PSpice available.

OrCAD®, PSpice®, and Cadence® are registered trademarks of the Cadence Design Systems, Inc.
MSWord is a registered trademark of the Microsoft Corporation.
Program Editor is a registered trademark of the WordPerfect Corporation.

Printed in the United States of America

10 9 8 7 6 5 4 3 2 1

ISBN 0-13-101988-0

Pearson Education Ltd., *London*
Pearson Education Australia Pty. Ltd., *Sydney*
Pearson Education Singapore, Pte. Ltd.
Pearson Education North Asia Ltd., *Hong Kong*
Pearson Education Canada, Inc., *Toronto*
Pearson Educación de Mexico, S.A. de C.V.
Pearson Education—Japan, *Tokyo*
Pearson Education Malaysia, Pte. Ltd.
Pearson Education, Inc., *Upper Saddle River, New Jersey*

To my parents,
my wife Fatema,
and
my family, Fa-eza, Farzana, Hasan, and Hussain

Contents

Preface

The Engineering Accreditation Commission of the Accreditation Board for Engineering and Technology (EAC/ ABET) requirements specify the integration of computer-aided analysis and design in electrical and computer engineering curricula. SPICE is very popular software for analyzing electrical and electronic circuits. The MicroSim Corporation first introduced the PSpice simulator, which can run on personal computers (PCs). It is similar to the University of California (UC) Berkeley SPICE. The student version of PSpice, which is available free to students, is ideal for classroom use and for assignments requiring computer-aided simulation and analysis. PSpice widens the scope for the integration of computer-aided simulation to circuits and electronics courses for undergraduate and graduate students.

It may not be possible to add a one-credit-hour course on SPICE to integrate computer-aided analysis in circuits and electronics courses. However, students need some basic knowledge of how to use SPICE. They are constantly under pressure with course loads and do not always have the free time to read the details of SPICE, PSpice, or OrCAD from manuals and books of a general nature.

This book is the outcome of the author's experience in integrating SPICE in circuits and electronics courses at the 200-, 300-, or 400-level. The objective is to introduce the SPICE simulator to the electrical and computer engineering curriculum at the sophomore or junior level with a minimum amount of time and effort. This book requires no prior knowledge of the SPICE simulator. A course on basic circuits should be a prerequisite or co-requisite. Once the student develops an interest in and an appreciation for the applications of circuit simulators like SPICE, he or she can move on to more advanced materials for the full utilization of SPICE, PSpice, or OrCAD in solving complex circuits and systems.

This book can be divided into six parts:

(1) introduction to SPICE simulation	Chapters 1 and 2;
(2) DC, transient and AC circuit analysis	Chapters 3, 4, and 5;
(3) advanced SPICE commands and analysis	Chapter 6;
(4) semiconductor devices modeling and circuits	Chapters 7, 8, and 9;
(5) op-amp circuits and differential amplifiers	Chapter 10, and
(6) difficulties	Chapter 11.

Chapters 8, 9, and 10 describe the simple equivalent circuits of transistors and op-amps, which are commonly used in analyzing electronic circuits. Although SPICE generates the parameters of complex transistor models, analysis with a simple circuit model exposes the students to the mechanism of computation by SPICE .MODEL commands. This approach has the advantage that the students can compare the results, which are obtained in a classroom environment with the simple circuit models of devices, to those obtained by using complex SPICE models.

The commands, models, and examples that are described for PSpice are also applicable to UC Berkeley SPICE with minor modifications. The changes for running a PSpice circuit file on SPICE and vice-versa are discussed in Chapter 11. The filenames for the circuit files in this book are named using all uppercase so that the same file can be run on either the PSpice or the SPICE simulator

Probe is a graphics post-processor and is very useful in plotting the results of simulation, especially with the capability of arithmetic operation. It can be used to plot impedance, power, and so on. Once students have experience programming in PSpice, they will really appreciate the advantages of .Probe command. *Probe* is an option on PSpice, available with the student version. Running *Probe* does not require a math co-processor. The students can also get the normal printer output or printer plotting. The prints and plots are very helpful to the students in their theoretical understanding and in making judgments on the merits of a circuit and its characteristics.

This book can be used as a textbook on SPICE with a course on basic circuits being the prerequisite or co-requisite. It can also be used as a supplement to any standard textbook on basic circuits or electronics. In the latter case, the following sequence is recommended for the integration of SPICE at the basic circuits level of the curriculum:

1. As a supplement to a basic circuits course with three hours of lectures (or equivalent lab hours) and self-study assignments from Chapters 1 to 6. Starting from Chapter 2, the students should work hands-on with PCs.
2. In an electronics course it should continue to be used, with two hours of lectures (or equivalent lab hours) and self-study assignments from Chapters 6 to 10.

For integrating SPICE at the electronics level, three hours of lectures (or equivalent lab hours) are recommended on Chapters 1 to 6. Chapters 7 to 10 could be left for self-study assignments. From the author's experience in the class, it has been observed that after three lectures of 50 minutes duration, all students could solve assignments independently without any difficulty. The class could progress in a normal manner with one assignment per week on electronic circuits simulation and analysis with SPICE. Although the materials of this book have been tested in a basic circuits course for engineering students and in two electronics courses for electrical and computer engineering students, the book is also recommended for electrical engineering technology students.

PSPICE SOFTWARE AND PROGRAM FILES

The Lite version of OrCAD Capture 9.2 is included on the enclosed CD-ROM. It can also be obtained from

> Cadence Design Systems, Inc.
> 2655 Seely Avenue
> San Jose, CA 95134

> Websites: http://www.cadence.com
> http://www.orcad.com
> http://www.pspice.com
> http://www.ema-eda.com

The enclosed CD also contains (a) the user-defined model library file Rashid_PCE3_MODEL.LIB, and (b) all PSpice circuit files (with the extension .CIR) in the folder Rashid_PCE3_AD Circuits, PSpice Schematics (with the extension .SCH) in the folder Rashid_PCE3_PSpice_Schematics, and OrCAD Capture files (with extensions .OPJ and. DSN) in the folder Rashid_PCE3_Orcad_Capture for use with the book.

Important Notes: The PSpice circuit files (with the extension .CIR) are self-contained—each file contains any necessary device or component models. However, the PSpice Schematics files (with the extension .SCH) need the user-defined model library file Rashid_PCE3_MODEL.LIB, which is included with the PSpice Schematic files. This file ***must be included*** from the Analysis menu of PSpice Schematics. Similarly, the OrCAD Capture files (with extensions .OPJ and .DSN) also need the user-defined model library file Rashid_PCE3_MODEL.LIB, which is included with the OrCAD Schematic files. It ***must be included*** from the PSpice Simulation settings menu of OrCAD Capture. If these files are not included while running the simulation, it will not run and will give errors.

When importing PSpice Schematics files into OrCAD Capture, OrCAD Capture will require specifying the location of the msim_evl.ini file. Your OrCAD Capture folder may not have the msim_evl.ini file and you must find the location of that file in your computer. If you cannot locate the file, you can copy the msim_evl.ini file from the enclosed CD to your Windows folder (i.e., C:\WINNT) so that its location becomes C:\WINNT\msim_evl.ini. Depending on the OrCAD version you are using, your PSPICEIV.ini file will be in either the WINNT directory or the WINDOWS directory.

ACKNOWLEDGMENTS

I would like to thank the following persons for their comments and suggestions:

> Frank H. Hielscher, *Lehigh University*
> A. Zielinski, *University of Victoria, Canada*

Emil C. Neu, *Stevens Institute of Technology*
Balachandran Ruthramurthy, *College TAFE, Seremban, Malaysia*

I gratefully acknowledge Cadence Design Systems, Inc. in making OrCAD and PSpice available. It has been a great pleasure working with the editor, Alice Dworkin, and production editor, Craig Little. Finally, I would thank my family for their love, patience, and understanding.

Any comments and suggestions regarding this book are welcomed and should be sent to the author.

Dr. Muhammad H. Rashid
Professor and Director
Electrical and Computer Engineering
University of West Florida
11000 University Parkway
Pensacola, FL 32514-5754

E-mail: mrashid@uwf.edu

MUHAMMAD H. RASHID
Pensacola, Florida

About the Author

Muhammad H. Rashid received the B.Sc. degree in electrical engineering from the Bangladesh University of Engineering and Technology and the M.Sc. and Ph.D. degrees from the University of Birmingham, UK.

Currently, he is a Professor of electrical engineering with the University of Florida and the Director of the UF/UWF Joint Program in Electrical and Computer Engineering. Previously, he was a Professor of electrical engineering and the Chair of the Engineering Department at Indiana University–Purdue University at Fort Wayne. In addition, he was a Visiting Assistant Professor of electrical engineering at the University of Connecticut, Associate Professor of electrical engineering at Concordia University (Montreal, Canada), Professor of electrical engineering at Purdue University, Calumet, and Visiting Professor of electrical engineering at King Fahd University of Petroleum and Minerals, Saudi Arabia. He has also been employed as a design and development engineer with Brush Electrical Machines Ltd. (UK), as a Research Engineer with Lucas Group Research Centre (UK), and as a Lecturer and Head of Control Engineering Department at the Higher Institute of Electronics (Malta). He is actively involved in teaching, researching, and lecturing in power electronics. He has published 14 books and more than 100 technical papers. His books have been adopted as textbooks all over the world. His book *Power Electronics* has been translated into Spanish, Portuguese, Indonesian, Korean and Persian. His book *Microelectronics* has been translated into Spanish in Mexico and Spain. He has had many invitations from foreign governments and agencies to be a keynote lecturer and consultant, from foreign universities to serve as an external Ph.D. examiner, and from funding agencies to serve as a research proposal reviewer. His contributions in education have been recognized by foreign governments and agencies. He has previously lectured and consulted for NATO for Turkey in 1994, UNDP for Bangladesh in 1989 and 1994, Saudi Arabia in 1993, Pakistan in 1993, Malaysia in 1995 and 2002, and Bangkok in 2002, and has been invited by foreign universities in Australia, Canada, Hong Kong, India, Malaysia, Singapore to serve as an external examiner for undergraduate, master's and Ph.D. degree examinations, by funding agencies in Australia, Canada, United States, and Hong Kong to review research proposals, and by U.S. and foreign universities to evaluate promotion cases for professorship. He has previously authored seven books published by Prentice Hall: *Power Electronics—Circuits, Devices, and Applications* (1988, 2/e 1993), *SPICE For Power Electronics* (1993), *SPICE for Circuits and Electronics Using PSpice*

(1990, 2/e 1995), *Electromechanical and Electrical Machinery* (1986), and *Engineering Design for Electrical Engineers* (1990). He has authored five IEEE self-study guides: *Self-Study Guide on Fundamentals of Power Electronics, Power Electronics Laboratory Using PSpice, Selected Readings on SPICE Simulation of Power Electronics*, and *Selected Readings on Power Electronics* (IEEE Press, 1996) and *Microelectronics Laboratory Using Electronics Workbench* (IEEE Press, 2000). He also wrote two books: *Electronic Circuit Design using Electronics Workbench* (January 1998), and *Microelectronic Circuits—Analysis and Design* (April 1999) by PWS Publishing). He is editor of *Power Electronics Handbook* published by Academic Press, 2001.

Dr. Rashid is a registered Professional Engineer in the Province of Ontario (Canada), a registered Chartered Engineer (UK), a Fellow of the Institution of Electrical Engineers (IEE, UK) and a Fellow of the Institute of Electrical and Electronics Engineers (IEEE, USA). He was elected as an IEEE Fellow with the citation *"Leadership in power electronics education and contributions to the analysis and design methodologies of solid-state power converters."* He was the recipient of the *1991 Outstanding Engineer Award* from The Institute of Electrical and Electronics Engineers (IEEE). He received the 2002 IEEE Educational Activity Award (EAB) Meritorious Achievement Award in Continuing Education with the citation *"for contributions to the design and delivery of continuing education in power electronics and computer-aided-simulation"*. He was also an ABET program evaluator for electrical engineering from 1995 to 2000 and he is currently an engineering evaluator for the Southern Association of Colleges and Schools (SACS, USA). He has been elected as an IEEE-Industry Applications Society (IAS) Distinguished Lecturer. He is the Editor-in-Chief of the *Power Electronics and Applications Series,* published by CRC Press.

CHAPTER 1

Introduction

The learning objective of this chapter is to develop an understanding of the following:

- General description and the types of the SPICE software
- Types of analysis that can be performed on electronic and electrical circuits
- Limitations of PSpice software
- Online resources on SPICE.

1.1 INTRODUCTION

Electronic circuit design requires accurate methods for evaluating circuit performance. Because of the enormous complexity of modern integrated circuits, computer-aided circuit analysis is essential and can provide information about circuit performance that is almost impossible to obtain with laboratory prototype measurements. Computer-aided analysis permits

1. Evaluation of the effects of variations in elements, such as resistors, transistors, transformers, and so on
2. The assessment of performance improvements or degradations
3. Evaluation of the effects of noise and signal distortion without the need of expensive measuring instruments
4. Sensitivity analysis to determine the permissible bounds due to tolerances on each and every element value or parameter of active elements
5. Fourier analysis without expensive wave analyzers
6. Evaluation of the effects of nonlinear elements on the circuit performance
7. Optimization of the design of electronic circuits in terms of circuit parameters.

SPICE is a general-purpose circuit program that simulates electronic circuits. SPICE can perform various analyses of electronic circuits: the operating (or the quiescent) points of transistors, a time-domain response, a small-signal frequency response, and so on. SPICE contains models for common circuit elements, active as

1

well as passive, and it is capable of simulating most electronic circuits. It is a versatile program and is widely used in both industries and universities. The acronym SPICE stands for *Simulation Program with Integrated Circuit Emphasis*.

Until recently, SPICE was available only on mainframe computers. In addition to the high initial cost of the computer system, such a machine can be inconvenient for classroom use. In 1984, MicroSim introduced the PSpice simulator, which is similar to the Berkeley SPICE and runs on an IBM-PC or compatible. It is available at no cost to students for classroom use. PSpice, therefore, widens the scope for the integration of computer-aided circuit analysis into electronic circuit courses at the undergraduate level. Other versions of PSpice that will run on computers such as the Macintosh II, VAX, SUN, and NEC are also available.

1.2 DESCRIPTIONS OF SPICE

The development of SPICE spans a period of about 30 years. During the mid-1960s, the program ECAP was developed at IBM [2]. Later, ECAP served as the starting point for the development of program CANCER at the University of California (UC), Berkeley in the late 1960s. Based on CANCER, SPICE was developed at Berkeley in the early 1970s. SPICE2, which is an improved version of SPICE, was then developed during the mid-1970s at UC—Berkeley.

The algorithms of SPICE2 are general in nature, but are robust and powerful for simulating electrical and electronics circuits, and SPICE2 has become a standard tool in the industry for circuit simulations. The development of SPICE2 was supported by public funds at UC—Berkeley, and the program is in the public domain. (SPICE3, which is a variation of SPICE2, is designed especially to support the computer-aided design (CAD) research program at UC—Berkeley.)

SPICE2 has become an industry standard and is now referred to simply as SPICE. The input syntax for SPICE is a free-format style; it does not require that data be entered in fixed column locations. SPICE assumes reasonable default values for un-specified circuit parameters. In addition, it performs a considerable amount of error checking to ensure that a circuit has been entered correctly.

PSpice, which uses the same algorithms as SPICE2 and is a member of the SPICE family, is equally useful for simulating all types of circuits in a wide range of applications. In both PSpice and SPICE, a circuit is described by statements that are stored in a file called the *circuit file*. The circuit file is read by the SPICE simulator. Each statement is self-contained and independent; the statements do not interact with each other. SPICE (or PSpice) statements are easy to learn and use. A schematic editor can be used to draw the circuit and create a schematic file, which can then be read by PSpice for running the simulation.

1.3 TYPES OF SPICE

The commercially supported versions of SPICE2 can be divided into two types: mainframe versions and PC-based versions. Their methods of computation may differ, but their features are almost identical to those of SPICE2. However, some versions may include such additions as a preprocessor or shell program to manage input and provide interactive control, as well as a postprocessor for refining the normal SPICE output. A person who is familiar with one SPICE version (e.g., PSpice) should be able to work with other versions.

The mainframe versions are as follows:

HSPICE (Meta-Software), which is designed for integrated circuit design with special device models

RAD-SPICE (Meta-Software), which simulates circuits subjected to ionizing radiation

IG-SPICE (A.B. Associates) and I-SPICE (NCSS Time Sharing), which are designed for interactive circuit simulation with graphic output

Precise (Electronic Engineering Software)

PSpice (MicroSim), which is similar to PSpice

AccuSim (Mentor Graphics)

Cadence-SPICE (Cadence Design)

SPICE-Plus (Valid Logic)

The following are PC versions:

AllSpice (Acotech)

IS-SPICE (Intusoft)

Z-SPICE (Z-Tech)

SPICE-Plus (Analog Design Tools)

DSPICE (Daisy Systems)

PSpice (MicroSim)

Spice (KEMET)

B2-Spice A/D (Beige Bag Software)

AIM-Spice (Automatic Integrated Circuit Modeling Spice)

Visual Spice (Island Logix)

Spice3f4 (Kiva Design)

MD-SPICE (Zeland Software, Inc)

Ivex Spice (Ivex Design)

1.4   TYPES OF ANALYSIS

PSpice allows various types of analyses. Each analysis is invoked by including its command statement. For example, a statement beginning with the .DC command invokes the DC sweep. The types of analyses and their corresponding .(dot) commands are described below:

Dc Analysis is used for circuits with time-invariant sources (e.g., steady-state dc sources). It calculates all node voltages and branch currents over a range of values, and their quiescent (dc) values are the outputs:

Dc sweep of an input voltage/current source, a model parameter, or temperature over a range of values (.DC)

Determination of the linearized model parameters of nonlinear devices (.OP)

Dc operating point to obtain all node voltages (.OP)

Small-signal transfer function with small-signal gain, input resistance, and output resistance (Thévenin's equivalent) (.TF)

Dc small-signal sensitivities (.SENS)

Transient Analysis is used for circuits with time-variant sources (e.g., ac sources and switched dc sources). It calculates all node voltages and branch currents over a time interval, and their instantaneous values are the outputs:

Circuit behavior in response to time-varying sources (.TRAN)

Dc and Fourier components of the transient analysis results (.FOUR)

Ac Analysis is used for small-signal analysis of circuits with sources of variable frequencies. It calculates all node voltages and branch currents over a range of frequencies, and their magnitudes and phase angles are the outputs:

Circuit response over a range of source frequencies (.AC)

Noise generation at an output node for every frequency (.NOISE)

In schematic versions, the commands are invoked from the setup menu as shown in Fig. 1.1.

FIGURE 1.1

Analysis setup in PSpice Schematics.

1.5   DESCRIPTIONS OF SIMULATION SOFTWARE TOOLS

There are many simulation software tools. Some example are as follows:

AIM SPICE Software: Integrated Circuit Modeling Spice (AIM-Spice) is a version of SPICE, the most popular analog circuit simulator.

AKNM Circuit Magic: Circuit Magic allows you to design, simulate, and learn about electrical circuits. Circuit Magic is an easy-to-use educational tool to allow simple direct (DC) and alternate current (AC) electrical circuits to be constructed and analyzed. The software shows circuit calculations using Kirchhoff's laws, node voltage, and mesh current methods and includes a schematic editor and a vector diagram editor.

Schematic CAD software and EDA software: Circuit Simulation, Schematic Entry, PCB Layout, Gerber Viewer.

Beige Bag Software: B2 Spice A/D 2000 is a fully featured mixed-mode simulator that combines powerful capabilities with a deceptively easy-to-use interface.

Electronics Workbench Software: Electronics Workbench offers a professional circuit design solution with a suite of integrated tools that includes schematic capture, simulation, layout, and autorouting for printed circuit boards (PCBs) and programmable logic devices (FPGAs/CPLDs).

Intusoft Software: This software includes SPICE simulation, analog and mixed-signal circuit design tools, magnetics transformer design, and test program development.

Island Logix Software: This program is part of the VisualSpice Electronic Circuit Design and Simulation Software, in which a completely integrated modern user interface circuit design environment allows you to quickly and easily capture your schematic designs, perform your simulation, and analyze the results.

Ivex Design Software: This software contains Windows-based EDA tools, schematic capture, spice tools, and printed circuit board layout.

MacroSim Digital Simulator: MacroSim Digital Simulator is a sophisticated software tool that integrates the process of designing and simulating the operation of digital electronic circuitry. MacroSim has been developed for professional engineers, hobbyists, and tertiary and late secondary students.

MDSPICE Software: MDSPICE is a mixed frequency and time-domain spice simulator for predicting the time-domain response of high-speed networks, high-frequency circuits, and nonlinear devices directly using s-parameters.

Micro-Cap (Spectrum) Software: This software is an analog/digital simulation that is compatible with SPICE and PSpice.

NOVA-686 Linear RF Circuit Simulation: NOVA-686 is a shareware, Rf circuit simulation program for the RF design engineer, radio amateur, and hobbyist.

SIMetrix Software: SIMetrix is an affordable mixed-mode circuit simulator designed for professional circuit designers.

SPICE Software: SPICE (Simulation Program with Integrated Circuit Emphasis) is one of the most widely used computer programs for analog circuit analysis. Spice software includes programs for DOS, Windows, OS/2, Linux, Macintosh, and UNIX.

1.6 PSPICE PLATFORM

The PSpice platform used depends on the SPICE version. The three platforms for PSpice are as follows:

> PSpice A/D or OrCAD PSpice A/D (version 9.1 or above)
> PSpice Schematics (version 9.1 or below)
> OrCAD Capture Lite (version 9.2 or above)

1.6.1 PSpice A/D

The platform for the PSpice A/D is shown in Fig. 1.2. The circuit that is described by statements and analysis commands is simulated by the run command from the platform. The output results can be displayed and viewed from platform menus.

1.6.2 PSpice Schematics

The platform for PSpice Schematics is shown in Fig. 1.3. The circuit that is drawn on the platform is run from the analysis menu. The simulation type and its settings are specified from the analysis menu. After the simulation run is completed, PSpice automatically opens the PSpice A/D platform for displaying and viewing the output results.

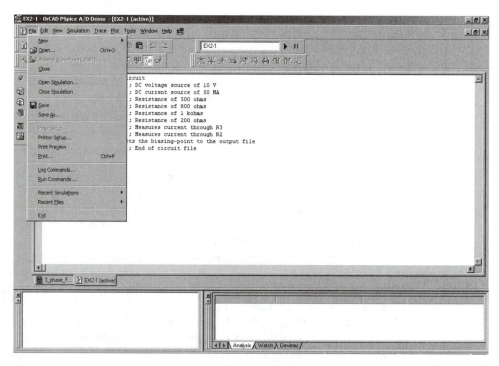

FIGURE 1.2

Platform for PSpice A/D (version 9.1).

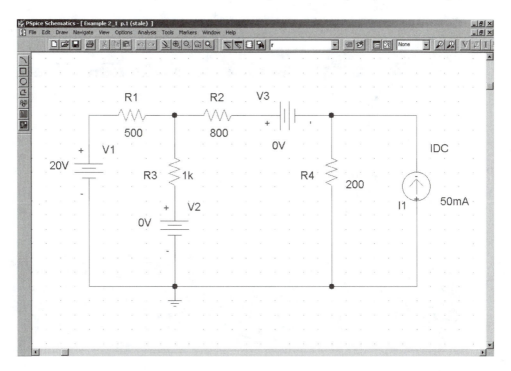

FIGURE 1.3

Platform for PSpice Schematics (version 9.1).

1.6.3 OrCAD Capture

The platform for the OrCAD Capture, which is similar to that of the PSpice Schematics, but has more features, is shown in Fig. 1.4. The circuit that is drawn on the platform is run from the PSpice menu. The simulation type and its settings are specified from the PSpice menu. After the simulation run is completed, the Capture automatically opens the PSpice A/D platform for displaying and viewing the output results.

1.7 PSPICE SCHEMATIC VERSUS OrCAD CAPTURE

OrCAD Capture has some new features and a platform similar to that of the PSpice Schematic. The Schematic files (with extension .SCH) can be imported to OrCAD capture (with extension .OPJ). However, the OrCAD file cannot be run on PSpice Schematic. Therefore, it is advisable to those readers familiar with PSpice Schematics to use PSpice Schematic version 9.1, which has the similar platform to version 8.0, but its PSpice A/D platform is similar to that of OrCAD capture. The PSpice Schematics (version 9.1) as shown in Fig. 1.5 can be copied from Cadence Design Systems at the website:

http://www.cadence.com

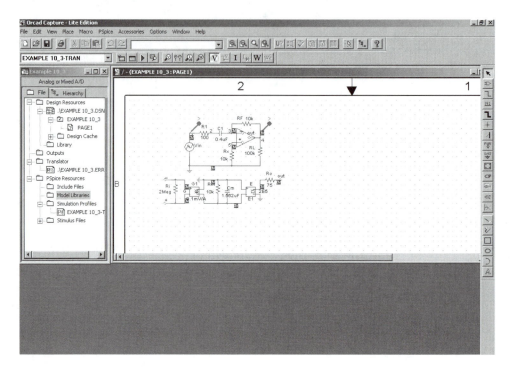

FIGURE 1.4

Platform for OrCAD Capture (version 9.2).

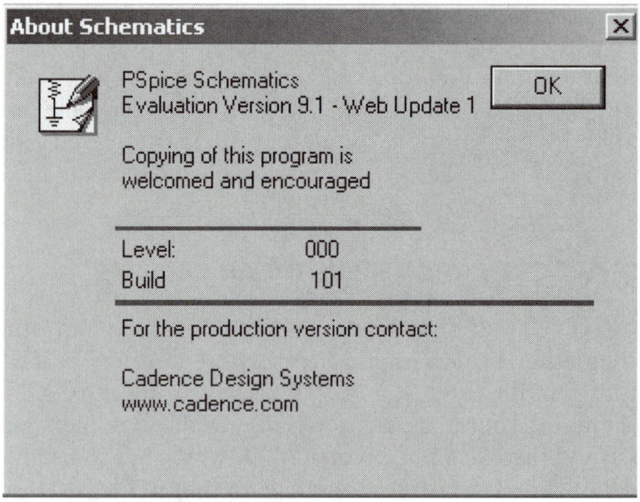

FIGURE 1.5

About Schematics.

1.8   LIMITATIONS OF PSPICE

As a circuit simulator, PSpice has the following limitations:

1. The student version of PSpice is restricted to circuits with a maximum of 10 transistors. However, the professional DOS (or production) version can simulate a circuit with up to 200 bipolar transistors (or 150 MOSFETs).

2. The program is not interactive; that is, the circuit cannot be analyzed for various component values without editing the program statements.

3. PSpice does not support an iterative method of solution. If the elements of a circuit are specified, the output can be predicted. On the other hand, if the output is specified, PSpice cannot be used to synthesize the circuit elements.

4. The input impedance cannot be determined directly without running the graphic postprocessor, Probe. The student version does not require a floating-point coprocessor for running Probe, but the professional version does require such a coprocessor.

5. The PC version requires 512 kilobytes of memory (RAM) to run.

6. Distortion analysis is not available in PSpice. (SPICE2 allows distortion analysis, but it gives wrong answers.)

7. The output impedance of a circuit cannot be printed or plotted directly.

8. The student version will run *with* or *without* the floating-point coprocessor. If the coprocessor is present, the program will run at full speed; otherwise, it will run 5 to 15 times slower. The professional version requires a coprocessor; it is not optional.

1.9   SPICE RESOURCES

There are many online resources. Some of them are as follows:

Websites with Free SPICE Models

Analog Devices
   http://products.analog.com/products_html/list_gen_spice.html

Apex Microtechnology
   http://eportal.apexmicrotech.com/mainsite/index.asp

Coilcraft
   http://www.coilcraft.com/models.cfm

Comlinear
   http://www.national.com/models

Elantec
   http://www.elantec.com/pages/products.html

Epcos Electronic Parts and Components
   http://www.epcos.de/web/home/html/home_d.html

Fairchild Semiconductor Models and Simulation Tools
   http://www.fairchildsemi.com/models/

Infineon Technologies AG
   http://www.infineon.com/

Intersil Simulation Models
   http://www.intersil.com/design/simulationModels.asp
International Rectifier
   http://www.irf.com/product-info/models/
Johanson Technology
   http://www.johansontechnology.com/
Linear Technology
   http://www.linear-tech.com/software/
Maxim
   http://www.maxim-ic.com/
Microchip
   http://www.microchip.com/index.asp
Motorola Semiconductor Products
   http://www1.motorola.com/
National Semiconductor
   http://www.national.com/models
Philips Semiconductors
   http://www.semiconductors.philips.com/
Polyfet
   http://www.polyfet.com/
Teccor
   http://www.teccor.com/asp/sitemap.asp?group=downloads
Texas Instruments
   http://www.ti.com/sc/docs/msp/tools/macromod.htm#comps
Zetex
   http://www.zetex.com/

Websites with SPICE Models

Analog & RF Models
   http://www.home.earthlink.net/~wksands/
Analog Innovations
   http://www.analog-innovations.com/
Duncan's Amp Pages
   http://www.duncanamps.com/
EDN Magazine
   http://www.e-insite.net/ednmag/
Intusoft Free SPICE Models
   http://www.intusoft.com/models.htm
MOSIS IC Design Models
   http://www.mosis.org/
Planet EE
   http://www.planetee.com/
PSpice.com
   http://www.pspice.com/

SPICE Models from Symmetry
http://www.symmetry.com/

SPICE Model Index
http://homepages.which.net/~paul.hills/Circuits/Spice/ModelIndex.html

SPICE and Circuit Simulation Information Sites

AboutSpice.com
http://www.aboutspice.com/

Artech House Publishers - Books and Software for High-Technology Professionals
http://www.artech-house.com/

EDTN Home Page
http://www.edtn.com/

E/J Bloom Associates Home Page—SMPS Books & Software
http://www.ejbloom.com/

MOSIS IC Foundry
http://www.mosis.org/

NCSU SPICE Benchmarks
http://www.cbl.ncsu.edu/pub/Benchmark_dirs/

NIST Modeling Validation Group
http://ray.eeel.nist.gov/modval.html

Norman Koren Vacuum Tube Audio Page
http://www.normankoren.com/Audio/

PSpice.com
http://www.pspice.com/

Ridley Engineering PWM Simulation
http://www.ridleyengineering.com/

SGS-Thomson
http://us.st.com/stonline/index.shtml

SPICE Simulations Dr Vincent G Bello
http://www.spicesim.com/

Temic (Siliconix)
http://www.temic.com/index_en.html?

University of Exeter's Online SPICE3 User's Manual
http://newton.ex.ac.uk/teaching/CDHW/Electronics2/userguide/

Virtual Library Electrical Engineering
http://webdiee.cem.itesm.mx/wwwvlee/

Yahoo Club Circuit Simulation Chat Room
http://login.yahoo.com/config/login?.intl=uk&.src=ygrp&.done
=http://uk.groups.yahoo.com%2Fclubs%2Felectroniccircuitsimulation

Engineering Magazines with SPICE Articles

EDN Home Page
http://www.e-insite.net/ednmag/

Electronic Design
http://www.elecdesign.com/

PCIM Home Page
   http://www.pcim.com/

Personal Engineering & Instrumentation
   http://www.pcim.com/

Planet EE
   http://www.planetee.com/

REFERENCES

[1] Attia, John Okyere. *PSPICE and MATLAB for Electronics: An Integrated Approach*. Boca Raton, FL: CRC Press, 2002.

[2] Walter Banzhaf, *Computer-Aided Circuit Analysis Using PSpice*. Englewood Cliffs, New Jersey: Prentice Hall, 1992 (1989).

[3] Edward Brumgnach, *PSpice for Windows*. Delmar Publishers, 1995.

[4] Rahul Chattergy, *Spicy Circuits: Elements of Computer-Aided Circuit Analysis*. Boca Raton, Florida: CRC Press, 1992.

[5] Roger Conant, *Engineering Circuit Analysis with PSpice and Probe*. New York: McGraw-Hill, 1993.

[6] Connelly and Choi, *Macromodeling with SPICE*. Englewood Cliffs, New Jersey: Prentice Hall, 1992.

[7] L. H. Fenical, *PSpice A Tutorial*. Englewood Cliffs, New Jersey: Prentice Hall, 1992.

[8] Roy W. Goody, *PSpice for Windows-A Circuit Simulation Primer*. Englewood Cliffs, New Jersey: Prentice Hall, 1995.

[9] Roy W. Goody, *PSpice for Windows* Vol II: Operational Amplifiers & Digital Circuits. Upper Saddle River, New Jersey: Prentice Hall, 1996.

[10] Roy W. Goody. *OrCAD PSpice for Windows Volume 1: DC and AC Circuits*, 3d ed. Upper Saddle River, New Jersey: Prentice Hall, 2000.

[11] Roy W. Goody. *OrCAD PSpice for Windows Volume II: Devices, Circuits, and Operational Amplifiers*, 3d ed. Upper Saddle River, New Jersey: Prentice Hall, 2000.

[12] James Gottling and Richard C. Dorf, *Introduction to Electric Circuits*, 3d ed. (along with *Hands on PSpice with 3.5 Exercise Diskette set*). New York: John Wiley & Sons, 1996.

[13] Bashir Al-Hashimi, *The Art of Simulation Using PSpice Analog and Digital*. Boca Raton, Florida: CRC Press, 1995.

[14] Marc E. Herniter, *Schematic Capture with PSpice*. New York: Merrill (Macmillan Publishing Company), 1994.

[15] Marc E. Herniter, *Schematic Capture with Cadence PSpice*. Upper Saddle River, New Jersey: Prentice Hall, 2001.

[16] R. W. Jensen and L. P. McNamee, *Handbook of Circuit Analysis Languages and Techniques*. Englewood Cliffs, New Jersey: Prentice Hall, 1976.

[17] R. W. Jensen and M. D. Liberman, *IBM Electronic Circuit Analysis Program and Applications*. Englewood Cliffs, New Jersey: Prentice Hall, 1968.

[18] John Keown, *PSpice and Circuit Analysis*, 3d ed. New York: Merrill (Macmillan Publishing Company), 1997.

[19] Paul G. Krol, *Inside OrCAD Capture*. Clifton Park, New York: OnWord Press, 1998.

[20] Robert Lamey, *The Illustrated Guide to PSpice for Windows*. New York: Delmar Publishers, 1995.

[21] Yim-Shu Lee, *Computer-Aided Analysis and Design of Switch-Mode Power Supplies*. New York: Marcel Dekker, 1993.

[22] Giuseppe Massobrio and Paolo Antognetti, *Semiconductor Device Modeling with SPICE*, 2d ed. New York: McGraw-Hill, 1993.

[23] Franz Monssen, *PSpice with Circuit Analysis*. New York: Merrill (Macmillan Publishing Company), 1993.

[24] Jack F. Morris, *Introduction to PSpice with Student Exercise Disk*. New York: Houghton Mifflin, 1991.

[25] Karl H. Muler, *A Spice Cookbook*. San Pedro, California: Intusoft, Inc., 1991.

[26] James W. Nilsson and Susan A. Riedel, *Introduction to PSpice*. New York: Addison-Wesley, 1993.

[27] T.E. Price, *Analog Electronics: An Integrated PSpice Approach*. Upper Saddle River, New Jersey: Prentice Hall, 1996.

[28] *PSpice Manual*. Irvine, California: MicroSim Corporation, 1992.

[29] R. Ramshaw and D. Schuurman, *PSpice Simulation of Power Electronics Circuits*. New York: Kluwer Academic Publishers, 1997.

[30] Muhammad H. Rashid, *SPICE for Circuits and Electronics Using PSpice*. Englewood Cliffs, New Jersey: Prentice Hall, 1995.

[31] Gordon W. Roberts and Adel S. Sedra, *SPICE*. New York: Oxford University Press, 1997.

[32] Chris Schroeder, *Inside OrCAD Capture for Windows*. Burlington, Massachusetts: Newnes, 1998.

[33] John A. Stuller, *Basic Introduction to PSpice*. New York: John Wiley & Sons, 1995.

[34] James A. Svoboda, *PSpice for Linear Circuits*. New York: John Wiley & Sons, 2002.

[35] Thomas W. Thorpe, *Computerized Circuit Analysis with SPICE*. New York: Wiley Interscience, 1992.

[36] Paul Tuinenga, *SPICE: A Guide to Circuit Simulation and Analysis Using PSpice*, 3d. Englewood Cliffs, New Jersey: Prentice Hall, 1995.

[37] Andrei Vladimirescu, *The Spice Book*. New York: John Wiley & Sons, 1994.

CHAPTER 2

Circuit Descriptions

After completing this chapter, students should be able to

- Describe circuits for PSpice simulation;
- Create input files, which can be read by PSpice;
- Obtain output results and plots of simulations;
- Perform analyses for finding dc operating voltages and currents, and plot dc sweep and nested sweeps of a dc source voltage and current.

2.1 INTRODUCTION

SPICE is a general-purpose circuit program that can be applied to simulate and calculate the performance of electrical and electronic circuits. A circuit is described to a computer by using a file called the *circuit file*, which is normally typed in from a keyboard. The circuit file contains the circuit details of components and elements, the information about the sources, and the commands for what to calculate and what to provide as output. The circuit file is the input file to the SPICE (or PSpice) program, which, after executing the commands, produces the results in another file called the *output file*.

2.2 INPUT FILES

Input files can be either schematic files or netlist files (also known as the circuit files). In circuit files, the user assigns the node numbers. The nodes connect the circuit elements, the semiconductor devices, and the sources. If a resistor R is connected between Nodes 1 and 2, SPICE relates the voltage across R, v_R, and the current through R, i, by

$$v_R = V(1) - V(2) = Ri.$$

If an inductor L is connected between Nodes 3 and 4, SPICE relates the voltage across L, v_L, and the current through L, i, by

$$v_L = V(3) - V(4) = R\frac{di}{dt}.$$

If a capacitor C is connected between Nodes 5 and 6, SPICE relates the voltage across C, v_C, and the current through C, i, by

$$v_C = V(5) - V(6) = C\int i\, dt.$$

From these descriptions of the circuit elements in the form of a netlist, PSpice develops a set of matrices and solves for all voltages and currents for specified input signals and supply voltages.

The main disadvantage of the circuit (or input) file is that one has to draw the circuit and generate the net list of components and devices. The PSpice Design Center has a *Schematics editor* for circuit drawings that uses part symbols to represent devices and wire symbols for connections. The circuit diagram is saved with a .SCH extension and is called the *Schematics input file*. One can set up simulation parameters, run PSpice, and run Probe from the Schematics menu. For a schematic input file, PSpice Schematics creates the netlist automatically and then solves for all voltages and currents. Probe processes the simulation results. One can also generate a circuit file directly from the circuit diagrams and simulation specifications. For many circuit simulations, working from a circuit diagram is much more convenient than typing in a circuit file. One could also interface *OrCADS*'s DRAFT Schematics editor to PSpice for generating the NETLIST.

For simulations requiring advanced commands and techniques, the generation of the circuit file is often convenient and necessary. One must often work from both the circuit file and the Schematics file, depending upon the applications and the types of simulation problems. However, the student version of PSpice allows only 20 components, so it is not complex enough for classroom examples in circuits and electronics courses.

Once the descriptions of the circuit and the type of analysis are specified, PSpice schematics can simulate the schematic file. The circuit file can be simulated by PSpice A/D. Table 2.1 shows the names and function of programs within PSpice. Probe is a graphical prostprocessor for displaying output variables such as voltages, currents, powers, and impedances.

A circuit must be specified in terms of element names, element values, nodes, variable parameters, and sources. Consider the circuit in Fig. 2.1 that is to be simulated for calculating all node voltages and currents through R_2 and R_3. We shall show (1) how to describe this circuit to PSpice, (2) how to specify the type of analysis to be

TABLE 2.1 Names and function of programs within PSpice

Program Name	Program Function
Schematics	Schematic circuit entry, symbol editor, and design management
PSpice A/D	Mixed analog and digital circuit simulation
Probe	Presentation and postprocessing of simulation results
Stimulus Editor	Graphic creation of input signals

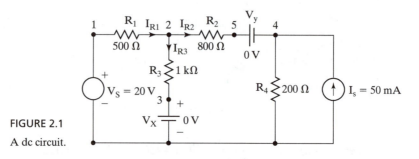

FIGURE 2.1

A dc circuit.

performed, and (3) how to define the required output variables. The description and analysis of a circuit require specifying the following:

Element values
Nodes
Circuit elements
Element models
Sources
Types of analysis
Output variables
PSpice output commands
Format of circuit files
Format of output files.

2.3   ELEMENT VALUES

The element values are written in standard floating-point notation with optional scale and units suffixes. Some values without suffixes that are allowable in PSpice are as follows:

> 5   5.   5.0   5E + 3   5.0E + 3   5.E3

There are two types of suffixes: the scale suffix and the units suffix. The scale suffix multiplies the number that it follows. The scale suffixes recognized by PSpice are

$F = 1E - 15$
$P = 1E - 12$
$N = 1E - 9$
$U = 1E - 6$
$MIL = 25.4E - 6$
$M = 1E - 3$
$K = 1E3$
$MEG = 1E6$
$G = 1E9$
$T = 1E12$

The unit suffixes that are normally used are

V = volt
A = amp
HZ = hertz
OHM = ohm (Ω)
H = henry
F = farad
DEG = degree

The scale suffix is always the first suffix, and the unit suffix follows the scale suffix. In the absence of a scale suffix, the first suffix may be a units suffix, provided it is not the symbol of a scale suffix. If the value of an inductor is 15 μH it is written as 15U or 15UH. In the absence of scale and units suffixes, the units of voltage, current, frequency, inductance, capacitance, and angle are, by default, volts, amps, hertz, henrys, farads, and degrees, respectively. PSpice ignores any units suffix, so the following values are equivalent:

25E–3   25.0E–3   25M   25MA   25MV   25MOHM   25MH

Notes

1. The scale suffixes are all uppercase, but PSpice allows lowercase.
2. M means "milli," not "mega." 2 MΩ is written as 2MEG or 2MEGOHM.

2.4   NODES

For PSpice A/D: The location of an element is identified by the node numbers. Each element is connected between two nodes. Node numbers are assigned to the circuit in Fig. 2.1. Node 0 is predefined as the ground. All nodes must be connected to at least two elements and should, therefore, appear at least twice. Node numbers must be integers from 0 to 9999 for SPICE2, but need not be sequential. PSpice allows any alphanumeric string up to 131 characters long. The node names shown in Table 2.2 are reserved and cannot be used.

The node numbers to which an element is connected are specified after the name of the element. All nodes must have a dc path to the ground node. This condition, which is not always satisfied in some circuits, is normally met by connecting very large resistors; it is discussed in Sec. 11.10.

TABLE 2.2   Reserved Node Names

Reserved Node Names	Value	Description
0	0 volts	Analog ground
$D_HI	1	Digital high level
$D_LO	0	Digital low level
$D_X	X	Digital unknown level

For PSpice schematics, node numbers are assigned by PSpice and are usually in an alphanumeric string such as $N_0002 and $N_0001 as follows:

First Node	Second Node
$N_0002	$N_0001
0	+
_	out
$N_0001	_
0	out

2.5   CIRCUIT ELEMENTS

For PSpice A/D: Circuit elements are identified by names. A name must start with a letter symbol corresponding to the element, but after that it can contain either letters or numbers. Names can be up to eight characters long for SPICE2, and up to 131 characters long for PSpice. However, names longer than eight characters are not normally necessary and not recommended. Table 2.3 shows the first letters of elements and sources. For example, the name of a resistor must start with R, an independent current source with I, and an independent voltage source with V.

The format for describing passive elements is

```
<element name> <positive node> <negative node> <value>
```

where positive current is assumed to flow into positive node $N+$ and out of negative node $N-$. If the nodes are interchanged, the direction of the current through the element will be reversed. The formats for passive elements are described in Chapters 4, 5, and 6. The active elements are described in Chapters 7, 8, and 9.

TABLE 2.3   Symbols of Circuit Elements and Sources

First Letter	Circuit Elements and Sources
B	GaAs MES field-effect transistor
C	Capacitor
D	Diode
E	Voltage-controlled voltage source
F	Current-controlled current source
G	Voltage-controlled current source
H	Current-controlled voltage source
I	Independent current source
J	Junction field-effect transistor
K	Mutual inductors (transformer)
L	Inductor
M	MOS field-effect transistor
Q	Bipolar junction transistor
R	Resistor
S	Voltage-controlled switch
T	Transmission line
V	Independent voltage source
W	Current-controlled switch

Note: Voltage-controlled switch, current-controlled switch, and GaAs are not available in SPICE2, but they are available in SPICE3.

The values of some circuit elements depend on other parameters, such as the resistance as a function of temperature and the capacitance as a function of voltage. Models may be used to assign values to the various parameters of circuit elements. The techniques for specifying models of resistors are described in Sections 3.2 and 3.3, and those of inductors and capacitors in Section 4.2.

The passive elements in Fig. 2.1 are described as follows:

- The statement that R_1 has a value of 500 Ω and is connected between nodes 1 and 2 is

    ```
    R1 1 2 500
    ```

- The statement that R_2 has a value of 800 Ω and is connected between nodes 2 and 5 is

    ```
    R2 2 5 800
    ```

- The statement that R_3 has a value of 1 kΩ and is connected between nodes 2 and 3 is

    ```
    R3 2 3 1K
    ```

- The statement that R_4 has a value of 200 Ω and is connected between nodes 4 and 0 is

    ```
    R4 4 0 200
    ```

For PSpice Schematics: The name of an element begins with a specific letter as shown in Table 2.3. The part symbols can be obtained from the Get New Part in Draw menu as shown in Fig. 2.2(a). The symbols for passive symbols are obtained from the

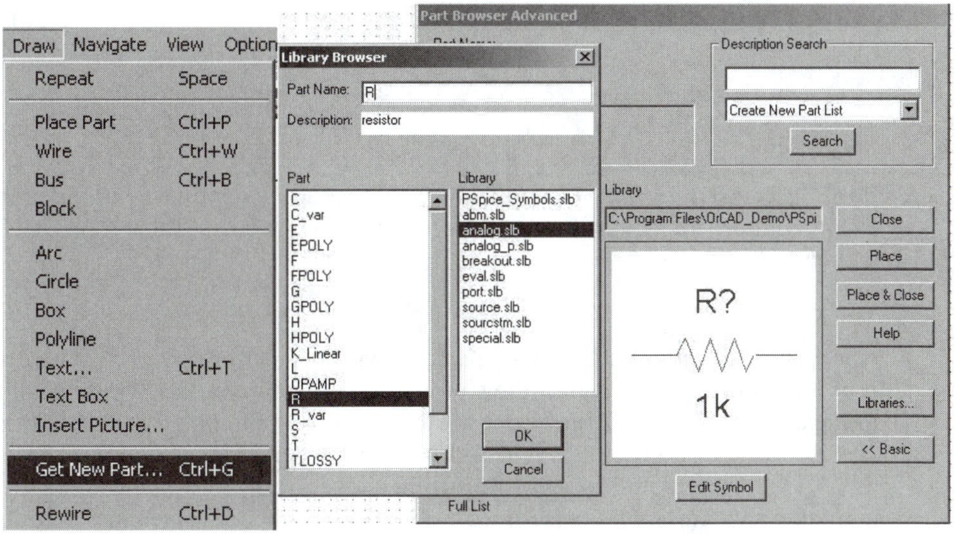

(a) Draw menu for getting parts   (b) Symbols from analog.slb file

FIGURE 2.2

Draw and part menus for analog.slb file.

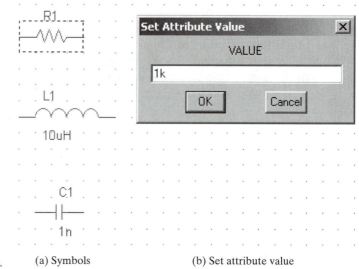

FIGURE 2.3

Symbols and names of typical elements.   (a) Symbols   (b) Set attribute value

analog.slb library as shown in Fig. 2.2(b). The element name and its value can be changed within the schematic as shown in Fig. 2.3. The netlist is created by PSpice automatically. A typical listing follows:

Element Name	First Node	Second Node	Value
R_R1	$N_0002	$N_0001	100
R_Rx	0	+	10k
R_RF	–	out	10k
C_C1	$N_0001	–	0.4uF
R_RL	0	out	100k

2.6   SOURCES

For PSpice A/D: Voltage (or current) sources can be dependent or independent. The letter symbols for the names of sources are also listed in Table 2.3. The format for sources is

```
<source name> <positive node> <negative node> <source model>
```

where the voltage of node $N+$ is specified with respect to node $N-$. The positive current is assumed to flow out of the source (from positive) node $N+$ through the circuit to negative node $N-$. If the nodes are interchanged, the polarity of the source will be reversed. The order of nodes $N+$ and $N-$ is important.

An independent voltage (or current) source can be dc, sinusoidal, pulse, exponential, polynomial, piecewise linear, or single-frequency frequency modulation. The techniques for specifying models of sources are described in Sections 3.5, 3.6, and 4.3.

The model for a simple dc source is

```
DC <value>
```

- For $V_S = 20$ V, assuming that node 1 is at a higher potential with respect to node 0, the SPICE statement to specify the voltage source V_S that is connected between nodes 1 and 0 is

```
VS 1 0 DC 20V
```

- For $I_S = 50$ mA, assuming that I_S flows from node 0 to node 4, the SPICE statement to specify the current source I_S that is connected between nodes 0 and 4 is

```
IS 0 4 DC 50MA
```

For PSpice Schematics: The name of a source with the specified letter is shown in Table 2.3. The source symbols can be obtained from the source.slb library as shown in Fig. 2.4(a). The symbols for sources are obtained from the source.slb library as shown in Fig. 2.4(b). The source name and the value can be changed within the schematic as shown in Fig. 2.5.

The typical netlist for voltage and current sources that are created by PSpice automatically is as follows:

Source Name	First Node	Second Node	Value	
V_Vs	$N_0001	0	20V	
V_Vx	$N_0002	0	0V	
I_Is	0	$N_0005	DC	50mA

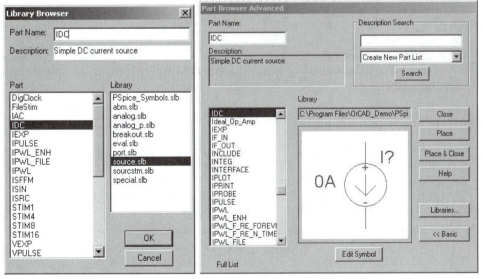

(a) Source.slb library   (b) Source symbols from source.slb library

FIGURE 2.4

Sources from the source.slb library.

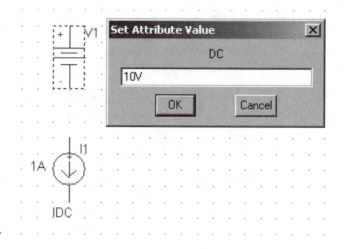

FIGURE 2.5

Symbols and names of typical dc sources.

2.7   TYPES OF ANALYSIS

For PSpice A/D: SPICE and PSpice (hereafter called SPICE/PSpice) can perform various analyses, discussed in Section 1.4. Each analysis is invoked by including its command statement. Since Fig. 2.1 is a dc circuit, we are concerned with dc analysis only. Whenever a circuit file is run, SPICE/PSpice always calculates the dc bias point, which consists of all node voltages and the currents through all voltage sources.

The details of all node voltages as well as the current and power dissipation of all voltage sources can be sent to the output file by the .OP command (discussed in Section 3.9.1), whose format is

```
.OP
```

For PSpice Schematics: The type analysis to be performed on a circuit is involved in the setup as shown in Fig. 2.6(b) within the analysis menu as shown in Fig. 2.6(a).

2.8   OUTPUT VARIABLES

For PSpice A/D: SPICE/PSpice has some unique features in printing or plotting output voltages or currents. The various types of output variables that are permitted by PSpice are discussed in Sections 3.7, 4.5, and 5.2. The voltage of node 4 with respect to node 0 is specified by $V(4,0)$ or $V(4)$, and the voltage of node 2 with respect to node 3 is specified by $V(2,3)$.

SPICE/PSpice can give the current of a voltage source as an output. A dc dummy voltage source of 0 V (say, $V_Z = 0$ V) is normally added and used as an ammeter to measure the current of that source, for example, $I(V_Z)$. Voltage sources V_X and V_Y in Fig. 2.1 act as ammeters and measure the currents of resistors R_2 and R_3, respectively. The statements for V_X and V_Y are as follows:

```
VX   3   0   DC   0V   ; Measures current through R3
VY   5   4   DC   0V   ; Measures current through R2
```

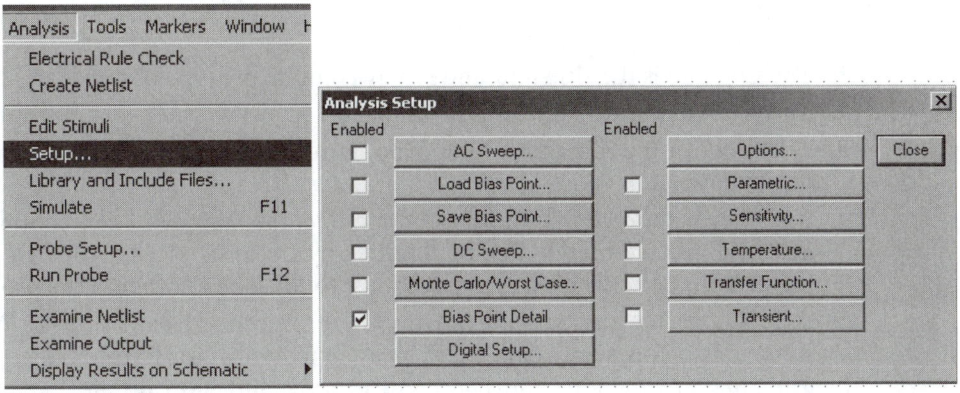

(a) Analysis menu   (b) Invoking the analysis type: Bias Point Detail

FIGURE 2.6

Analysis setup menu.

For PSpice Schematics: PSpice assigns the node numbers. Thus, the user does not know the node numbers and cannot refer the output voltages to the node numbers. The output voltage is rather referred to one of two terminals of an element. For example, the out voltage at the terminal 1 of a resistor R1 defined to V(R1:1), at the terminal 2 as V(R1:2) and between terminals 1 and 2 as V(R1:1,R1:2).

2.9   PSPICE OUTPUT COMMANDS

The most common forms of output are print tables and plots, which require output commands. However, with the .OP command, SPICE/PSpice automatically directs all node voltages and the current and power dissipation of all voltage sources to the output file, and therefore does not require any output command. Note that Probe is normally used for graphical outputs rather than the print command.

2.10   FORMAT OF CIRCUIT FILES

For PSpice A/D: A circuit file that can be read by SPICE/PSpice may be divided into five parts: (1) the title, which describes the type of circuit or any comments; (2) the circuit description, which defines the circuit elements and the set of model parameters; (3) the analysis description, which defines the type of analysis; (4) the output description, which defines the way the output is to be presented; and (5) the end of the program (the .END command). The format for a circuit file is as follows:

```
Title
Circuit description
Analysis description
Output description
.END    (end-of-file statement)
```

Notes

1. The first line is the title line, and it may contain any type of text.

2. The last line must be the .END command.

3. The order of the remaining lines is not important and does not affect the results of simulations.

4. If a PSpice statement is more than one line, the statement can continue on the next line. A continuation line is identified by a plus sign ($+$) in the first column of the next line. The continuation lines must follow one another in the proper order.

5. A comment line may be included anywhere, preceded by an asterisk (*). Within a statement, a comment is preceded by a semicolon (;) for PSpice only.

6. The number of blanks between items is not significant (except in the title line). Tabs and commas are equivalent to blanks. For example, " " and " " and "," and " , " are both equivalent.

7. PSpice statements or comments can be in either upper- or lowercase.

8. SPICE2 statements must be in uppercase only. *It is advisable to type PSpice statements in uppercase, so that the same circuit file can also be run on SPICE2.*

9. If you are not sure of any command or statement, the best thing is to run the circuit file by using that command or statement to determine what will happen. SPICE/PSpice is user-friendly software; it gives an error message in the output file that identifies a problem.

10. In electrical circuits, subscripts are normally assigned to symbols for voltages, currents, and circuit elements. However, in SPICE the symbols are represented without subscripts. For example, V_s, I_s, and R_1 are represented by VS, IS, and R1, respectively. As a result, the SPICE symbols in a circuit description of voltages, currents, and circuit elements are often different from the normal circuit symbols.

The menu is shown in Fig. 2.7. The new circuit file can be created or an existing file can be opened by using the PSpice file menu as shown in Fig. 2.8(a). The file can

FIGURE 2.7

The PSpice A/D menu.

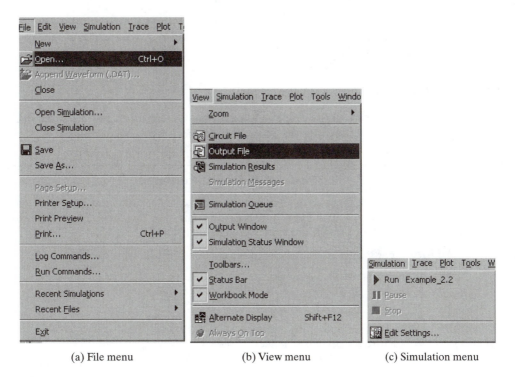

(a) File menu   (b) View menu   (c) Simulation menu

FIGURE 2.8

Submenus of the PSpice A/D.

also be created by using any text file editor (e.g., Notepad, a word processor). The circuit file can be run from the PSpice simulation menu as shown in Fig. 2.8(b). The output file can be viewed from the PSpice view menu as shown in Fig. 2.8(c) and the graphical plot can be viewed by selecting "Simulation Results" from the view menu as shown in Fig. 2.8(b).

2.11   FORMAT OF OUTPUT FILES

The results of simulation by SPICE/PSpice are stored in an output file. It is possible to control the type and amount of output by various commands. If there is any error in the circuit file, SPICE/PSpice will display a message on the screen indicating that there is an error and will suggest looking at the output file for details. The output falls into four types:

1. A description of the circuit itself that includes the net list, the device list, the model parameter list, and so on

2. Direct output from some of the analyses without the .PLOT and .PRINT commands (including the output from .OP, .TF, .SENS, .NOISE, and .FOUR analyses)

3. Prints and plots by .PLOT and .PRINT commands (including the output from the .DC, .AC, and .TRAN analyses)

4. Run statistics (i.e., the various kinds of summary information about the whole run, including times required by various analyses and the amount of memory used).

2.12 EXAMPLES OF SPICE SIMULATIONS

We have discussed all the details of describing the circuit of Fig. 2.1 as a SPICE/PSpice input file. We will illustrate the PSpice simulations by two examples.

Example 2.1: Finding node voltages and currents

The circuit of Fig. 2.1 is to be simulated on PSpice to calculate and print all node voltages and the current and power of all voltage sources (V_S, V_X, and V_Y). The circuit file is to be stored in file EX2-1.CIR, and the outputs are to be stored in file EX2-1.OUT.

Solution **For PSpice A/D:** The circuit file contains the following statements.

```
Example 2-1  Simple dc circuit
▲ VS   1   0   DC   20V    ; Dc voltage source of 10 V
  IS   0   4   DC   50MA   ; Dc current source of 50 mA
▲ ▲ R1 1   2   500        ; Resistance of 500 ohms
    R2   2   5   800        ; Resistance of 800 ohms
    R3   2   3   1KOHM      ; Resistance of 1 kohms
    R4   4   0   200        ; Resistance of 200 ohms
    VX   3   0   DC   0V    ; Measures current through R3
    VY   5   4   DC   0V    ; Measures current through R2
▲ ▲ ▲.OP               ; Directs the bias point to the output file
.END                   ; End of circuit file
```

Performing a simulation with PSpice requires that you know how to install the PSpice software, create a circuit file, and run the circuit file. Running PSpice on PCs is described in Appendix A. Later versions of PSpice are menu driven and do not require typing separate run commands. PSpice has a built-in text editor. The circuit file Ex2-1.CIR can be typed in from the File menu and then can be run from the Analysis menu. The menus of the PSpice control shell are shown in Fig. 2.7.

The results are stored by default in an output file that has the same name as the input file and is on the same drive, but has the extension .OUT. The output file EX2-1.OUT contains the results, which can be seen by using the Browse menu of PSpice A/D. The results that appear in the output file EX2-1.OUT are as follows:

```
****     SMALL-SIGNAL BIAS SOLUTION        TEMPERATURE =   27.000 DEG C
NODE    VOLTAGE     NODE   VOLTAGE    NODE   VOLTAGE     NODE   VOLTAGE
(   1)   20.0000  (   2)   12.5000  (   3)    0.0000  (   4)   10.5000
(   5)   10.5000
VOLTAGE SOURCE CURRENTS
NAME          CURRENT
VS            -1.500E-02                        IR1 = 15 mA
VX             1.250E-02                        IR3 = 12.5 mA
VY             2.500E-03                        IR2 = 2.5 mA
TOTAL POWER DISSIPATION 3.00E-01 WATTS
       JOB CONCLUDED
       TOTAL JOB TIME          1.04
```

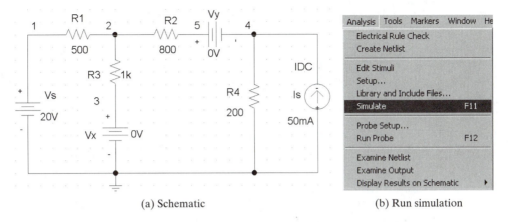

(a) Schematic   (b) Run simulation

FIGURE 2.9

PSpice Schematic for Example 2.1.

For PSpice Schematics: The PSpice schematic is shown in Fig. 2.9(a). (See Appendix A for instructions on drawing a schematic.) Selecting "Simulation" from the analysis menu will run the schematic. After the simulation is completed, the PSpice A/D menu is opened automatically. The output file and the simulation results can be viewed from the view menu as shown in Fig. 2.9(b).

Note. Since the type of analysis is not specified, the .OP command will not change the information about the bias point on the output file. Students are encouraged to run the circuit file EX2-1.CIR with and without the .OP command and to compare the output files.

Example 2.2   Finding the node voltage and specific currents

The circuit of Fig. 2.1 is to be simulated on PSpice to calculate and print the voltage at node 4, the current I_{R_2} and I_{R_3} for $V_S = 10$ V, 20 V, and 30 V. The circuit file is to be stored in the file EX2-2.CIR, and the outputs are to be stored in the file EX2-2.OUT.

Solution   The circuit file is similar to that of Example 2.1, except that the statements for the analysis and the output are different.

For PSpice A/D: Since the dc voltage V_S is swept from 10 V to 30 V with a 10-V increment, the dc analysis (discussed in Section 3.9.3) can be invoked by the .DC command, whose format is

```
.DC SWNAME SSTART SEND SINC
```

where SWNAME = the sweep variable name, either voltage or current source
        SSTART = the sweep start value
          SEND = the sweep end value
          SINC = the sweep increment value

For SSTART = 10 V, SEND = 30 V, and SINC = 10 V, the statement for the dc sweep is

```
.DC VS 10V 30V 10V
```

The output variables for the dc analysis are discussed in Section 3.7, and the .PRINT command is discussed in Section 3.8.1. The statement to print V(4), I(VX), and I(VY) for the results of the .DC sweep is

```
.PRINT DC V(4) I(VX) I(VY)
```

If the .DC and .PRINT statements are included in the circuit file EX2-1.CIR, we will get the circuit file for EX2-2.CIR. The circuit file contains the following statements.

Example 2-2   Dc sweep of VS

```
▲ VS   1   0   DC   20V    ; Dc voltage source of 10 V
  IS   0   4   DC   50MA   ; Dc current source of 50 mA
▲ ▲ R1  1   2   500       ; Resistance of 500 ohms
    R2   2   5   800       ; Resistance of 800 ohms
    R3   2   3   1KOHM     ; Resistance of 1 kohms
    R4   4   0   200       ; Resistance of 200 ohms
    VX   3   0   DC   0V   ; Measures current through R3
    VY   5   4   DC   0V   ; Measures current through R2
▲ ▲ ▲  *   Dc sweep of VS from 10 V to 30 V in 10 V increment
       .DC  VS  10V  30V  10V
       .PRINT  DC  V(4)  I(VX)  I(VY)  ; Prints the results of dc sweep
.END                                  ; End of circuit file
```

The output file EX2-2.OUT contains the results, which can be seen by using the browse menu of PSpice. The results that appear in the output file EX2-2.OUT are as follows:

```
****       DC TRANSFER CURVES              TEMPERATURE =  27.000 DEG C
VS            V(4)         I(VX)        I(VY)
  1.000E+01   9.500E+00    7.500E-03  -2.500E-03
  2.000E+01   1.050E+01    1.250E-02   2.500E-03
  3.000E+01   1.150E+01    1.750E-02   7.500E-03
             JOB CONCLUDED
             TOTAL JOB TIME              1.21
```

For PSpice Schematics: The PSpice schematic shown in Fig. 2.10(a) is the same as that of Fig. 2.9(a), with the exception of one voltage marker and two current markers. The current markers eliminate the needs for dc voltages V_x and V_y that are used for measuring currents. Selecting DC sweep in the analysis menu opens the menu for Dc sweep parameters as shown in Fig. 2.10(b). Selecting "Simulation" from the analysis menu will run the schematic. After the simulation is completed, the PSpice A/D menu is opened up automatically. Selecting "Simulation

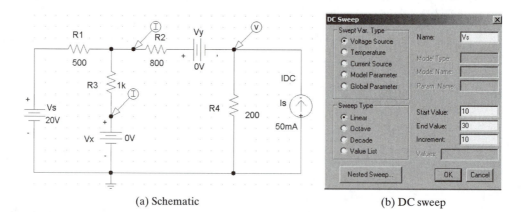

(a) Schematic   (b) DC sweep

FIGURE 2.10

PSpice Schematic for Example 2.2.

Results" in the view menu, as shown in Fig. 2.8(b), will automatically display the plots of the marker voltages and currents. The PSpice Schematic does give only plots rather than print outputs.

Notes

1. For the dc sweep in Example 2.2, PSpice calculates the dc bias point with $V_S = 20$ V, but will not direct the details of the bias point to the output file unless the .OP command is included.

2. Students are encouraged to run the circuit file EX2-2.CIR with and without the .OP command and to compare the output files.

Example 2.3   Finding node voltages and plot a specific current

The circuit of Fig. 2.1 is to be simulated on PSpice to calculate and print the voltage at node 4; the current I_{R_2} and I_{R_3} for $V_S = 5$ V, 20 V, and 30 V for each value of $I_S = 50$ mA, 100 mA, and 150 mA. The current I_{R_2} is to be plotted. The circuit file is to be stored in file EX2-3.CIR, and the outputs are to be stored in file EX2-3.OUT. The results should also be available for display and hard copy by the .PROBE command.

Solution   The circuit file is similar to that of Example 2.1, but the current source I_S has three values, and a plot of I_{R_2} versus V_S is required. For $V_S = 5$ V, 20 V, and 30 V, V_S cannot be swept linearly with an increment. We can simply list the values by using the key word LIST (discussed in Section 3.9.3).

   For PSpice A/D: The format for the dc sweep with a list of values is

```
.DC SWNAME LIST <value>*
```

where SWNAME = the sweep variable name, either voltage or current source;
   LIST = a key word
   $\langle$value$\rangle^*$ = list of sweep values

For $V_S = 5$ V, 20 V, and 30 V, the statement for the dc list sweep becomes

```
.DC VS LIST 5V 20V 30V
```

   Since I_S has three values, there will be three sets of results for every value of V_S, for a total of nine sets. I_S can be regarded as the nested sweep within the sweep of V_S and can be swept from 50 mA to 150 mA with a 50-mA increment. The statement (discussed in Section 3.9.3) for the nested dc sweeps of V_S and I_S is

```
.DC VS LIST 5V 20V 30V IS 50MA 150MA 50MA
```

which sweeps the current I_S linearly within the list sweep of V_S.

   The command for output in the form of plots is .PLOT. The statement for the plots of I(VY) from the results of the .DC sweep is

```
.PLOT DC I(VY)
```

The outputs of .PRINT and .PLOT commands are stored in an output file created automatically by PSpice.

   Probe is a *graphical waveform analyzer* for PSpice. The statement to invoke Probe is

```
.PROBE
```

This command makes the results of simulation available for graphical outputs on the display and on the hard copy. After executing the .PROBE command, Probe puts a menu on the screen that allows you to obtain graphical output. It is very easy to use Probe. The .PRINT command gives a table of data, .PLOT generates the plot on the output file, and the .PROBE command provides graphical output on the monitor screen that can be dumped directly into a plotter or a printer. The .PLOT and .PRINT commands could generate a large amount of data in the output file and should be avoided if graphical output is available, as it is in PSpice with .PROBE. With the .PROBE command, there is *no* need for the .PLOT or .PRINT commands.

The circuit file contains the following statements:

Example 2-3   Dc sweep of VS and IS
```
▲ VS    1   0   DC   20V       ; Dc voltage source of 10 V
  IS    0   4   DC   50MA      ; Dc current source of 50 mA
▲ ▲ R1  1   2   500            ; Resistance of 500 ohms
    R2  2   5   800            ; Resistance of 800 ohms
    R3  2   3   1KOHM          ; Resistance of 1 kohms
    R4  4   0   200            ; Resistance of 200 ohms
    VX  3   0   DC   0V        ; Measures current through R3
    VY  5   4   DC   0V        ; Measures current through R2
▲ ▲ ▲ * Dc sweep of VS from 10 V to 50 V in 20 V increment and
      * Dc sweep of IS from 0 to 100 mA in 50 mA increment
      .DC    VS    LIST  5V 20V 30V      IS  50MA 150MA 50MA
      .PRINT DC  V(4)   I(VX)  I(VY) ; Prints the results of dc sweep
      .PLOT  DC  I(VY)                ; Plots I(VY) on the output file
      .PROBE                         ; Graphical waveform analyzer
  .END                               ; End of circuit file
```

The output file EX2-3.OUT contains the results, which can be seen by using the browse menu of PSpice. The results that appear in the output file EX2-3.OUT are as follows:

```
****        DC TRANSFER CURVES              TEMPERATURE = 27.000 DEG C
  VS            V(4)                   I(VX)              I(VY)
   5.000E+00   9.000E+00   5.000E-03  −5.000E-03
   2.000E+01   1.050E+01   1.250E-02   2.500E-03
   3.000E+01   1.150E+01   1.750E-02   7.500E-03
   5.000E+00   1.750E+01   7.500E-03  −1.250E-02
   2.000E+01   1.900E+01   1.500E-02  −5.000E-03
   3.000E+01   2.000E+01   2.000E-02   3.000E-12
   5.000E+00   2.600E+01   1.000E-02  −2.000E-02
   2.000E+01   2.750E+01   1.750E-02  −1.250E-02
   3.000E+01   2.850E+01   2.250E-02  −7.500E-03
```

The plots of the dc sweep I_{R_2} versus V_S that are displayed on the monitor by .PROBE command and are dumped directly into a plotter are shown in Fig. 2.11. Figure 2.12 shows the plots of the dc sweep I_{R_2} versus V_S that are stored in the output file EX2-3.OUT by the .PLOT command and are printed on a printer (by pressing Print-Screen and then typing TYPE EX2-3.OUT).

For PSpice Schematics: The PSpice schematic is the same as that of Fig. 2.10(a). The dc sweep values of V_s are listed rather than an increment change as shown in Fig. 2.13(a). For each value of V_s, the value I_S is increased as a nested sweep with an increment as shown in Fig. 2.13(b). Selecting "Simulation" from the analysis menu will run the schematic. After the simulation is completed, the PSpice A/D menu is opened automatically. Selecting "Simulation Results" in the view menu, as shown in Fig. 2.8(b), will display automatically the plots of the marker voltage and currents.

Note. Notice the quality of plots obtained with the .PROBE command compared to those obtained with the .PLOT command. The .PLOT statement generates graphical plots in the

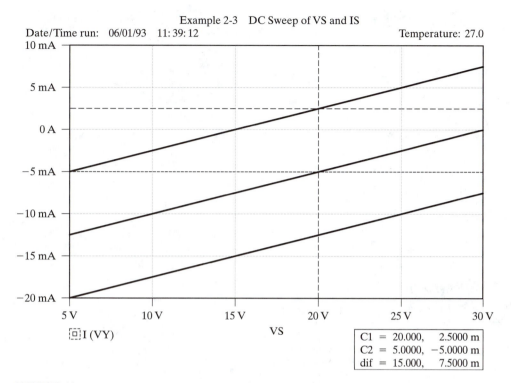

FIGURE 2.11

Plots of dc sweeps obtained by the .PROBE command.

Example 2-3   DC Sweep of VS and IS

★★★★   DC TRANSFER CURVES   TEMPERATURE   =   27.000 DEG C

★★★

VS	I (VY)					
(★) − − − − − − − −	−2.0000E−02	−1.0000E−02	0.0000E+00	1.0000E−02	2.0000E−02	
5.000E+00	−5.000E−03	.	.	★	.	.
2.000E+01	2.500E−03	.	.	.	★	.
3.000E+01	7.500E−03	.	.	.	.	★
5.000E+00	−1.250E−02	.	★	.	.	.
2.000E+01	−5.000E−03	.	.	★	.	.
3.000E+01	3.000E−12	.	.	.	★	.
5.000E+00	−2.000E−02	★	.	.	.	.
2.000E+01	−1.250E−02	.	★	.	.	.
3.000E+01	−7.500E−03	.	.	★	.	.

JOB CONCLUDED

TOTAL JOB TIME   1.43

FIGURE 2.12

Plots of dc sweeps obtained by the .PLOT command.

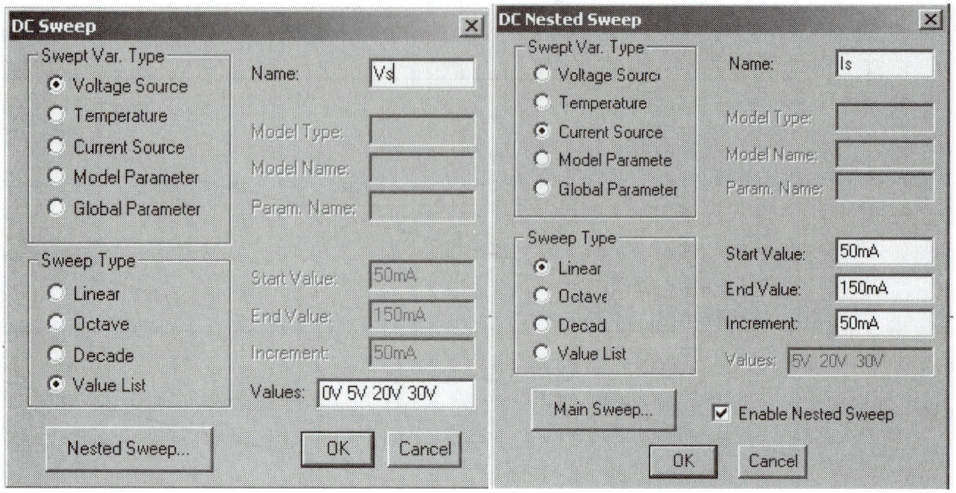

(a) Main sweep

(b) Nested sweep

FIGURE 2.13

PSpice Schematic for Example 2.3.

output file. If the .PROBE command is included, there is no need for the .PLOT command. With the availability of Probe in PSpice schematics and A/D, the plot output as shown in Fig. 2.12 with the .PLOT command is not normally used anymore.

2.13   OrCAD CAPTURE

If the OrCAD Lite Software is properly installed, it will have the following portions:

- Capture—The schematic capture as the from-end input interface that has replaced Microsim Schematics. Its menu layout has slightly changed, but the basic elements have not.
- PSpice A/D—The mixed-signal simulation tool. This tool is similar to Microsim Design Center, and it is relatively unchanged.
- Probe—The graphical postprocessor for viewing the simulation results. This tool is similar to Microsim Design Center and is relatively unchanged as well.

2.13.1   OrCAD Capture Layout

Figure 2.14 shows the layout of the OrCAD Capture. The top menu shows all of the main menus. The right-side menu shows the schematic "drawing" menu for selecting and placing parts. For example, a part can be placed either from the "Place" menu or from the drawing menu as shown in Fig. 2.15. Figure 2.16(a) shows the "Place" menu for selecting

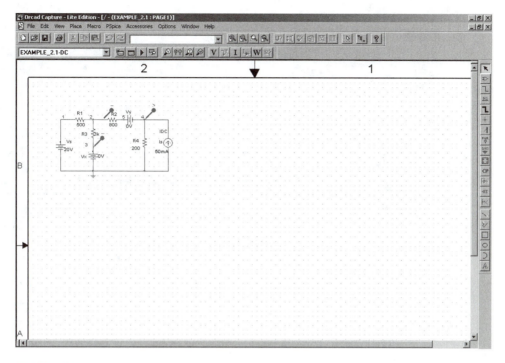

FIGURE 2.14

OrCAD Capture layout.

Place parts	Place wires	Place net alias	Place bus	Place junction	Place bus entry	Pace power	Pace ground	Hierarchical Block	Place port	Place pin	Off-pageconnector	Place no-connect	Place line	Place polyline	Place rectangle	Place ellipse	Place arc	Place text

FIGURE 2.15

Draw menu for placing parts.

a part, and Figure 2.16(b) shows the menu for placing a part (e.g., R) from the schematic library (e.g., Analog).

Figure 2.17(a) shows the PSpice menu for selecting the simulation profile, and Fig. 2.17(b) shows the menu for selecting analysis type (e.g., DC sweep, time domain, AC sweep, or bias point) and options such as parametric sweep, temperature sweep, or Monte Carlo/worst case.

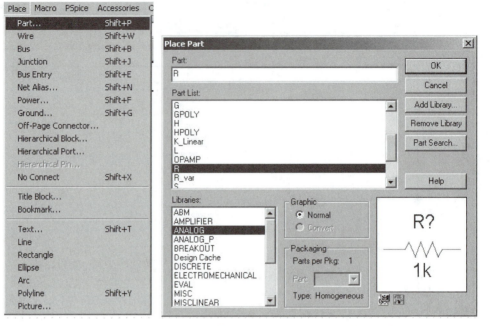

(a) Place menu   (b) Selecting a part from a schematic library

FIGURE 2.16

Place and part menus.

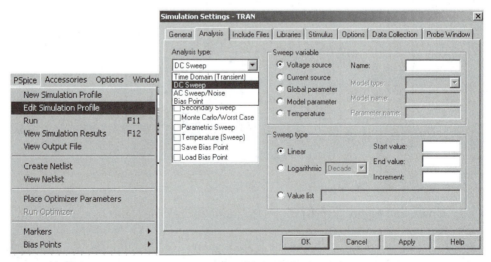

(a) PSpice menu   (b) Analysis type and set up menu

FIGURE 2.17

PSpice menu and analysis setup.

2.13.2   PSpice A/D

PSpice A/D combines the PSpice and Probe for editing and running a simulation file, viewing output and simulation results, and setting the simulation profile. Its menu is shown in Fig. 2.18(a). The left-hand side menu, as shown in Fig. 2.18(b), allows viewing of the simulation results and setting of simulation profiles.

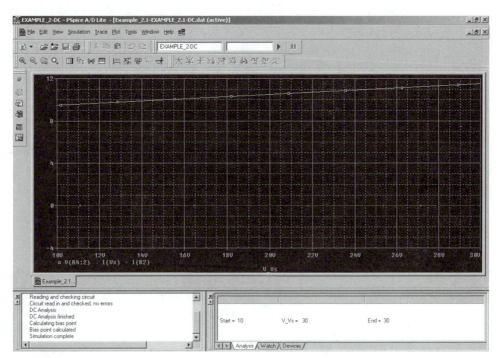

(a) PSpice A/D menu

(b) PSpice A/D viewing and setting menus

FIGURE 2.18

PSpice menu.

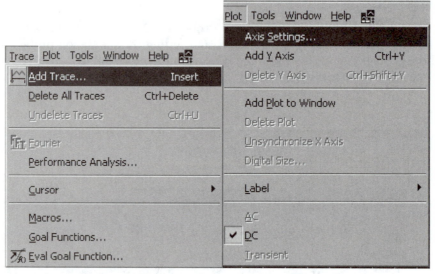

(a) Trace menu of Probe   (b) Plot menu of Probe

FIGURE 2.19

Probe menu of PSpice A/D.

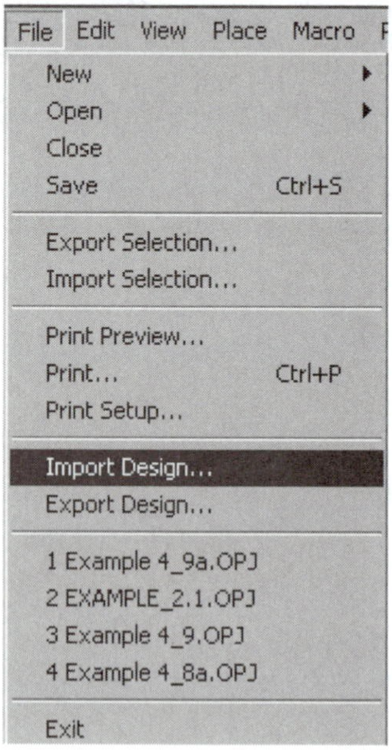

(a) File menu 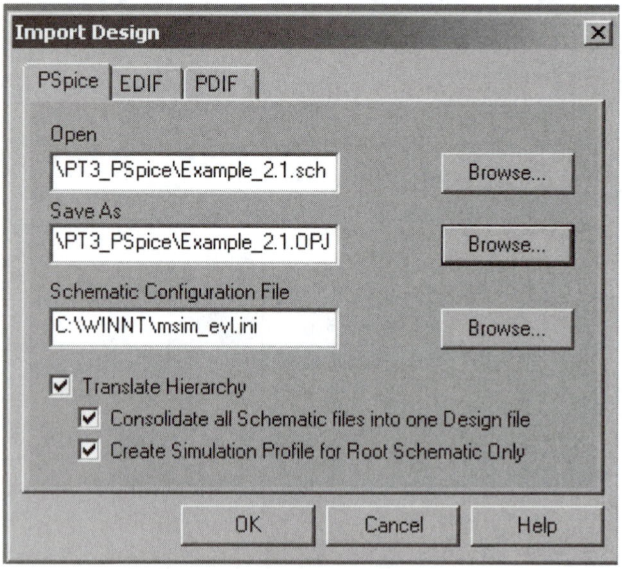   (b) Design import

FIGURE 2.20

Importing Microsim schematic file.

2.13.3 Probe

The Probe menu is combined within PSpice A/D. Figure 2.19(a) shows the trace menu for plotting the output variables, and Fig. 2.19(b) shows the menu for setting the plot axes and the number of plots.

2.14   IMPORTING MICROSIM SCHEMATIC IN OrCAD CAPTURE

OrCAD Capture has a file extension of .OPJ. However, Microsim Schematic version 9.1 or below uses the file extension of .SCH. Therefore, a schematic file (.sch) can be run directly on OrCAD Capture without conversion and can be imported to OrCAD Capture from its file menu as shown in Fig. 2.20(a). It requires the identifying the location of the following files:

- The name and the location of the schematic file with .SCH extension
- The name and the location of the Capture file with .OPJ extension, where the imported file will be saved
- The name and the location of the schematic configuration file with .ini extension.

Once the conversion is completed, the imported file with .OPJ extension can be simulated on OrCAD Capture.

Note: When importing PSpice Schematics files into OrCAD Capture, Capture will require you to specify the location of the `msim_evl.ini` file. If you cannot locate the file on your computer, you can copy the `msim_evl.ini` file from the enclosed CD-ROM to your Windows directory (i.e., `C:\WINNT`) so that its location becomes `C:\WINNT\msim_evl.ini`.

REFERENCES

[1]   Boylestad and Nashelsky, *Electronic Devices and Circuits Theory*. Upper Saddle River, New Jersey: Prentice Hall, 2002.

[2]   Robert Boylestad, *Introductory Circuit Analysis* (10th ed). Upper Saddle River, New Jersey: Prentice Hall, 2003.

[3]   L. Richard Carley, *Introduction to Electrical and Computer Engineering* (2d ed). New York: McGraw-Hill, 1997.

[4]   J.R. Cogdell, *Fundamentals of Electric Circuits*. Upper Saddle River, New Jersey: Prentice Hall, 1999.

[5]   Richard C. Dorf and James A. Svoboda, *Introduction to Electric Circuits*. New York: John Wiley & Sons, 2001.

[6]   William H. Hayt Jr., Jack E. Kemmerly, and Steven M. Durbin, *Engineering Circuit Analysis* (6th ed). Upper Saddle River, New Jersey: Prentice Hall, 2003.

[7]   J. David Irwin and David V. Kerns, Jr., *Introduction to Electrical Engineering*. Englewood Cliffs, New Jersey: Prentice Hall, 1995.

[8]   J.D. Irwin, *Basic Engineering Circuit Analysis* (5th ed). Upper Saddle River, New Jersey: Prentice Hall, 1996.

[9]   R.W. Jensen and M.D. Liberman, *IBM Electronic Circuit Analysis Program and Applications*. Englewood Cliffs, New Jersey: Prentice Hall, 1968.

[10]   R.W. Jensen and L.P. McNamee, *Handbook of Circuit Analysis Languages and Techniques*. Englewood Cliffs, New Jersey: Prentice Hall, 1976.

[11]   Jack F. Morris, *Basic Circuit Analysis*. New York: Houghton Mifflin, 1991.

[12]   James W. Nilsson and Susan A. Riedel, *Electric Circuits* (6th ed). Upper Saddle River, New Jersey: Prentice Hall, 2000.

[13] James W. Nilsson, *Electric Circuits*. New York, NY: Addison-Wesley Publishing Company, 4th edition, 1993.

[14] Dr. John O'Malley, *Schaum's Outline of Basic Circuit Analysis* (2d ed). New York: Mc-Graw-Hill, 1992.

[15] *PSpice Manual*. Irvine, California: MicroSim Corporation, 1992.

[16] M. H. Rashid, *SPICE for Power Electronics and Electric Power* (2d ed). Englewood Cliffs, New Jersey: Prentice Hall, 1993.

[17] Giorgio Rizzoni, *Principles and Applications of Electrical Engineering* (4th ed). New York: McGraw-Hill, 2003.

[18] John A. Stuller, *Basic Circuit Analysis* (2d ed). New York: John Wiley & Sons, 1995.

[19] Paul W. Tuinenga, *SPICE: A Guide to Circuit Simulation and Analysis Using PSpice* (3rd ed). Englewood Cliffs, New Jersey: Prentice Hall, 1995.

PROBLEMS

2.1 The circuit of Fig. P2.1 is to be simulated on PSpice to calculate and print **(a)** all node voltages, **(b)** the current and power dissipation of voltage sources V_A and V_B, and **(c)** the currents through all resistors.

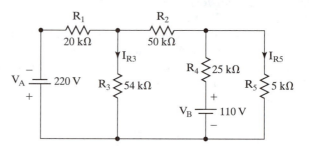

FIGURE P2.1

2.2 Repeat Problem 2.1 for $V_A = 100$ V, 120 V, and 150 V.
2.3 Repeat Problem 2.1 for $V_A = 100$ V, 120 V, and 150 V and for $V_B = 80$ V, 110 V, and 140 V.
2.4 The circuit of Fig. P2.4 is to be simulated on PSpice to calculate and print **(a)** all node voltages, **(b)** the current and power dissipation of voltage sources V_A and V_B, and **(c)** the

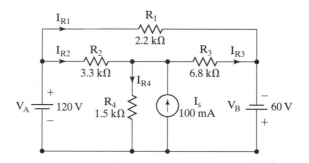

FIGURE P2.4

currents through all resistors.
2.5 Repeat Problem 2.4 for $V_A = 100$ V, 120 V, and 140 V.
2.6 Repeat Problem 2.4 for $V_A = 100$ V, 120 V, and 140 V and for $V_B = 40$ V, 60 V, and 80 V.
2.7 Repeat Problem 2.4 for $V_A = 100$ V, 120 V, and 140 V; for $V_B = 40$ V, 60 V, and 80 V; and $I_S = 50$ mA, 100 mA, and 150 mA.

C H A P T E R 3

DC Circuit Analysis

After completing this chapter, students should be able to

- Specify the operating temperature;
- Model temperature-dependent resistors;
- Model dependent and independent voltage and current sources;
- Understand the types of dc analysis and their output variables;
- Perform the dc analysis of a circuit, such as dc operating point, small-signal transfer function, and dc sweep.

3.1   INTRODUCTION

Sources in dc (direct current) circuits are constant voltages or currents; that is, voltages and currents are invariant with time. In a circuit analysis course, dc circuits are covered first in order to introduce the basic circuit laws and techniques, for example, Kirchhoff's voltage and current laws, node-voltage and mesh-current methods, and Thévenin and Norton equivalents. Students can simulate dc circuits in SPICE to verify and reinforce their theoretical knowledge and to study the effects of parameter variation on voltages and currents in dc circuits. The simulation of dc circuits with passive elements requires the modeling of

> Resistors
> Model parameters
> Operating temperature
> Dc sources
> Dc output variables
> Types of dc analysis.

We shall illustrate the SPICE simulations of dc circuits by examples. Students are encouraged to apply the basic circuit laws and to verify the SPICE results by hand calculation.

3.2 RESISTORS

The voltage and current relationships of resistors are shown in Fig. 3.1(a). The symbol for a resistor is R. The name of a resistor must start with R. The PSpice schematic is shown in Fig. 3.1(b), and the parameters of resistors are shown in Fig. 3.1(c). The resistor's name and its nominal value can be changed. Also, a tolerance value can be assigned to it. The netlist takes the general form of

```
R <name> N+ N−  RNAME RVALUE
```

A resistor does have a polarity, and the order of the nodes does matter. However, by defining $N+$ as the positive node and $N-$ as the negative node, the current is assumed to flow from node $N+$ through the resistor to node $N-$. RNAME is the model name that defines the parameters of the resistor and is discussed in Section 3.3. RVALUE is the nominal value of the resistance.

Note. Some versions of PSpice or SPICE do not recognize the polarity of resistors and do not allow reference to currents through resistors, for example, $I(R_1)$.

The model parameters are shown in Table 3.1. It should be noted that the model parameter R is a resistance multiplier, rather than the value of the resistance. If RNAME is omitted, RVALUE is the resistance in ohms and can be positive or negative, but *must* not be zero.

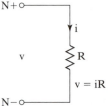

(a) Voltage and current relationships.

(b) PSpice schematic.

(c) Resistor's parameters.

FIGURE 3.1

Voltage/current relationship and parameters of resistors.

TABLE 3.1 Model Parameters for Resistors

Name	Meaning	Units	Default
R	Resistance multiplier		1
TC1	Linear temperature coefficient	$°C^{-1}$	0
TC2	Quadratic temperature coefficient	$°C^{-2}$	0
TCE	Exponential temperature coefficient	$\%\ °C$	0

If RNAME is included and TCE is specified, the resistance as a function of temperature is calculated from

```
                           TCE * (T - T0)
     RES = RVALUE * R * 1.01
```

If RNAME is included and TCE is not specified, the resistance as a function of temperature is calculated from

```
     RES = RVALUE * R * [1 + TC1 * (T - T0) + TC2 * (T - T0)²]
```

where T and T0 are the operating temperature and the room temperature, respectively, in degrees Celsius.

In the PSpice schematic, the user can assign the model name of the breakout devices in the library *breakout.slb* shown in Fig. 3.2(a) and can also edit the model parameters shown in Fig. 3.2(b) and (c).

Some resistor statements

```
R1        6     5      10K
RLOAD     12    11     ARES    2MEG
.MODEL    ARES    RES  (R=1    TC1=0.02   TC2=0.005)
RINPUT    15    14     BRES    5K
.MODEL    BRES    RES  (R=0.8  TCE=2.5)
```

3.3 MODELING OF ELEMENTS

The models are necessary to take into account the parameter variations; that is, the value of a resistor depends on the operating temperature. A model that specifies a set of parameters for an element is specified in PSpice by the .MODEL command. The .(dot) is an integral part of the command. The same model can be used by one or more elements in the same circuit. The general form of the model statement is

```
     .MODEL MNAME TYPE (P1=A1 P2=A2 P3=A2 . . . PN=AN)
```

MNAME is the name of the model and must start with a letter. Although not necessary, it is advisable to make the first letter the symbol of the element (e.g., *R* for resistor and *L* for inductor). The list of symbols for elements is shown in Table 2.1.

$P_1, P_2, \ldots$ are the element parameters (Table 3.1 for resistors) and $A_1, A_2, \ldots$ are their values, respectively. TYPE is the type name of the elements and must have the correct type, as shown in Table 3.2. An element must have the correct model type name. That is, a resistor must have the type name RES, not the type IND or CAP. However, there can be more than one model of the same type in a circuit with different

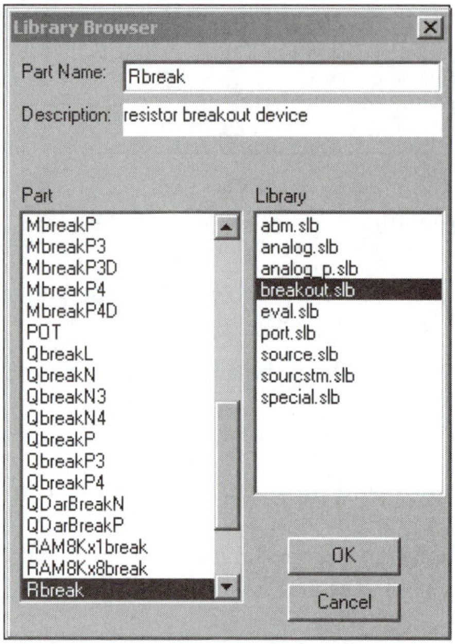

(a) Breakout devices

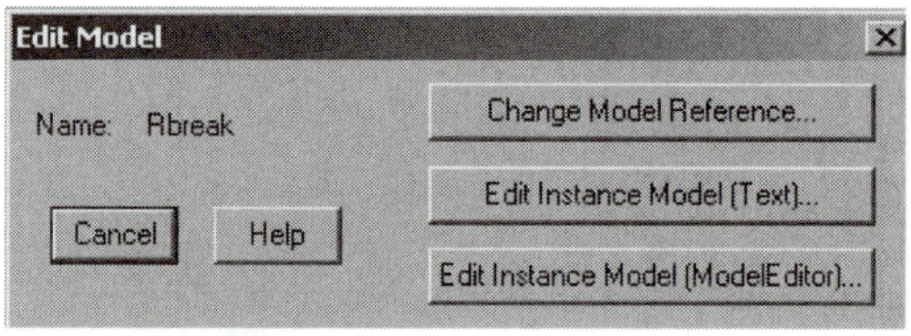

(b) Editing model parameters

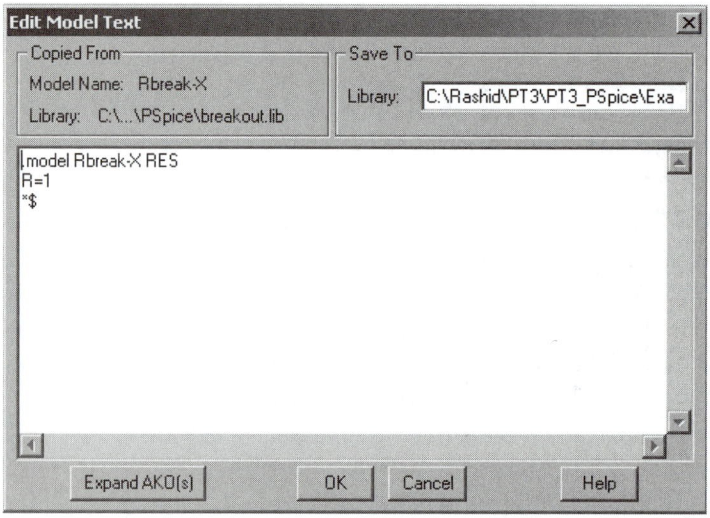

(c) Model parameters

FIGURE 3.2

Breakout devices.

TABLE 3.2   Type Names of Elements

Breakout Device Name	Type Name	Resistor
Rbreak	RES	Resistor
Cbreak	CAP	Capacitor
Dcbreak	D	Diode
Lbreak	IND	Inductor
QbreakN	NPN	Bipolar junction transistor
QbreakP	PNP	Bipolar junction transistor
JbreakN	NJF	N-channel junction FET
JbreakP	PJF	P-channel junction FET
MbreakN	NMOS	N-channel MOSFET
MbreakP	PMOS	P-channel MOSFET
Bbreak	GASFET	N-channel GaAs MOSFET
Sbreak	VSWITCH	Voltage-controlled switch
Wbreak	ISWITCH	Current-controlled switch
XFRM_LINEAR	None	Linear magnetic core (transformer)
XFRM_NONLINEAR	CORE	Nonlinear magnetic core (transformer)
ZbreakN	NIGBT	N-channel IGBT

model names. In the PSpice schematic, there is no need to identify the model name, and PSpice opens up the menu with the device model name as shown in Fig. 3.2(c).

Some Model Statements

```
.MODEL   RMOD     RES    (R=1.1  TCE=0.001)
.MODEL   RLOAD    RES    (R=1     TC1=0.02   TC2=0.005)
.MODEL   CPASS    CAP    (C=1     VC1=0.01   VC2=0.002  TC1=0.02  TC2=0.005)
.MODEL   LFILTER  IND    (L=1     IL1=0.1    IL2=0.002  TC1=0.02  TC2=0.005)
.MODEL   DNOM     D      (IS=1E-6)
.MODEL   QOUT     NPN    (BF=50  IS=1E-9)
```

Note that the model parameter R is a resistance multiplier and scales the actual resistance value, RVALUE. $R = 1.1$ means that RVALUE is multiplied by 1.1, *not* that RVALUE is 1.1 Ω.

3.4   OPERATING TEMPERATURE

The operating temperature of an analysis can be set to any desired value by the .TEMP command. The general form of the statement is

```
.TEMP <(one or more temperature) values>
```

The temperatures are in degrees Celsius. If more than one temperature is specified, then the analysis is performed for each temperature.

The model parameters are assumed to be measured at a nominal temperature, which, by default, is 27°C. The default nominal temperature of 27°C can be changed by

Analysis Setup ×

Enabled		Enabled		
☐	AC Sweep...		Options...	Close
☐	Load Bias Point...	☐	Parametric...	
☐	Save Bias Point...	☐	Sensitivity...	
☐	DC Sweep...	☑	Temperature...	
☐	Monte Carlo/Worst Case...	☐	Transfer Function...	
☐	Bias Point Detail	☐	Transient...	
	Digital Setup...			

FIGURE 3.3

Setting up operating temperature.

the TNOM option in the .OPTIONS statements that are discussed in Section 6.10. The operating temperature can be set from the analysis setup as shown in Fig. 3.3.

Some Temperature Statements

```
.TEMP   50
.TEMP   25   50
.TEMP   0    25   50   100
```

3.5 INDEPENDENT DC SOURCES

The independent sources can be time invariant or time variant. They can be currents or voltages, as shown in Fig. 3.4.

3.5.1 Independent Dc Voltage Source

The symbol for an independent voltage source is V, and the general form is

```
V<name> N+ N- [DC <value>]
```

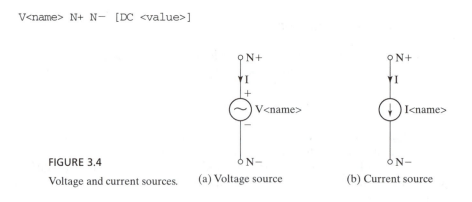

FIGURE 3.4

Voltage and current sources.   (a) Voltage source   (b) Current source

where $N+$ is the positive node, and $N-$ is the negative node, as shown in Fig. 3.4. Positive current flows from node $N+$ through the voltage source to the negative node $N-$. The voltage source need not be grounded.

The source is set to the dc value in dc analysis. A voltage source may be used as an *ammeter* in PSpice by inserting a zero-valued voltage source into the circuit for the purpose of measuring current. Since a zero-valued source behaves as a short circuit, there will be no effect on circuit operation.

Typical Statements

```
V1   15   0   6V          ; Dc voltage of 6 V, by default
V2   15   0   DC   6V   ; Dc voltage of 6 V
```

3.5.2 Independent Dc Current Source

The symbol of an independent current source is I, and the general form is

```
I<name> N+ N- [DC <value>]
```

where $N+$ is the positive node, and $N-$ is the negative node, as shown in Fig. 3.4(b). Positive current flows from node $N+$ through the current source to the negative node $N-$. The current source needs not be grounded. The source specifications are similar to those of independent voltage sources.

Typical Statements

```
I1   15   0   2.5MA          ; By default, dc current of 2.5 mA
I2   15   0   DC   2.5MA   ; Dc current of 2.5 mA
```

3.5.3 Schematic Independent Sources

The PSpice source library *source.slb* is shown in Fig. 3.5(a). Dc voltage and current sources are shown in Fig. 3.5(b) and (c). The user can change the values of the sources.

3.6 DEPENDENT SOURCES

The four types of dependent sources that follow are shown in Fig. 3.6:

> Voltage-controlled voltage source
> Voltage-controlled current source
> Current-controlled current source
> Current-controlled voltage source.

These sources can have either a fixed value or a polynomial expression.

3.6.1 Polynomial Source

The symbol for a polynomial or nonlinear source is POLY(n), where n is the number of dimensions of the polynomial. The default value of n is 1. The dimensions depend on the number of controlling sources. The general form is

```
POLY (n) <(controlling) nodes> <(coefficients) values>
```

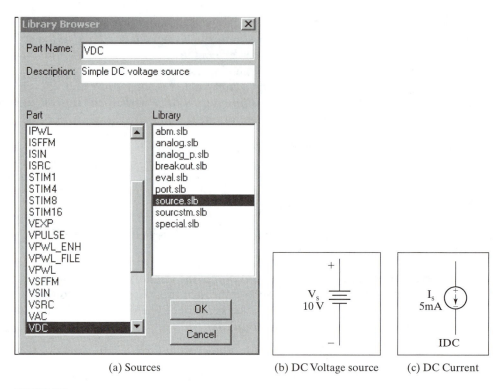

(a) Sources   (b) DC Voltage source   (c) DC Current

FIGURE 3.5

Independent DC sources.

The output sources or the controlling sources can be voltages or currents. For voltage-controlled sources, the number of controlling nodes must be twice the number of dimensions. For current-controlled sources, the number of controlling sources must be equal to the number of dimensions. The number of dimensions and the number of coefficients are arbitrary.

Let us call A, B, and C the three controlling variables, and Y the output source. Fig. 3.6 shows a source that is controlled by A, B, and C. The output source takes the form of

$$Y = f(A, B, C, \dots)$$

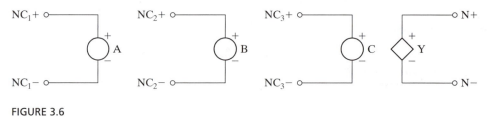

FIGURE 3.6

Polynomial source.

where

Y can be a voltage or current

A, B, and C can be a voltage or current or any combination thereof.

For a polynomial of $n = 1$ with A as the only controlling variable, the source function takes the form of

$$Y = P_0 + P_1A + P_2A^2 + P_3A^3 + P_4A^4 + \cdots + P_nA^n$$

where $P_0, P_1, \ldots, P_n$ are the coefficient values; this is written in PSpice as

```
POLY NC1+ NC1−  P₀ P₁ P₂ P₃ P₄ P₅ . . .  Pₙ
```

where NC1+ and NC1− are the positive and negative nodes, respectively, of controlling source A.

For a polynomial of degree $n = 2$ with A and B as the controlling sources, the source function takes the form of

$$Y = P_0 + P_1A + P_2B + P_3A^2 + P_4AB + P_5B^2 + P_6A^3$$
$$+ P_7A^2B + P_8AB^2 + P_9B^3 + \cdots$$

This is described in PSpice as

```
POLY(2) NC1+ NC1−  NC2+ NC2−  P₀ P₁ P₂ P₃ P₄ P₅ . . .  Pₙ
```

where NC1+, NC2+ and NC1−, NC2− are the positive and negative nodes, respectively, of the controlling sources.

For a polynomial of degree $n = 3$ with A, B, and C as the controlling sources, the source function takes the form of

$$Y = P_0 + P_1A + P_2B + P_3C + P_4A^2 + P_5AB + P_6AC + P_7B^2 + P_8BC + P_9C^2$$
$$+ P_{10}A^3 + P_{11}A^2B + P_{12}A^2C + P_{13}AB^2 + P_{14}ABC + P_{15}AC^2 + P_{16}B^3$$
$$+ P_{17}B^2C + P_{18}BC^2 + P_{19}C^3 + P_{20}A^4 + \cdots$$

This is written in PSpice as

```
POLY(3) NC1+ NC1−  NC2+ NC2−  NC3+ NC3−  P₀ P₁ P₂ P₃ P₄ P₅ . . .  Pₙ
```

where NC1+, NC2+, NC3+ and NC1−, NC2−, NC3− are the positive and negative nodes, respectively, of the controlling sources.

Typical model statements.   For $Y = 2V(10)$, the model is

```
POLY 10 0 2.0
```

For $Y = V(5) + 2[V(5)]^2 + 3[V(5)]^3 + 4[V(5)]^4$, the model is

```
POLY 5 0 0.0 1.0 2.0 3.0 4.0
```

For $Y = 0.5 + V(3) + 2V(5) + 3[V(3)]^2 + 4V(3)\,V(5)$, the model is

```
POLY(2) 3 0 5 0 0.5 1.0 2.0 3.0 4.0
```

For $Y = V(3) + 2V(5) + 3V(10) + 4[V(3)]^2$, the model is

```
POLY(3) 3 0 5 0 10 0 0.0 1.0 2.0 3.0 4.0
```

If $I(\text{VN})$ is the controlling current through voltage source V_N and $Y = I(\text{VN}) + 2[I(\text{VN})]^2 + 3[I(\text{VN})]^3 + 4[I(\text{VN})]^4$, the model is

```
POLY VN 0.0 1.0 2.0 3.0 4.0
```

If $I(\text{VN})$ and $I(\text{VX})$ are the controlling currents and $Y = I(\text{VN}) + 2I(\text{VX}) + 3[I(\text{VN})]^2 + 4I(\text{VN})I(\text{VX})$, the model is

```
POLY(2) VN VX 0.0 1.0 2.0 3.0 4.0
```

Note. If the source is of one dimension and only one coefficient is specified, as in the first example, PSpice assumes that $P_0 = 0$ and takes the specified value to be P_1. That is, $Y = 2A$.

3.6.2 Voltage-Controlled Voltage Source

The symbol of a voltage-controlled voltage source in Fig. 3.7(a) is E; it takes a linear form, as in

```
E<name> N+ N- NC+ NC- <(voltage gain) value>
```

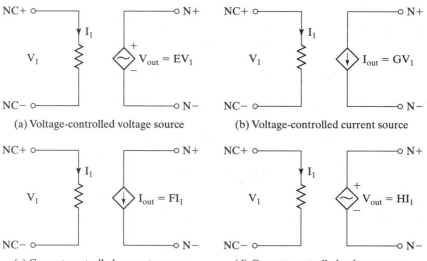

(a) Voltage-controlled voltage source

(b) Voltage-controlled current source

(c) Current-controlled current source

(d) Current-controlled voltage source

FIGURE 3.7

Dependent sources.

where $N+$ and $N-$ are the positive and negative output nodes, respectively, and NC+ and NC− are the positive and negative nodes, respectively, of the controlling voltage. The nonlinear form is

```
E<name>    N+ N- [POLY(<value>)
+              <<(+ controlling) node> <(- controlling) node>> (pairs)
+              <(polynomial coefficients) values>]
```

The POLY source was described in Section 3.6.1. The number of controlling nodes is twice the number of dimensions. A particular node may appear more than once, and the output and controlling nodes could be the same.

Typical Statements

```
EAB        1   2    4    6    1.0
EVOLT      4   7    20   22   2E5
ENONLIN    25  40   POLY(2)  3  0  5  0  0.0  1.0  1.5  1.2  1.7
E2         10  12   POLY  5  0  0.0  1.0  1.5  1.2  1.7
```

Notes

1. The source ENONLIN that specifies a polynomial voltage source between nodes 25 and 40 is controlled by $V(3)$ and $V(5)$. Its value is given by

$$Y = V(3) + 1.5V(5) + 1.2[V(3)]^2 + 1.7V(3)V(5)$$

2. The source $E2$ that specifies a polynomial voltage source between nodes 10 and 12 is controlled by $V(5, 0)$ and is given by

$$Y = V(5) + 1.5[V(5)]^2 + 1.2[V(5)]^3 + 1.7[V(5)]^4$$

3.6.3 Voltage-Controlled Current Source

The symbol of a voltage-controlled current source, as shown in Fig. 3.7(b), is G: its linear form is

```
G<name> N+ N- NC+ NC- <(transconductance) value>
```

where $N+$ and $N-$ are the positive and negative output nodes, respectively. NC+ and NC− are the positive and negative nodes, respectively, of the controlling voltage. The nonlinear form is

```
G<name>    N+ N- [POLY(<value>)
+              <<(+ controlling) node> <(- controlling) node>> (pairs)
+              <(polynomial coefficients) values>]
```

Typical Statements

```
GAB        1   2    4    6    1.0
GVOLT      4   7    20   22   2E5
GNONLIN    25  40 POLY(2)  3  0  5  0  0.0  1.0  1.5  1.2  1.7
G2         10  12 POLY  5  0  0.0  1.0  1.5  1.2  1.7
```

Notes

1. The source GNONLIN that specifies a polynomial current source from node 25 to node 40 is controlled by $V(3)$ and $V(5)$ and is given by

$$I = V(3) + 1.5V(5) + 1.2[V(3)]^2 + 1.7V(3)V(5)$$

2. The source $G2$ that specifies a polynomial current source from node 10 to node 12 is controlled by $V(5)$; it is given by

$$I = V(5) + 1.5[V(5)]^2 + 1.2[V(5)]^3 + 1.7[V(5)]^4$$

3. A nonlinear conductance can be simulated by a voltage-controlled current source. A linear voltage-controlled current source is the same as a conductance if the controlling nodes are the same as the output nodes.

   ```
   GRES 4 6 4 6 0.1
   ```

   is a conductance of 0.1 mhos with a resistance of $1/0.1 = 10\ \Omega$.

   ```
   GHMO 1 2 POLY 1 2 0.0 1.5M 1.7M
   ```

   represents

$$I = 1.5 \times 1^{-3}\, V(1,2) + 1.7 \times 10^{-3}\, [V(1,2)]^2$$

   and is a nonlinear conductance in mhos.

3.6.4   Current-Controlled Current Source

The symbol of the current-controlled current source, as shown in Fig. 3.7(c), is F, and its linear form is

```
F<name> N+ N− VN <(current gain) value>
```

where $N+$ and $N-$ are the positive and negative nodes, respectively, of the current source. VN is a voltage source through which the controlling current flows. The controlling current is assumed to flow from the positive node of VN through the voltage source VN to the negative node of VN. The current through the controlling voltage source I(VN) determines the output current. The voltage source VN that monitors the controlling current must be an independent voltage source, and it can have a **zero** or finite value. If the current through a resistor controls the source, a dummy voltage source of 0 V should be connected in series with the resistor to monitor the controlling current.

The nonlinear form is

```
F<name>    N+ N− [POLY(<value>
+          VN1, VN2, VN3.....
+          <(polynomial coefficients) values>]
```

The POLY source was described in Section 3.6.1. The number of controlling current sources must be equal to the number of dimensions.

Typical Statements

```
FAB        1    2    VIN   10
FAMP       13   4    VCC   50
FNONLIN    25   40   POLY  VN   0.0   1.0   1.5   1.2   1.7
```

Note. The source FNONLIN that specifies a polynomial current source from node 25 to node 40 is given by

$$I = I(\text{VN}) + 1.5[I(\text{VN})]^2 + 1.2[I(\text{VN})]^3 + 1.7[I(\text{VN})]^4$$

3.6.5 Current-Controlled Voltage Source

The symbol of a current-controlled voltage source, as shown in Fig. 3.7(d), is H, and its linear form is

```
H<name> N+ N− VN <(transresistance) value>
```

where $N+$ and $N-$ are the positive and negative nodes, respectively, of the voltage source. VN is a voltage source through which the controlling current flows, and its specification is similar to that for a current-controlled current source.

The nonlinear form is

```
H<name>    N+ N−  [POLY(<value>)
+          VN1, VN2, VN3 ...
+          <(polynomial coefficients) values>]
```

Typical Statements

```
HAB        1    2    VIN   10
HAMP       13   4    VCC   50
HNONLIN    25   40   POLY  VN   0.0   1.0   1.5   1.2   1.7
```

Notes

1. The source HNONLIN that specifies a polynomial voltage source between nodes 25 and 40 is controlled by $I(\text{VN})$ and is given by

$$V = I(\text{VN}) + 1.5[I(\text{VN})]^2 + 1.2[(\text{VN})]^3 + 1.7[I(\text{VN})]^4$$

2. A nonlinear resistance can be simulated by a current-controlled voltage source. A linear current-controlled voltage source is the same as a resistor if the controlling current is the same as the current through the voltage between output nodes.

```
HRES 4 6 VN 10
```

is a resistance of 10 Ω.

```
HMHO 1 2 POLY VN 0.0 1.5M 1.7M
```

represents

$$H = 1.5 \times 1^{-3} I(\text{VN}) + 1.7 \times 10^{-3} [I(\text{VN})]^2$$

and is a nonlinear resistance in ohms.

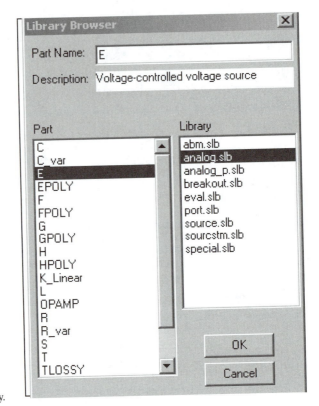

FIGURE 3.8

Analog library.

3.6.6   Schematic Dependent Sources

The PSpice analog library *analog.slb* is shown in Fig. 3.8. The voltage-controlled voltage source (E), the current-controlled current source (F), the voltage-controlled current source (G), and the current-controlled voltage source (H) of the PSpice library are shown in Fig. 3.9(a), (b), (c), and (d).

3.7   DC OUTPUT VARIABLES

PSpice has some unique features for printing or plotting output voltages or currents. The output variables can be divided into two types: voltage output and current output. An output variable can be assigned the symbol of a device (or element) to identify whether the output is the voltage across the device or current through the device (or element). Table 3.3 shows the symbols of two-terminal elements. Table 3.4 shows the symbols and terminal symbols of three- or four-terminal devices.

3.7.1   Voltage Output

For dc sweep and transient analysis (discussed in Chapter 4), the output voltages can be obtained by the following statements:

V($\langle$node$\rangle$)	Voltage at $\langle$node$\rangle$ with respect to ground
V(N1, N2)	Voltage at node N_1 with respect to node N_2

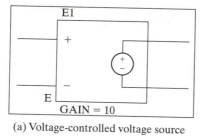

(a) Voltage-controlled voltage source   (b) Current-controlled current source

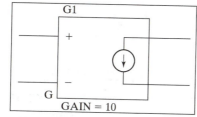

(c) Voltage-controlled current source   (d) Current-controlled voltage source

FIGURE 3.9

PSpice-dependent sources.

V(⟨name⟩)	Voltage across two-terminal device, ⟨name⟩
Vx(⟨name⟩)	Voltage at terminal x of three-terminal device, ⟨name⟩
Vxy(⟨name⟩)	Voltage across terminals x and y of three-terminal device, ⟨name⟩
Vz(⟨name⟩)	Voltage at port z of transmission line, ⟨name⟩

Variables	**Meaning**
V(5)	Voltage at node 5 with respect to ground
V(4,2)	Voltage of node 4 with respect to node 2
V(R1)	Voltage of resistor R_1, where the first node (as defined in the circuit file) is positive with respect to the second node
V(L1)	Voltage of inductor L_1, where the first node (as defined in the circuit file) is positive with respect to the second node
V(C1)	Voltage of capacitor C_1, where the first node (as defined in the circuit file) is positive with respect to the second node
V(D1)	Voltage across diode D_1 where the anode is positive with respect to cathode
VC(Q3)	Voltage at the collector of transistor Q_3 with respect to ground
VDS(M6)	Drain-source voltage of MOSFET M_6
VB(T1)	Voltage at port B of transmission line T_1

Note. SPICE and some versions of PSpice do not permit measuring voltage across a resistor, an inductor, and a capacitor, for example, V(R1), V(L1), and V(C1). This type of statement is used only for outputs by .PLOT and .PRINT commands.

TABLE 3.3   Symbols for Two-Terminal Elements

First letter	Element
C	Capacitor
D	Diode
E	Voltage-controlled voltage source
F	Current-controlled current source
G	Voltage-controlled current source
H	Current-controlled voltage source
I	Independent current source
L	Inductor
R	Resistor
V	Independent voltage source

TABLE 3.4   Symbols and Terminal Symbols for Three- or Four-Terminal Devices

First letter	Device	Terminals
B	GaAs MESFET	D (Drain)
		G (Gate)
		S (Source)
J	JFET	D (Drain)
		G (Gate)
		S (Source)
M	MOSFET	D (Drain)
		G (Gate)
		S (Source)
		B (Bulk, substrate)
Q	BJT	C (Collector)
		B (Base)
		E (Emitter)
		S (Substrate)
T	Transmission line	A (Input port)
		B Output port
Z	IGBT	G(Gate)
		C(Collector)
		E(Emitter)

3.7.2   Current Output

For dc sweep and transient analysis (discussed in Chapter 4), the output currents can be obtained by the following statements:

I($\langle$name$\rangle$)	Current through $\langle$name$\rangle$
Ix($\langle$name$\rangle$)	Current into terminal x of $\langle$name$\rangle$
Iz($\langle$name$\rangle$)	Current at port z of transmission line $\langle$name$\rangle$

Variables	**Meaning**
I(VS)	Current flowing into dc source V_S
I(R5)	Current flowing into resistor R_5, where the current is assumed to flow from the first node (as defined in the circuit file) through R_5 to the second node

I(D1)	Current into diode D_1
IC(Q4)	Current into the collector of transistor Q_4
IG(J1)	Current into gate of JFET J_1
ID(M5)	Current into drain of MOSFET M_5
IA(T1)	Current at port A of transmission line T_1

Note. SPICE and some versions of PSpice do not permit measuring the current through a resistor, for example, I(R5). The easiest way is to add a dummy voltage source of 0 V (say, $V_X = 0$ V) and to measure the current through that source, for example, I(VX).

Example 3.1:   Defining the output variables of a BJT circuit

For the bipolar junction transistor (BJT) circuit in Fig. 3.10, write the various currents and voltages in forms that are allowed by PSpice. The dc sources of 0 V are introduced to measure currents I_1 and I_2.

Solution

PSpice	VARIABLES	
I_B	IB(Q1)	Base current of transistor Q_1
I_C	IC(Q1)	Collector current of transistor Q_1
I_E	IE(Q1)	Emitter current of transistor Q_1
I_S	I(VCC)	Current through voltage source V_{CC}
I_1	I(VX)	Current through voltage source V_X
I_2	I(VY)	Current through voltage source V_Y
V_B	VB(Q1)	Voltage at the base of transistor Q_1
V_C	VC(Q1)	Voltage at the collector of transistor Q_1
V_E	VE(Q1)	Voltage at the emitter of transistor Q_1
V_{CE}	VCE(Q1)	Collector-emitter voltage of transistor Q_1
V_{BE}	VBE(Q1)	Base-emitter voltage of transistor Q_1

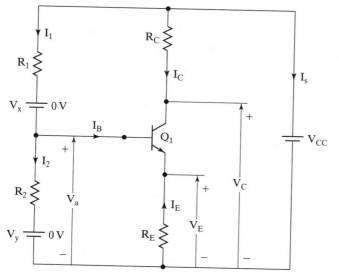

FIGURE 3.10

Bipolar junction transistor circuit.

3.8   TYPES OF OUTPUT

The commands that are available to get output from the results of simulations are as follows:

```
.PRINT      Print
.PLOT       Plot
.PROBE      Probe
Probe Output
.WIDTH      Width
```

3.8.1   .PRINT (Print Statement)

The results from dc analysis can be obtained in the form of tables. The print statement for dc outputs takes the form

```
.PRINT DC [output variables]
```

The maximum number of output variables is eight in any .PRINT statement. However, more than one .PRINT statement can be used to print all of the desired output variables.

The values of the output variables are printed as a table with each column corresponding to one output variable. The number of digits for output values can be changed by the NUMDGT option on the .OPTIONS statement. (See Section 6.10.) The results of the .PRINT statement are stored in the output file. An example of a print statement is

```
.PRINT DC V(2), V(3,5), V(R1), VCE(Q2), I(VIN), I(R1), IC(Q2)
```

Note. Having two .PRINT statements for the same variables will not produce two tables. PSpice will ignore the first statement and produce output for the second statement.

3.8.2   .PLOT (Plot Statement)

The results from dc analysis can also be obtained in the form of line printer plots. The plots are drawn by using characters, and the results can be obtained from any kind of printer. The plot statement for dc outputs takes the following form:

```
.PLOT DC <output variables>
+ [<(lower limit) value>, <(upper limit) value>]
```

The maximum number of output variables is eight in any .PLOT statement. More than one .PLOT statement can be used to plot all the desired output variables.

The range and increment of the *x*-axis is fixed by the dc analysis command. The range of the *y*-axis is set by adding [⟨(lower limit) value⟩, ⟨(upper limit) value⟩] at the end of a .PLOT statement. The *y*-axis range [⟨(lower limit) value⟩, ⟨(upper limit) value⟩] can be placed in the middle of a set of output variables. The output variables will follow the specified range that comes immediately to the right.

If the *y*-axis range is omitted, PSpice assigns a default range determined by the range of the output variable. If the ranges of output variables vary widely, PSpice assigns the ranges corresponding to the different output variables.

Plot Statements

```
.PLOT DC V(2), V(3,5), V(R1), VCE(Q2), I(VIN),  I(R1), IC(Q2)
.PLOT DC V(5) V(4,7) (0, 10V) IB(Q1) (0, 50MA) IC(Q1) (-50MA, 50MA)
```

Notes. In the first statement, the *y*-axis is by default. In the second statement, the range for voltages V(5) and V(4,7) is 0 V to 10 V, that for current IB(Q1) is 0 MA to 50 MA, and that for the current IC(Q1) is −50 MA to 50 MA.

3.8.3   .PROBE (Probe Statement)

Probe is a graphics postprocessor/waveform analyzer for PSpice and is available as an option for the professional version of PSpice. However, Probe comes with the student version of PSpice. The simulation results cannot be used directly by Probe. First, the results have to be processed by the .PROBE command, which writes the processed data on a file, PROBE.DAT, for use by Probe. The command takes one of these forms:

```
.PROBE
.PROBE <one or more output variables)
```

In the first form, where no output variable is specified, the .PROBE command writes all the node voltages and all the element currents into the PROBE.DAT file. The element currents are written in the forms that are permitted as output variables (discussed in Section 3.7).

In the second form, where the output variables are specified, PSpice writes only the specified output variables to the PROBE.DAT file. This form is suitable for users without a fixed disk to limit the size of the PROBE.DAT file.

Probe Statements

```
.PROBE
.PROBE V(5), V(4, 3), V(C1), VM(2), I(R2), IB(Q1), VBE(Q1)
```

3.8.4   Probe Output

Once the results of the simulations are processed by the .PROBE command, the results are available for graphical displays and can be further manipulated through expressions. Probe comes with a first menu, as shown in Fig. 3.11, that allows one to choose the type of analysis. After the first choice, the second level is the choice for the plots and coordinates of output variables, as shown in Fig. 3.12. After the choices are made, the output is displayed as shown in Fig. 3.13. As just illustrated, it is very easy to use Probe.

Probe disregards case for letters; that is, V(8) and v(8) are equivalent. However, there is an exception with the scale suffix *m*, and one should be careful about "milli" and "mega"; *m* means "milli" (1E − 3), but *M* means "mega" (1E + 6). The suffixes MEG and MIL are not acceptable. The following units are recognized by Probe:

V	Volts
A	Amps
W	Watts
d	Degrees (of phase)
s	Seconds
H	Hertz

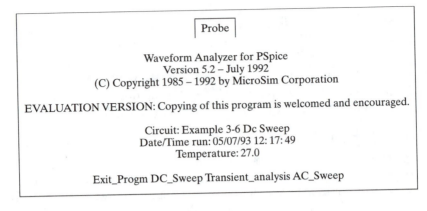

FIGURE 3.11

Select analysis menu for Probe.

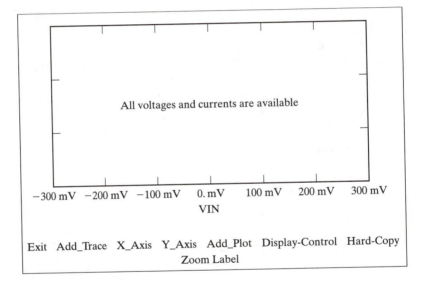

FIGURE 3.12

Select plot/graphics output.

If units are omitted, Probe recognizes that W = V * A, V = W/A, and A = W/V and gives the results in appropriate units.

The arithmetic expressions of common output variables use the following functions and the operators +, −, *, /, along with parentheses:

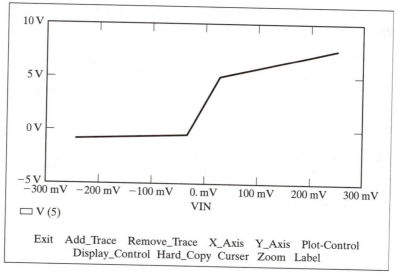

FIGURE 3.13

Output display.

PROBE FUNCTION	MEANING
ABS(x)	$\|x\|$ (absolute value)
B(Kxy)	Flux density of coupled inductor Kxy
H(Kxy)	Magnetization of coupled inductor Kxy
SGN(x)	$+1$ (if $x > 0$), 0 (if $x = 0$), -1 (if $x < 0$)
EXP(x)	e^x
DB(x)	$20 \log(\|x\|)$ (log of base 10)
LOG(x)	$\ln(x)$ (log of base e)
LOG10(x)	$\log(x)$ (log of base 10)
PWR(x,y)	$\|x\|^y$
SQRT(x)	$x^{1/2}$
SIN(x)	$\sin(x)$ (x in radians)
COS(x)	$\cos(x)$ (x in radians)
TAN(x)	$\tan(x)$ (x in radians)
ARCTAN(x)	$\tan^{-1}(x)$ (result in radians)
d(y)	Derivative of y with respect to the x-axis variable
s(y)	Integral of y over the x-axis variable
AVG(x)	Running average of x
RMS(x)	Running RMS of x

If a dc voltage source VS is connected between nodes 2 and 3, the plot of the trace (or curve)

```
V(2, 3) * I(VS)
```

will give the power delivered by the source VS.

Two or more traces can be added to the same plot by typing the variables, for example,

```
V(8) dV(8) s(V(8)) AVG(V(8)) RMS(V(8)) V(2,3) * I(VS)
```

This is faster than adding one plot at a time.

If the analysis is completed for more than one parameter or sweep variable, the PROBE.DAT file can contain all the results. If V(8) is calculated for three parameter values, the expression

```
V(8))
```

will give three curves instead of the usual one curve. One could specify a particular trace by the expression

```
V(8) @n
```

which will give the curve of V(8) for the nth transient analysis. The difference between mth curve and nth curve can be obtained by the expression

```
v(8) @m-V(8) @n
```

Notes

1. The .PROBE command requires a math coprocessor for the professional version of PSpice, but not for the student version.
2. Probe is not available on SPICE. However, the newest version of SPICE (SPICE3) has a postprocessor similar to Probe called Nutmeg.
3. It is required that the type of display and the type of hard-copy devices in the PROBE.DEV file be specified as follows:

   ```
   Display = <display name>

   Hard copy = <port name>, <device name>
   ```

   The details of names for display, port, and device (printer) can be found in the README.DOC file that comes with the PSpice program or in the PSpice manual.
4. The display and hard-copy devices can be set from the Display/Printer Setup menu.

3.8.5   .WIDTH (Width Statement)

The width of the output in columns can be set by the .WIDTH statement, which has the general form of

```
.WIDTH OUT=<value>
```

The ⟨value⟩ is in columns and must be either 80 or 132. The default value is 80.

3.9   TYPES OF DC ANALYSIS

In dc analysis, all the independent and dependent sources are dc types. If inductors and capacitors are present in a circuit, they are considered short circuits and open circuits, respectively, because at zero frequency, the impedance represented by an inductor is zero and that of a capacitor is infinite. The commands that are commonly used for dc analysis are

.OP	Dc operating point
.TF	Small-signal transfer function
.DC	Dc sweep
.PARAM	Dc parametric sweep

In the PSpice schematic, these commands are invoked in the analysis setup by the setup menu as shown in Fig. 3.14.

3.9.1   .OP (Operating Point)

Electronic and electrical circuits contain nonlinear devices (e.g., diodes, transistors), whose parameters depend on the *operating point*. The operating point is also known as a *bias point* or *quiescent point*.

Let us consider the circuit of Fig. 3.15(a), where the voltage across the resistor R_2 depends upon its current. Using Kirchhoff's voltage law (KVL), the current I_S can be expressed as

$$V_S = R_1 I_S + V_2$$

or

$$I_S = -\frac{V_2}{R_1} + \frac{V_S}{R_1}, \qquad (3.1)$$

which is the equation of a straight line.

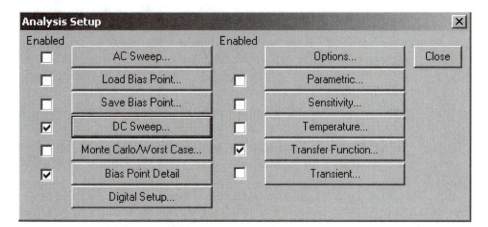

FIGURE 3.14

Setup for .OP, .TF, and .DC commands.

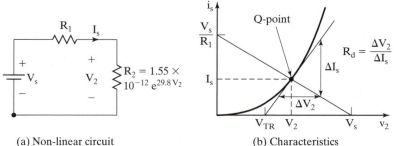

(a) Non-linear circuit (b) Characteristics

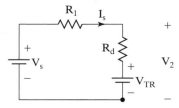

(c) Small-signal representation

FIGURE 3.15

Dc circuit with a nonlinear resistance.

Let us assume that the current I_S is related to the voltage V_2 by

$$I_S = 1.55 \times 10^{-12} e^{29.8V_2} \quad \text{for} \quad V_2 > 0. \tag{3.2}$$

I_S depends on V_2, which in turn depends on I_S. The intersection of the two curves defined by Eqs. (3.1) and (3.2) gives the operating point or quiescent point (Q-point) as shown in Fig. 3.15(b). If V_S changes by a small amount, then the operating point will also change. The slope of the tangent at the Q-point will be a measure of this change, and will represent a small-signal resistance R_d, so that $R_d = \Delta V_2 / \Delta I_S$. R_2 can be represented by a small-signal resistance R_d and a resistance that corresponds to the fixed threshold voltage V_{TR}. This type of representation is known as a *piecewise linear model* and is shown in Fig. 3.15(c).

The operating point is always calculated by PSpice in calculating the small-signal parameters of nonlinear devices during the dc sweep and transfer-function analysis. The command takes the form

.OP

The .OP command controls the output of the bias point, but not the method of bias analysis. If the .OP command is omitted, PSpice prints only a list of the node voltages.

If the .OP command is present, PSpice prints the currents and power dissipations of all the voltage sources. The small-signal parameters of all nonlinear controlled sources and all the semiconductor devices are also printed.

The .OP command has no effect in circuits with linear elements. However, in circuits with nonlinear devices (e.g., diodes and transistors), the .OP command prints the small-signal model parameters of the devices.

Example 3.2:   Varying the model parameter of resistors

Repeat Example 2.1 with the values of R_1 and R_2 increased by +5% and those of R_3 and R_4 decreased by 10%.

Solution   Since there are two types of resistances, we will use two models: RMOD1 with $R - 1.05$ and RMOD2 with $R - 0.9$. The PSpice schematic is shown in Fig. 3.16(a). The dc bias detail is set at the analysis setup and the dc values can be displayed by setting the display option of the analysis menu as shown in Fig. 3.16(b). A library file that contains the model statements of RMOD1 and RMOD2 is created and saved in a library file called Rashid_Model.LIB:

.MODEL	RMOD1	RES $(R = 1.05)$
.MODEL	RMOD2	RES $(R = 0.90)$

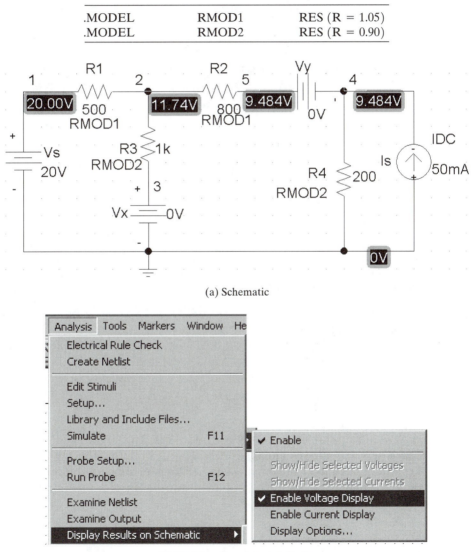

(a) Schematic

(b) Enable voltage/current display

FIGURE 3.16

PSpice schematic for Example 3.2.

The listing of the circuit file follows.

Example 3.2 Simple dc circuit

```
▲ VS   1   0   DC   20V      ;   Dc voltage source of 20 V
  IS   0   4   DC   50MA     ;   Dc current source of 2 mA
▲ ▲ R1   1   2      RMOD1  500     ; Resistance of 500 ohms with model RMOD1
  R2   2   5      RMOD1  800     ; Resistance of 800 ohms with model RMOD1
  .MODEL  RMOD1   RES (R = 1.05) ; Model for R1 and R2
  R3   2   3      RMOD2 1KOHM    ; Resistance of 1 kilohms with model RMOD2
  R4   4   0      RMOD2 200      ; Resistance of 200 ohms with model RMOD2
  .MODEL  RMOD2   RES (R = .9)   ; Model for R1 and R2
  VX   3   0      DC   0V        ; Measures current through R3
  VY   5   4      DC   0V        ; Measures current through R2
▲ ▲ ▲ *. OP                     ; Prints small-signal model parameters
  .END                          ;End of circuit file
```

The results that are obtained by printing the contents of output file EX3-2.OUT follow:

```
****      SMALL-SIGNAL BIAS SOLUTION        TEMPERATURE = 27.000 DEG C
NODE    VOLTAGE     NODE    VOLTAGE     NODE    VOLTAGE     NODE    VOLTAGE
(   1)    20.0000  (   2)   11.7410  (   3)    0.0000  (   4)    9.4836
(   5)    9.4836
     VOLTAGE SOURCE CURRENTS
     NAME      CURRENT        CALCULATED
     VS      -1.573E-02      -15.73 mA
     VX       1.305E-02       13.05 mA
     VY       2.687E-03       2.6869 mA
     TOTAL POWER DISSIPATION   3.15E-01 WATTS
```

Note. Students are encouraged to run Example 3.2 with the .OP command in order to see its effect.

3.9.2   .TF (Small-Signal Transfer Function)

The small-signal transfer function capability of PSpice can be used to compute the small-signal dc gain, the input resistance, and the output resistance of a circuit. If $V(1)$ and $V(4)$ are the input and output variables, respectively, PSpice will calculate the small-signal dc gain between nodes 1 and 4, defined by

$$A_v = \frac{\Delta V_{\text{out}}}{\Delta V_{\text{in}}} = \frac{V(4)}{V(1)}$$

as well as the input resistance between nodes 1 and 0 and the small-signal dc output resistance between nodes 4 and 0.

PSpice calculates the small-signal dc transfer function by linearizing the circuit around the operating point. The statement for the transfer function has one of the following forms:

```
.TF   VOUT   VIN
.TF   IOUT   IIN
```

where VIN is the input voltage and VOUT (or IOUT) is the output voltage (or current). If the output is a current, then that current must be through a voltage source. The output variable, VOUT or IOUT, has the same format and meaning as in a .PRINT statement. If there are inductors and capacitors in a circuit, the inductors are treated as short circuits and the capacitors as open circuits.

The .TF command calculates the parameters of Thévenin's (or Norton's) equivalent circuit for the circuit file. It automatically prints the output and does not require .PRINT, .PLOT, or .PROBE statements.

Statements for Transfer-Function Analysis

```
.TF V(2,4) VIN ; VIN is the input, and V(2, 4) is the output.
.TF V(10)  IIN ; IIN is the input, and V(10) is the output.
.TF I(VX)  IIN ; IIN is the input, and the current through VX is the output.
.TF I(VX)  VIN ; VIN is the input, and the current through VX is the output.
```

Example 3.3: Finding the Thevenin's equivalent

A dc circuit is shown in Fig. 3.17. Use PSpice to calculate and print (a) the voltage gain $A_v = V(2,4)/V_{in}$, (b) the input resistance $R_{in} = V_{in}/I_{in}$, (c) Thévenin's (output) resistance $R_{out} = R_{Th}$ between nodes 2 and 4, and (d) Thévenin's voltage V_{Th} between nodes 2 and 4.

Solution The output voltage $V(2,4)$ is between nodes 2 and 4. The .TF command can calculate and print the dc gain, the input resistance, and the output resistance. The voltage source V_x in Fig. 3.17 acts as an ammeter, and an independent source of 0 V is normally connected to measure a current. The PSpice schematic is shown in Fig. 3.18(a). The .TF command is set in the analysis setup. The output variable and the input source are identified in the TF set menu as shown in Fig. 3.18(b). The listing of the circuit file follows.

```
Example 3.3 Thévenin's analysis
▲ VIN  1   0  DC  10V ; Voltage source of 10 V dc
  IS   4   3  DC  2A  ; Current source of 2 A dc
▲ ▲ VX  4  5  DC  OV  ; Measures the current through R5
    R1  1  2  5
    R2  2  3 10
    R3  2  0 20
    R4  3  4 40
    R5  5  0 10
▲ ▲ ▲ .TF  V(2,4)  VIN ; Transfer-function analysis
 .END                  ; End of circuit file
```

The results are obtained by printing the contents of output file EX3-3.OUT. PSpice always prints the small-signal bias solutions to the output file, and they are as follows:

```
****    SMALL-SIGNAL BIAS SOLUTION        TEMPERATURE = 27.000 DEG C
NODE   VOLTAGE     NODE  VOLTAGE       NODE   VOLTAGE     NODE  VOLTAGE
(  1)   10.0000 (    2)  12.5000     (   3)    23.7500 (    4) −11.2500
(  5)  −11.2500
      VOLTAGE SOURCE CURRENTS
      NAME    CURRENT
      VIN    5.000E−01
      VX     −1.125E+00
      TOTAL POWER DISSIPATION −5.00E+00 WATTS
```

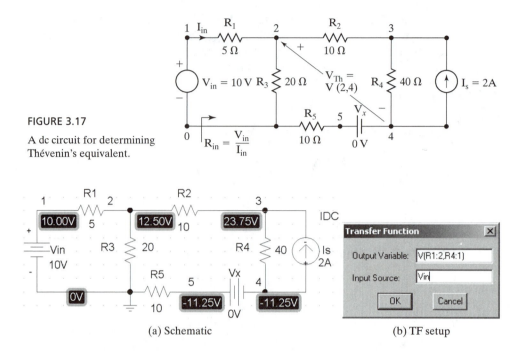

FIGURE 3.17

A dc circuit for determining
Thévenin's equivalent.

(a) Schematic (b) TF setup

FIGURE 3.18

PSpice Schematic for Example 3.3.

PSpice prints the results of .TF command to the output file, and they are as follows:

```
****      SMALL-SIGNAL CHARACTERISTICS
     V(2, 4)/VIN = 6.250E-01                    A_v   = 0.625
     INPUT RESISTANCE AT VIN = 2.000E+01        R_in  = 20 Ω
     OUTPUT RESISTANCE AT V(2, 4) = 1.094E+01   R_Th  = 10.94 Ω
     JOB CONCLUDED
     TOTAL JOB TIME           1.05
```

Thus, Thévenin's voltage V_{Th_1} is $A_v V_{in} = 0.625 \times 10 = 6.25$ V for V_{in} only.

Thevenin's output voltage for I_S only is $V_{Th2} = 17.5$ V. Thus, the total output voltage is $V_{Th} = V_{Th1} + V_{Th2} = 6.25 + 17.5$ V $= 23.75$ V when both V_{in} and I_S are applied at the same time. The Thevenin's output resistance is given by

$$R_{Th} = (R_2 + R_4) \| [(R_1 \| R_3) + R_5]$$
$$= (10 + 40) \| [(5 \| 20) + 10] = 10.94 \text{ k}\Omega.$$

Note

1. The transfer function analysis gives the gain with respect to the only input (e.g., the specified input V_{in}). However, PSpice calculates the node voltages with all sources. For example, $V(2) = 12.5$ V, $V(4) = -11.25$ V, and $V_{Th} = V(2) - V(4) = 12.5 - (-11.25) = 23.75$ V.

2. One should be very careful about identifying the output variable in PSpice Schematics. During the rotations of the components, their orientations are also changed. When a component such as a resistor R is placed on the schematic drawing, the left hand is the terminal "1" (R:1), and right-hand side is the terminal "2" (R:2). The positive current flows from the terminal 1 to the terminal 2.

Example 3.4:   Transfer function analysis of a BJT amplifier

An amplifier circuit is shown in Fig. 3.19. Calculate and print (a) the voltage gain $A_v = V(5)/V_{in}$; (b) the input resistance, R_{in}; and (c) the output resistance, R_{out}.

Solution   The .TF command can calculate and print the dc gain, the input resistance, and the output resistance. The PSpice schematic is shown in Fig. 3.20(a). The .TF command is set in the analysis setup. The output variable and the input source are identified in the TF set menu as shown in Fig. 3.20(b). The listing of the circuit file follows.

Example 3.4 Transfer-function analysis

```
▲ VIN  1 0 DC IV ; Dc input voltage of 1 V
▲ ▲ R1 1 2 1K
    R2 2 0 20K
    RP 2 6 1.5K
    RE 3 0 250
    F1 4 3 VX     40  ; Current-controlled current source
    RO 4 3 100K
    RC 4 5 2K
    VX 6 3 DC     OV  ; Measures the current through R_p
    VY 5 0 DC     OV  ; Measures the current through R_L
▲ ▲ ▲ .TF V(4) VIN   ; Transfer-function analysis
.END                 ; End of circuit file
```

The results are obtained by printing the contents of output file EX3-4.OUT. PSpice always prints automatically the small-signal bias solutions to the output file, which are as follows:

```
****       SMALL-SIGNAL BIAS SOLUTION       TEMPERATURE =   27.000  DEG C
NODE     VOLTAGE       NODE    VOLTAGE    NODE    VOLTAGE    NODE   VOLTAGE
(   1)    1.0000   (    2)     .8797  (    3)    .7653  (    4)   -5.9695
(   5)    0.0000   (    6)     .7653
    VOLTAGE SOURCE CURRENTS
    NAME          CURRENT
    VIN         -1.203E-04
    VX           7.630E-05
    VY          -2.985E-03
    TOTAL POWER DISSIPATION 1.20E-04 WATTS
```

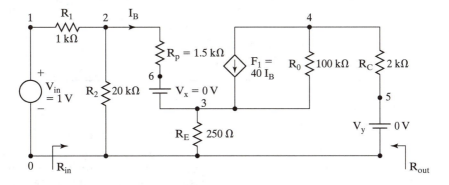

FIGURE 3.19

An amplifier circuit for determining Thévenin's equivalent.

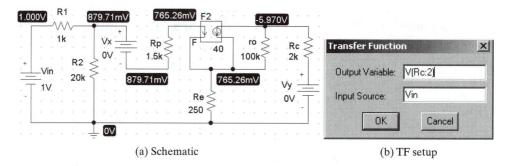

(a) Schematic   (b) TF setup

FIGURE 3.20

PSpice Schematic for Example 3.4.

The .TF command calculates the small-signal characteristics and sends the results automatically to the output file. The results of .TF command are as follows:

```
****      SMALL-SIGNAL CHARACTERISTICS
          V(4)/VIN = -5.969E+00                    Aᵥ  = -5.969
          INPUT RESISTANCE AT VIN = 8.313E+03      Rᵢₙ = 8.313 kΩ
          OUTPUT RESISTANCE AT V(4) = 1.992E+03    R_Th = 1.992 kΩ
          JOB CONCLUDED
          TOTAL JOB TIME              .99
```

Thus, Thévenin's voltage V_{Th} is $A_v V_{\text{in}} = -5.969 \times 1 = -5.969$ V.

The input resistance is given by

$$R_{\text{in}} = R_1 + R_2 \| [R_p + (1 + \beta_f)R_E]$$

$$= 1k + 20k \| [1.5k + (1 + 40) \times 250] = 8.4 \text{ k}\Omega.$$

The voltage gain is given by

$$A_v = \frac{-\beta_f R_C}{R_p + (1 + \beta_f)R_E} \times \frac{R_2 \| [R_p + (1 + \beta_f)R_E]}{R_{\text{in}}}$$

$$= \frac{-40 \times 2k}{1.5k + (1 + 40) \times 250} \times \frac{2k \| [1.5k + (1 + 40) \times 250]}{8.4k} = -5.998 \text{ V/V}.$$

The output resistance is approximately given by

$$R_o \approx R_C = 2 \text{ k}\Omega.$$

3.9.3   .DC (Dc Sweep)

The dc sweep is also known as the *dc transfer characteristic*. The input variable is varied over a range of values. For each value of the input variable, the dc operating point and the small-signal dc gain are computed by calling the small-signal transfer function capability of PSpice. The dc sweep (or dc transfer characteristic) is obtained by repeating the calculations of the small-signal transfer function for a set of values. An example of

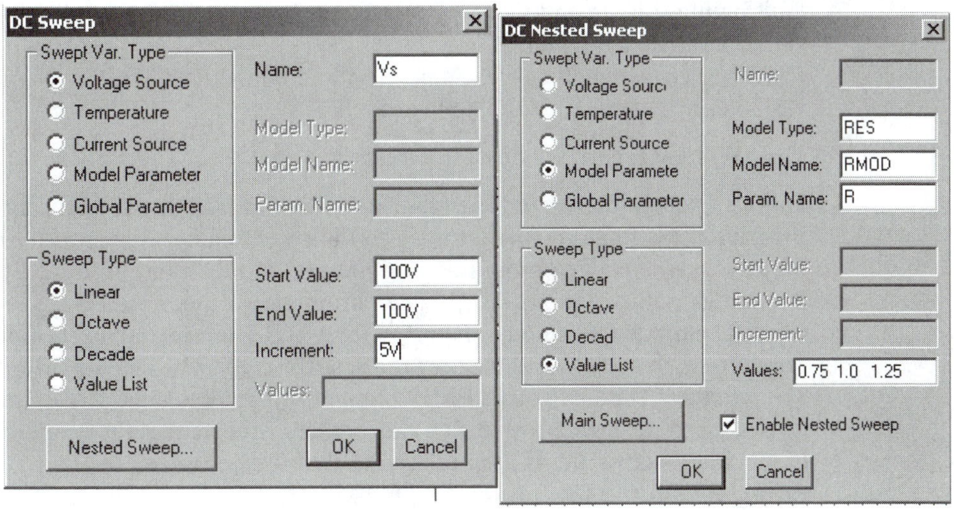

(a) Main DC sweep   (b) Nested sweep

FIGURE 3.21

DC sweep in PSpice schematic.

the dc sweep setup in PSpice schematic is shown in Fig. 3.21(a), and the nested sweep is shown in Fig. 3.21(b). The statement for performing the dc sweep takes one of the following general forms:

```
.DC     LIN     SWNAME SSTART SEND SINC
+               [(nested sweep specification)]
.DC     OCT     SWNAME  SSTART  SEND  NP
+               [(nested sweep specification)]
.DC     DEC     SWNAME  SSTART  SEND  NP
+               [(nested sweep specification)]
.DC     SWNAME     LIST <value>
+               [(nested sweep specification)]
```

SWNAME is the sweep variable name and could be either a voltage or a current source. SSTART, SEND, and SINC are the start value, the end value, and the increment value of the sweep variable, respectively. The sweep increment SINC must be positive; it must not be zero or negative. NP is the number of steps. LIN, OCT, or DEC specifies the type of sweep, as follows:

LIN (*Linear sweep*): SWNAME is swept linearly from SSTART to SEND. SINC is the step size. If LIN is omitted, SPICE assumes a linear sweep by default.

OCT (*Sweep by octave*): SWNAME is swept logarithmically by octave, and NP becomes the number of steps per octave. The next variable is generated by multiplying the present value by a constant larger than unity. OCT is used if the variable range is wide.

DEC (*Sweep by decade*): SWNAME is swept logarithmically by decade, and NP becomes the number of steps per decade. The next variable is generated by multiplying the present value by a constant larger than unity. DEC is used if the variable range is the widest.

LIST (*List of values*): There are no start and end values. The values of the sweep variables are listed after the keyword LIST.

The SWNAME can be one of the following types:

Source: The name of an independent voltage or current source. During the sweep, the source's voltage or current is set to the sweep value.

Model Parameter: The model name type and model name followed by a model parameter name in parentheses. The parameter in the model is set to the sweep value. The model parameters L and W for a MOS device and any temperature parameters such as TC1 and TC2 for the resistor *cannot* be swept.

Temperature: The keyword TEMP followed by the keyword LIST. The temperature is set to the sweep value. For each value of the sweep, the model parameters of all circuit components are updated to that temperature.

Global Parameter: The keyword PARAM followed by a parameter name. The parameter is swept. During the sweep, the global parameter's value is set to the sweep value, and all expressions are evaluated.

The dc sweep can be nested, similar to a DO loop within a DO loop in FORTRAN programming. The first sweep is the inner loop, and the second sweep is the outer loop. The first sweep is completed for each value of the second sweep. The nested sweep specification follows the same rules as the main sweep variable.

PSpice does not print or plot any output by itself for the dc sweep; the results of dc sweep are obtained by .PRINT, .PLOT, or .PROBE statements. Probe allows nested sweeps to be displayed as a family of curves.

Statements for the Dc Sweep

```
.DC VIN -5V 10V 0.25V              ; Sweeps the voltage VIN linearly.
.DC LIN IIN 50MA -50MA 1MA         ; Sweeps the current IIN linearly.
.DC VA 0 15V 0.5V IA 0 1MA 0.05MA  ; Sweeps the current IA linearly
                                     within the linear sweep of VA.
.DC RES RMOD(R) 0.9 1.1 0.001      ; Sweeps linearly the model parameter
                                     R of the resistor model RMOD.
.DC DEC NPN QM(IS) 1E-18 1E-14 10  ; Sweeps with a decade increment the
                                     parameter IS of the NPN transistor.
.DC TEMP LIST 0 50 80 100 150      ; Sweeps the temperature TEMP using
                                     the listed values.
.DC PARAM Vsupply -15V 15V 0.5V    ; Sweeps linearly the parameter PARAM
                                     Vsupply.
```

Notes

1. If the source has a dc value, its value is set by the sweep, overriding the dc value.
2. In the third statement, the current source IA is the inner loop, and the voltage source VA is the outer loop. PSpice will vary the value of the current source IA from 0 to 1 mA with an increment of 0.05 mA for each value of voltage source VA and generate an entire print table or plot for each value of voltage sweep.

3. The sweep-start value SSTART may be greater than or less than the sweep-end value SEND.
4. The sweep increment SINC must be greater than zero.
5. The number of points NP must be greater than zero.
6. After the dc sweep is finished, the sweep variable is set back to the value it had before the sweep started.

Example 3.5:   Transfer function analysis for finding the Thevenin's equivalent

A dc circuit with controlled sources is shown in Fig. 3.22. Use PSpice to calculate all node voltages and branch currents and Thévenin's equivalent circuit between nodes 2 and 5.

Solution   We can use (1) the .DC command to calculate all voltages and currents and (2) the .PRINT command to print the results of the dc analysis. The .TF command will calculate and print Thévenin's equivalent voltage and resistances along with all node voltages and currents through all independent voltage sources. The PSpice schematic is shown in Fig. 3.23(a). By enabling the voltage and current displays, the node voltages and branch currents are also displayed. The .TF command is set in the analysis setup. The output variable and the input source are identified in the TF setup menu as shown in Fig. 3.23(b). The circuit file follows:

```
Example 3.5  Dc circuit with controlled dc sources
▲ VS    1  0  DC   100V     ; Voltage source of 100 V dc
  IS    0  3  DC   5A       ; Current source of 5 A dc
▲ ▲ R1  1  2    10
    R2  2  5    20
    R3  3  0    50
    R4  3  4    40
    VX  5  3  DC  OV        ; Measures current through R2
    E1  4  0  1    0   0.5  ; Voltage-controlled voltage source
    F1  0  2  VS       0.5  ; Current-controlled current source
    G1  4  3  1    0   0.1  ; Voltage-controlled current source
    H1  1  3  VX       2    ; Current-controlled voltage source
▲ ▲ ▲ .DC  VS     100V    100V    5V         ; Only one dc sweep value
      .PRINT  DC   I(R1)   I(R2)   I(R3)   I(R4)  ; Prints branch currents
      .PRINT  DC   V(1)    V(2)    V(3)    V(4)  V(5) ; Prints node voltages
      .TF    V(2, 5)    VS      ; Transfer-function analysis
  .END                         ; End of circuit file
```

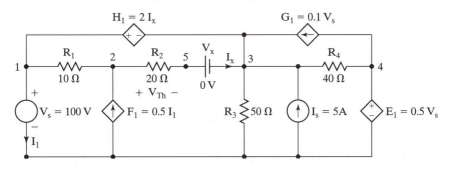

FIGURE 3.22

Dc circuit with controlled dc sources.

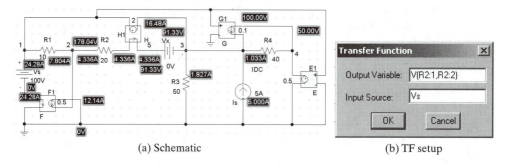

(a) Schematic (b) TF setup

FIGURE 3.23

PSpice schematic for Example 3.5.

Note that PSpice will sound a warning because there is only one item of data from dc sweep, and Probe can't plot the transfer characteristic. You can ignore this warning and look for the results on the output file.

PSpice always prints automatically the small-signal bias solutions to the output file, and they are as follows:

```
****       SMALL-SIGNAL BIAS SOLUTION       TEMPERATURE =   27.000 DEG C
  NODE      VOLTAGE       NODE    VOLTAGE      NODE    VOLTAGE      NODE    VOLTAGE
(    1)     100.0000   (    2)   178.0400  (    3)    91.3280  (    4)    50.0000
(    5)      91.3280
          VOLTAGE SOURCE CURRENTS
          NAME           CURRENT
          VS             2.428E+01
          VX             4.336E+00
          TOTAL POWER DISSIPATION    −2.43E+03 WATTS
```

The results of the dc sweep, which are obtained from the output file EX3-5.OUT, are shown in the following:

```
****      DC TRANSFER CURVES                TEMPERATURE =   27.000 DEG C
  VS           I(R1)          I(R2)          I(R3)         I(R4)
  1.000E+02   −7.804E+00     4.336E+00     1.827E+00     1.033E+00
  VS           V(1)           V(2)           V(3)         V(4)         V(5)
  1.000E+02    1.000E+02     1.780E+02     9.133E+01     5.000E+01    9.133E+01
```

The results of the .TF command, which are also obtained from the output file EX3-5.OUT, are as follows:

```
****       SMALL-SIGNAL CHARACTERISTICS
      V(2,5)/VS =  4.982E-01
      INPUT RESISTANCE AT VS = −7.169E+00
      OUTPUT RESISTANCE AT V(2,5) =  5.240E+00
          JOB CONCLUDED
          TOTAL JOB TIME             1.27
```

Note. The .TF command calculates the small-signal characteristics and automatically sends the results to the output file. The dc sweep calculates the various voltages and currents for dc transfer curves, but printing them requires a PRINT statement.

Example 3.6:   DC transfer characteristic with varying resistor's model parameter

For the amplifier circuit of Fig. 3.19, calculate and plot the dc transfer characteristics, V_{out} versus V_{in}. The input voltage is varied from 0 to 1 V with an increment of 0.5 V. The resistance R_E changes by ±25%. Plot the results with the .PLOT command, print with the .PRINT command, and display them by using Probe.

Solution   The .DC command will sweep the input voltage V_{in}. We can vary the resistance R_E by varying the model parameter R by ±25%. The PSpice schematic is shown in Fig. 3.24(a). By enabling the voltage and current displays, the node voltages and branch currents are also displayed. The .DC command is set in the analysis setup. Changing its model parameter R of the resistance R_E as shown in Fig. 3.24(b) changes the values of the resistance.
   The circuit file follows:

Example 3.6 Dc sweep
```
▲ VIN     1     0     DC    IV  ; Dc input voltage of 1 V
▲ ▲ R1     1     2     1K
      R2     2     0     20K
      RP     2     6     1.5K
      RE     3     0     RMOD   250        ; Resistance with model RMOD
      .MODEL    RMOD    RES (R = 1.0)      ; Model statement for RE
      F1     4     3     VX      40        ; Current-controlled current source
      RO     4     3     100K
      RL     4     5     2K
      VX     6     3     DC    OV          ; Measures the current through Rp
      VY     5     0     DC    OV          ; Measures the current through RL
▲ ▲ ▲ *    Dc sweep for VIN from 0 to 1 V with 0.5 V increment and
       *    use the listed values of parameter R in model RMOD
      .DC VIN    0     1.5   0.5 RES RMOD (R) LIST 0.75 1.0 1.25
      .PRINT   DC   V(1)   V(4)    ; Prints a table on the output file
      .PLOT    DC   V(0,4)         ; Plots V(0,4) on the output file
      .PROBE                       ; Graphical waveform analyzer
  .END                             ; End of circuit file
```

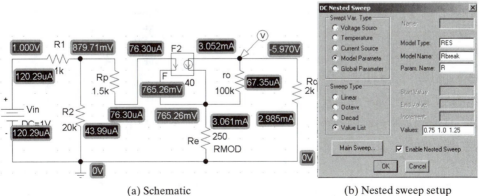

(a) Schematic   (b) Nested sweep setup

FIGURE 3.24

PSpice schematic for Example 3.6.

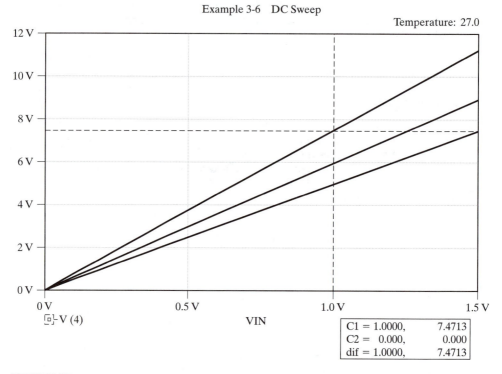

FIGURE 3.25

Dc transfer characteristics obtained by Probe for Example 3.6.

The transfer characteristics that are obtained by Probe are shown in Fig. 3.25. The plot produced by the .PLOT command is shown in Fig. 3.26. The results of the .PRINT command, which are obtained from the output file EX3-6.OUT, follow:

```
▲ ▲ ▲ ****       DC TRANSFER CURVES              TEMPERATURE = 27.000 DEG C
           VIN          V(1)           V(4)
        0.000E+00     0.000E+00       0.000E+00
        5.000E-01     5.000E-01      -3.736E+00
        1.000E+00     1.000E+00      -7.471E+00

        0.000E+00     0.000E+00       0.000E+00
        5.000E-01     5.000E-01      -2.985E+00
        1.000E+00     1.000E+00      -5.969E+00

        0.000E+00     0.000E+00       0.000E+00
        5.000E-01     5.000E-01      -2.485E+00
        1.000E+00     1.000E+00      -4.970E+00
           JOB CONCLUDED
           TOTAL JOB TIME               1.70
```

Note. The .PRINT command gives a table of data, and .PLOT generates the plot on the output file, while the .PROBE command gives graphical output on the monitor screen that can be dumped directly into a plotter or a printer. The .PLOT and .PRINT commands could generate a large amount of data in the output file and should be avoided if graphical output is available, as in PSpice, by .PROBE. With the .PROBE command, there is *no* need for the .PLOT and .PRINT commands.

Example 3-6   DC Sweep

★★★★   DC TRANSFER CURVES   TEMPERATURE   =   27.000 DEG C

★★★

VIN	V(0, 4)					
(★) --------	0.0000E+00	5.0000E+00	1.0000E+01	1.5000E+01	2.0000E+01
0.000E+00	0.000E+00	★	.	.	.	.
5.000E−01	3.736E+00	.	★	.	.	.
1.000E+00	7.471E+00	.	.	★	.	.
1.500E+00	1.121E+01	.	.	.	★	.
0.000E+00	0.000E+00	★	.	.	.	.
5.000E−01	2.985E+00	.	★	. ★	.	.
1.000E+00	5.969E+00	.	.	★	.	.
1.500E+00	8.954E+00	.	.	.	.	.
0.000E+00	0.000E+00	★	.	.	.	.
5.000E−01	2.485E+00	.	★	★	.	.
1.000E+00	4.970E+00	.	.	.	.	.
1.500E+00	7.456E+00	.	.	★	.	.

FIGURE 3.26

Dc transfer characteristics obtained by the .PLOT command for Example 3.6.

3.9.4   .PARAM (Dc Parametric sweep)

PSpice allows one to vary a parameter and to evaluate its effects on the dc analysis. (See Section 6.16.) The parameters could be a circuit element such as R, L, C, and their model parameters. The general statement for the dc sweep is given by

```
.DC PARAM SWNAME SWMIN SWMAX SWINC
```

where   SWNAME is the sweep parameter's name
        SWMIN is the minimum value of sweep parameters
        SWMAX is the maximum value of sweep parameters
        SWINC is the sweep increment

Example 3.7:   Plot the load power against the load resistance

The dc circuit shown in Fig. 3.27 has $R = 100\ \Omega$ and $V_S = 10$ V. Plot the load power against the load resistance R_L from 1 mΩ to 500 Ω with an increment of 1 Ω.

Solution   We will use the parametric sweep to vary the load resistance. The PSpice schematic is the same as shown in Fig. 3.27. The PARAM symbol can be selected from either the library special.slb as shown in Fig. 3.28(a) or the part browser as shown in Fig. 3.28(b).

   The values of the parameter RVAL are set from the menu as shown in Fig. 3.29. Up to three variables can be assigned to one PARAM command.

   The .PARAM command is selected from the analysis setup as shown in Fig. 3.30(a). The sweep parameters for RVAL are set from the dialog box as shown in Fig. 3.30(b). The plot of the load power against the load resistance RVAL $\equiv R_L$ is shown in Fig. 3.31. The maximum load

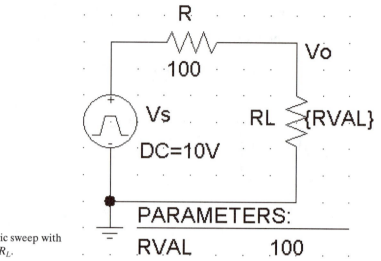

FIGURE 3.27

Dc circuit with parametric sweep with variable load resistance R_L.

power $P_{L(\max)} = 250$ mW occurs at 100 Ω, which is expected by the maximum power transfer rule when $R_L = R = 100$ Ω. The expected value of the maximum load power is given by

$$P_{L(\max)} = \frac{V_s^2}{4R} \quad \text{for } R_L = R$$

$$= \frac{10^2}{4 \times 100} = 25 \text{ mW}$$

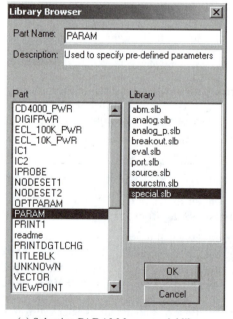

FIGURE 3.28

Selecting PARAM.

(a) Selecting PARAM from special library

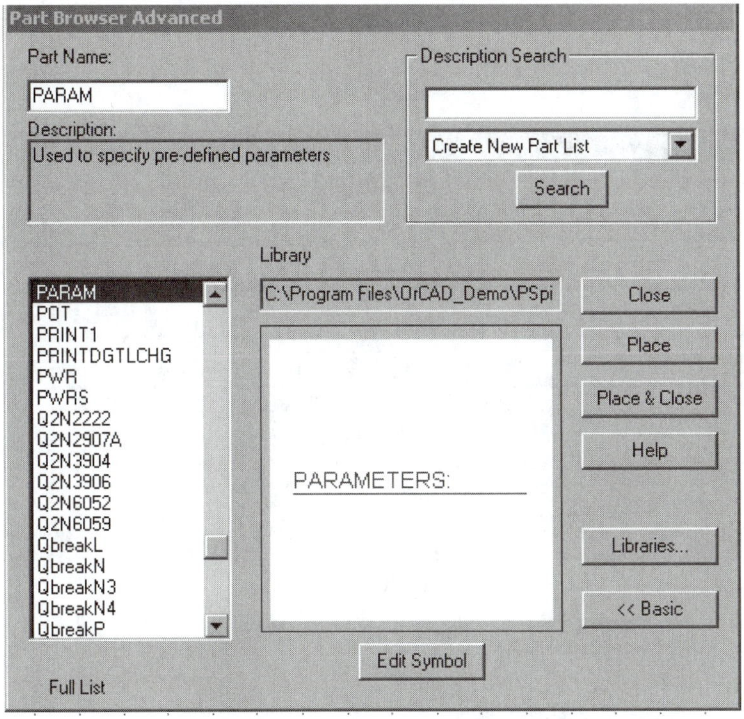

(b) Selecting PARAM from part browser

FIGURE 3.28
(*Continued*)

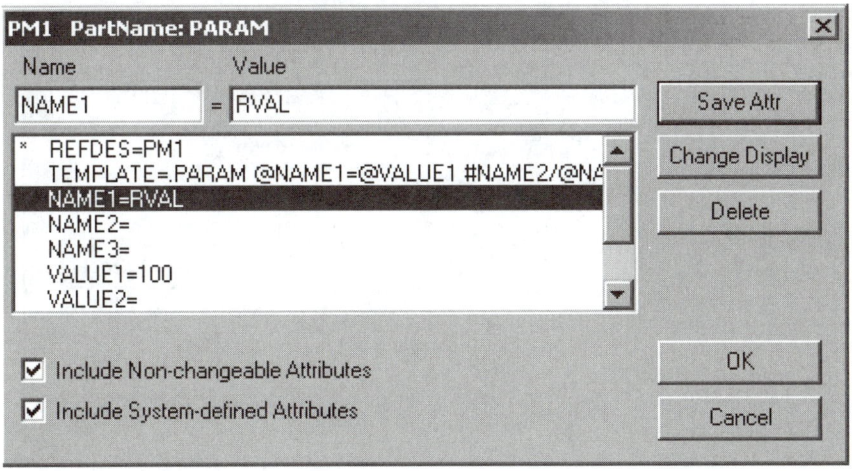

FIGURE 3.29

Setting the parameters of the PARAM.

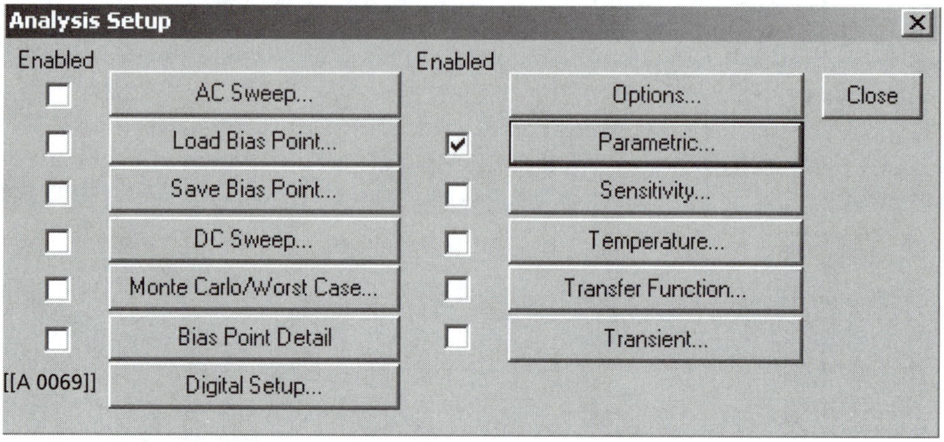

Analysis Setup

Enabled		Enabled	
☐	AC Sweep...		Options...
☐	Load Bias Point...	☑	Parametric...
☐	Save Bias Point...	☐	Sensitivity...
☐	DC Sweep...	☐	Temperature...
☐	Monte Carlo/Worst Case...	☐	Transfer Function...
☐	Bias Point Detail	☐	Transient...
[[A 0069]]	Digital Setup...		

Close

(a) Analysis setup

DC Sweep

Swept Var. Type
- ○ Voltage Source
- ○ Temperature
- ○ Current Source
- ○ Model Parameter
- ◉ Global Parameter

Name: RVAL
Model Type: RVAL
Model Name:
Param. Name:

Sweep Type
- ◉ Linear
- ○ Octave
- ○ Decade
- ○ Value List

Start Value: 1m
End Value: 500
Increment: 1
Values:

Nested Sweep...   OK   Cancel

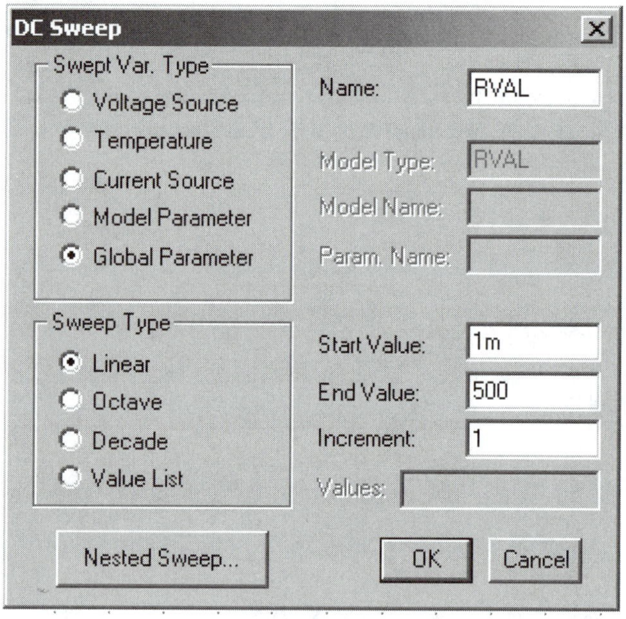

(b) Dc sweep for parameter RVAL

FIGURE 3.30

Setup for parametric sweep.

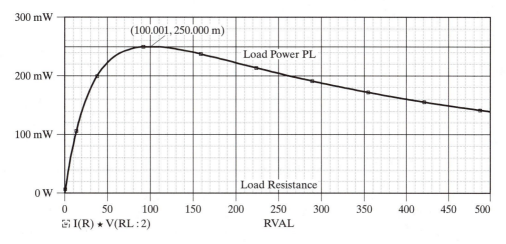

FIGURE 3.31

PSpice plot of the maximum load power against the load resistance R_L.

SUMMARY

The symbol and statement for resistors are

    R Resistor
             R⟨name⟩ N+ N−   RNAME   RVALUE

The symbols and statements for dc sources are as follows:

    E Voltage-controlled voltage source
             E⟨name⟩ N+ N− NC+ NC− ⟨(voltage gain) value⟩
    F Current-controlled current source
             F⟨name⟩ N+ N− VN ⟨(current gain) value⟩
    G Voltage-controlled current source
             G⟨name⟩ N+ N− NC+ NC− ⟨(transconductance) value⟩
    H Current-controlled voltage source
             H⟨name⟩ N+ N− VN ⟨(transresistance) value⟩
    I Independent current source
             I⟨name⟩ N+ N− [dc ⟨value⟩]
    V Independent voltage source
             V⟨name⟩ N+ N− [dc ⟨value⟩]

The following commands are generally used for dc analysis:

    .DC   Dc analysis
    .END   End of circuit
    .MODEL   Model
    .OP   Operating point

.PARAM	Parametric
.PLOT	Plot
.PRINT	Print
.PROBE	Probe
.TEMP	Temperature
.TF	Transfer function
.WIDTH	Width

REFERENCES

[1]   Robert Boylestad, *Introductory Circuit Analysis*, 10th ed. Upper Saddle River, New Jersey: Prentice Hall, 2003.

[2]   J.R. Cogdell, *Fundamentals of Electric Circuits*. Upper Saddle River, New Jersey: Prentice Hall, 1999.

[3]   R.C. Dorf and J.A. Svoboda, *Introduction to Electric Circuits*, 5th ed. New York: John Wiley & Sons, 2001.

[4]   A.E. Fitzerald, D.E. Higginbotham, and A. Grabel, *Basic Electrical Engineering*. New York: McGraw Hill, 1981.

[5]   Thomas L. Floyd, *Principles of Electric Circuits*, 6th ed. Upper Saddle River, New Jersey: Prentice Hall, 2000.

[6]   William H. Hayt, Jr., Jack E. Kemmerly, and Steven M. Durbin, *Engineering Circuit Analysis*, 6th ed. Upper Saddle River, New Jersey: Prentice Hall, 2003.

[7]   D. E. Johnson, J. Hilburn, and J. R. Johnson, *Basic Electric Circuit Analysis*. Englewood Cliffs, New Jersey: Prentice Hall, 1990.

[8]   Charles J. Monier, *Electric Circuit Analysis*. Upper Saddle River, New Jersey: Prentice Hall, 2001.

[9]   J.W. Nilsson and S. A. Riedel, *Electric Circuits*, 6th ed. Upper Saddle River, New Jersey: Prentice Hall, 2000.

[10]   G. Rizzoni, *Principles and Applications of Electrical Engineering*, 3d ed. New York: McGraw-Hill, 2000.

PROBLEMS

Write PSpice statements for Problems 3.1 to 3.10. Assume that the first node is the positive terminal and the second node is the negative terminal.

3.1 A resistor R_1 is connected between nodes 3 and 4 and has a nominal value of $R = 10\text{ k}\Omega$. The operating temperature is 55°C, and it has the form

```
R₁ = R * [1 + 0.2 * (T − T0) + 0.002 * (T − T0)²]
```

3.2 A resistor R_1 is connected between nodes 3 and 4 and has a nominal value of $R = 10\text{ k}\Omega$. The operating temperature is 55°C, and it has the form

```
R1 = R * 1.01^{4.5* (T-T0)}
```

3.3 A polynomial voltage source Y that is connected between nodes 1 and 2 is controlled by a voltage source V_1 connected between nodes 4 and 5. The source is given by

$$Y = 0.1V_1 + 0.2V_1^2 + 0.05V_1^3$$

3.4 A polynomial current source *I* that is connected between nodes 1 and 2 is controlled by a voltage source V_1 connected between nodes 4 and 5. The source is given by

$$Y = 0.1V_1 + 0.2V_1^2 + 0.05V_1^3$$

3.5 A voltage source V_{out} that is connected between nodes 5 and 6 is controlled by a voltage source V_1 and has a voltage gain of 25. The controlling voltage is connected between nodes 10 and 12. The source is expressed as

$$V_{out} = 25V_1$$

3.6 A current source I_{out} that is connected between nodes 5 and 6 is controlled by a current source I_1 and has a current gain of 10. The voltage through which the controlling current flows is V_C. The current source is given by

$$I_{out} = 10I_1$$

3.7 A current source I_{out} that is connected between nodes 5 and 6 is controlled by a voltage source V_1 between nodes 8 and 9. The transconductance is 0.05 mhos. The current source is given by

$$I_{out} = 0.05V_i$$

3.8 A voltage source V_{out} that is connected between nodes 5 and 6 is controlled by a current source I_1 and has a transresistance of 150 Ω. The voltage through which the controlling current flows is V_C. The voltage source is expressed as

$$V_{out} = 150I_1$$

3.9 A nonlinear resistance *R* that is connected between nodes 4 and 6 is controlled by a voltage source V_1 and has a resistance of the form

$$R = V_1 + 0.2V_1^2$$

3.10 A nonlinear transconductance G_m that is connected between nodes 4 and 6 is controlled by a current source. The voltage through which the controlling current flows is V_1. The transconductance has the form

$$G_m = V_1 + 0.2V_1^2$$

3.11 For the circuit in Fig. P3.11, calculate and print **(a)** the voltage gain, $A_v = V_{out}/V_{in}$; **(b)** the input resistance, R_{in}; and **(c)** the output resistance, R_{out}.

3.12 For the circuit in Fig. P3.11, calculate and plot the dc transfer characteristic V_{out} versus V_{in}. The input voltage is varied from 0 to 10 V with an increment of 0.5 V.

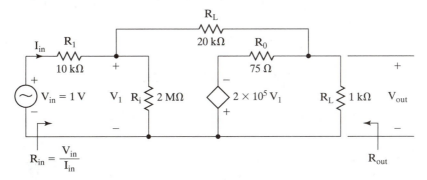

FIGURE P3.11

3.13 A dc circuit is shown in Fig. P3.13. Use PSpice to calculate and print Thévenin's equivalent circuit with respect to terminals *a* and *b*.

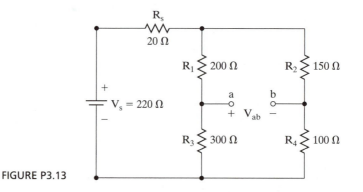

FIGURE P3.13

3.14 Use PSpice to calculate all node voltages and branch currents of Fig. P3.13. Assume an operating temperature of 55°C.

3.15 For Fig. P3.13, use PSpice to calculate and plot the dc transfer characteristics, V_{ab} versus V_S. The input voltage V_S is varied from 0 to 20 V with an increment of 5 V. The resistance R_1 changes by ±20%. Print the results with the .PRINT command, and display the transfer characteristics with Probe.

3.16 Repeat Example 3.3 with a .TF command **(a)** if the input is I_s and the output is the voltage V(2,4) between nodes 2 and 4, **(b)** if the input is I_s and the output is the current through V_{in}, and **(c)** if the input is V_{in} and the output is the current through V_x.

3.17 Repeat Problem 3.13 for the circuit of Fig. P3.17.

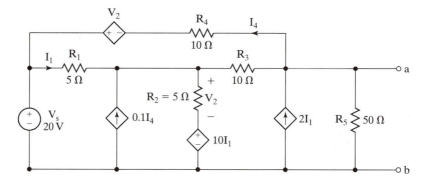

FIGURE P3.17

3.18 Repeat Problem 3.14 for the circuit of Fig. P3.17.

3.19 Repeat Problem 3.15 for the circuit of Fig. P3.17.

3.20 Use PSpice to find voltages at nodes *a*, *b*, *c*, and *d* for the circuit shown in Fig. P3.20. Assume an operating temperature of 25°C. Also find the input resistance $R_{in} = \dfrac{v_s}{i_s}$.

3.21 Use PSpice to find the Thevenin's equivalent voltage $V_{Th} = V_{ab}$ across nodes *a* and *b* and the equivalent resistance R_{Th} for the circuit as shown in Fig. P3.21. Assume an operating temperature of 25°C.

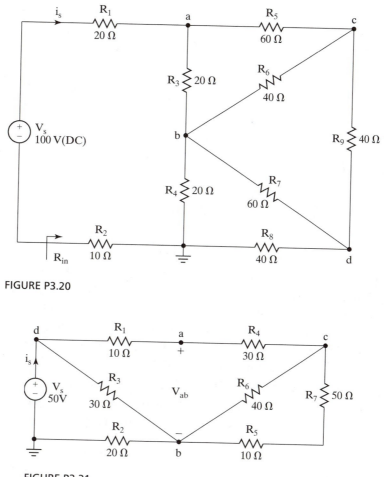

FIGURE P3.20

FIGURE P3.21

3.22 Use PSpice to find the value of load resistance R_L for the circuit in Fig. P3.22 so that the maximum power is delivered to the load. Assume an operating temperature of 25°C.

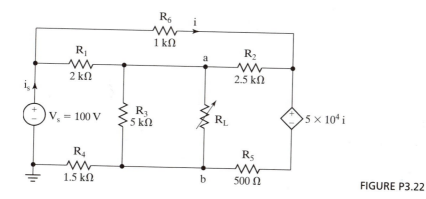

FIGURE P3.22

CHAPTER 4

Transient Analysis

After completing this chapter, students should be able to

- Specify the operating temperature
- Model temperature-dependent and -independent resistances, current-dependent and -independent inductors, and voltage-dependent and -independent capacitors
- Model transient voltage and current sources and specify their parameters
- Understand the response of transient analysis and obtain their output variables
- Assign initial conditions for transient analysis
- Understand the response of transient analysis and obtain their output variables
- Perform the transient analysis of a circuit and set the transient parameters
- Model transient voltage and current sources and specify their parameters.

4.1 INTRODUCTION

A transient analysis deals with the behavior of an electric circuit as a function of time. If a circuit contains an energy storage element(s), a transient can also occur in a dc circuit after a sudden change due to switches opening or closing. SPICE allows simulating transient behaviors, by assigning initial conditions to circuit elements, generating sources, and the opening and closing of switches. The simulation of transients in circuits with linear elements requires modeling of

> Resistors
> Capacitors and inductors
> Model parameters of elements
> Operating temperature
> Modeling of transient sources
> Transient sources
> Transient output variables

Output commands

Transient analysis.

The SPICE simulations of transient circuits are illustrated by examples. Students are encouraged to apply the techniques for transient analysis of simple circuit laws and to verify the SPICE results by hand calculations. We have already discussed the statements for resistors in Section 3.2, assigning model parameters of elements in Section 3.3, and specifying the operating temperature in Section 3.4.

4.2 CAPACITORS AND INDUCTORS

The voltage and current relationships of inductors and capacitors are shown in Fig. 4.1.

4.2.1 Capacitor

The symbol for a capacitor is C. The name of a capacitor must start with C, and it takes the general form

```
C<name> N+ N- CNAME CVALUE IC=V0
```

where $N+$ is the positive node and $N-$ is the negative node. The voltage of node $N+$ is assumed positive with respect to node $N-$, and the current flows from node $N+$ through the capacitor to node $N-$. CNAME is the model name, and CVALUE is the nominal value of the capacitor. IC defines the initial (time-zero) voltage of the capacitor, V_0.

The PSpice schematic for the capacitor is shown in Fig. 4.2(a), and its parameters are shown in Fig. 4.2(b). The capacitor's name and its nominal value can be changed. Also, a tolerance value can be assigned to it.

The model parameters are shown in Table 4.1. The model command is discussed in Section 3.3. If CNAME is omitted, CVALUE is the capacitance in farads, and the CVALUE can be positive or negative, but *must* not be zero. If CNAME is included, the capacitance that depends on voltage and temperature is calculated from

```
CAP = CVALUE * C * (1 + VC1 * V + VC2 * V²)
          * [1 + TC1 * (T - T0) + TC2 * (T - T0)²]
```

where T is the operating temperature in degrees Celsius and $T0$ is the room temperature in degrees Celsius.

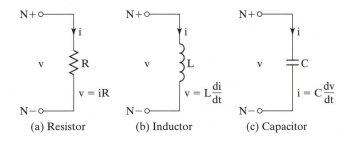

(a) Resistor   (b) Inductor   (c) Capacitor

FIGURE 4.1

Voltage and current relationships.

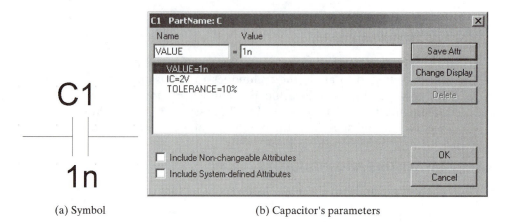

(a) Symbol   (b) Capacitor's parameters

FIGURE 4.2

Voltage/current relationship and parameters of capacitors.

TABLE 4.1   Model Parameters for Capacitors

Name	Meaning	Units	Default
C	Capacitance multiplier		1
VC1	Linear voltage coefficient	Volt^{-1}	0
VC2	Quadratic voltage coefficient	Volt^{-2}	0
TC1	Linear temperature coefficient	$°\text{C}^{-1}$	0
TC2	Quadratic temperature coefficient	$°\text{C}^{-2}$	0

In the PSpice schematic, the user can assign the model name of the breakout devices in the library *breakout.slb* shown in Fig. 4.3(a) and can also edit the model parameters shown in Fig. 4.3(c).

Some Capacitor Statements

```
C1        6     5      10UF
CLOAD     12    11     5PF      IC=2.5V
CINPUT    15    14     ACAP     10PF
C2        20    19     ACAP     20NF    IC=1.5V
.MODEL    ACAP   CAP (C=1    VC1=0.01   VC2=0.002 TC1=0.02 TC2=0.005)
```

Notes

1. The model parameter C is a capacitance multiplier, rather than the value of the capacitance. It scales the actual capacitance value, CVALUE. Thus, $C = 1.1$ means that CVALUE is multiplied by 1.1, *not* that CVALUE is 1.1 F.

2. The initial condition (if any) applies only if the UIC (use initial condition) option is specified in the .TRAN command statement, which is described in Section 4.7.2.

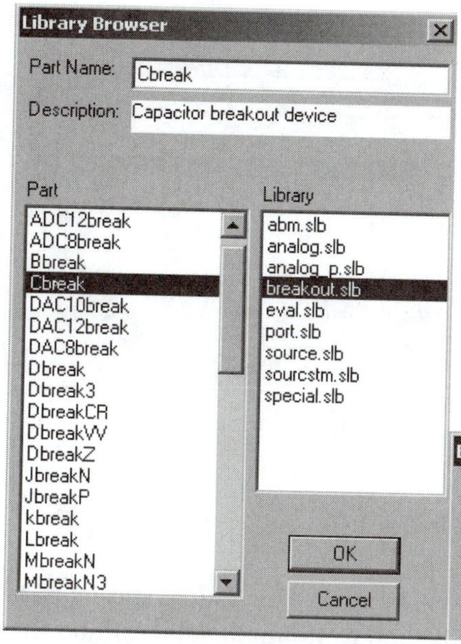

(a) Breakout devices

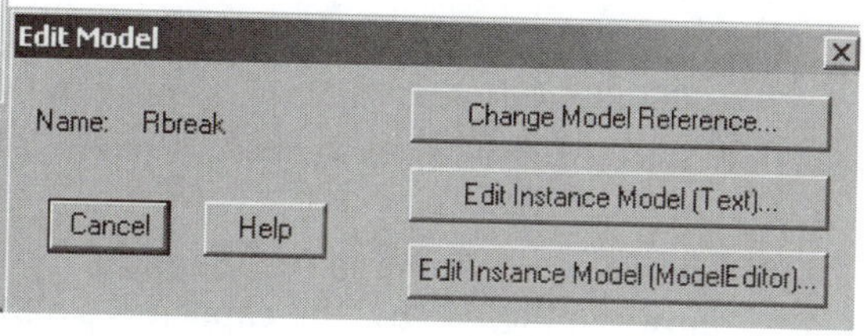

(b) Editing model parameters

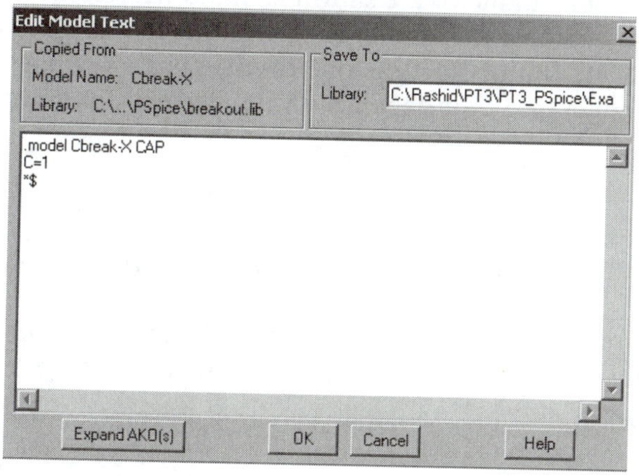

(c) Model parameters for capacitors

FIGURE 4.3

Breakout devices.

4.2.2   Inductor

The symbol for an inductor is L. The name of an inductor must start with L, and it takes the general form

```
L<name> N+ N- LNAME LVALUE IC=I0
```

where $N+$ is the positive node and $N-$ is the negative node. The voltage of $N+$ is assumed positive with respect to node $N-$, and the current flows from node $N+$ through the inductor to node $N-$. LNAME is the model name, and LVALUE is the nominal value of the inductor. IC defines the initial (time-zero) current of the inductor, I_0.

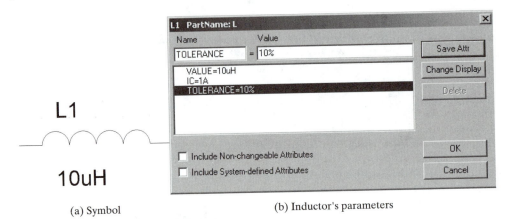

L1

10uH

(a) Symbol   (b) Inductor's parameters

FIGURE 4.4

Voltage/current relationship and parameters of inductors.

The PSpice schematic for the inductor is shown in Fig. 4.4(a), and its parameters are shown in Fig. 4.4(b). The capacitor's name and its nominal value can be changed. Also, a tolerance value can be assigned to it.

The model parameters of an inductor are shown in Table 4.2. Refer to Section 3.3 for a discussion of the model statement. If LNAME is omitted, LVALUE is the inductance in henrys, and LVALUE can be positive or negative, but *must* not be zero. If LNAME is included, the inductance that depends on the current and temperature is calculated from

$$
\begin{aligned}
\text{IND} = \text{LVALUE} \; * \; \text{L} \; * \; (1 \; + \; \text{IL1} \; * \; \text{I} \; + \; \text{IL2} \; * \; \text{I}^2) \\
* \; [1 \; + \; \text{TC1} \; * \; (\text{T} \; - \; \text{T0}) \; + \; \text{TC2} \; * \; (\text{T} \; - \; \text{T0})^2]
\end{aligned}
$$

where T is the operating temperature in degrees Celsius and $T0$ is the room temperature in degrees Celsius.

In the PSpice schematic, the user can assign the model name of the breakout devices in the library *breakout.slb* shown in Fig. 4.5(a) and can also edit the model parameters shown in Fig. 4.5(c).

Notes

1. The model parameter L is an inductance multiplier, rather than the value of the inductance. It scales the actual inductance value, LVALUE. Thus, $L = 1.1$ means that LVALUE is multiplied by 1.1, *not* that LVALUE is 1.1 F.

TABLE 4.2   Model Parameters for Inductors

Name	Meaning	Units	Default
L	Inductance multiplier		1
IL1	Linear current coefficient	Amps^{-1}	0
IL2	Quadratic current coefficient	Amps^{-2}	0
TC1	Linear temperature coefficient	$°\text{C}^{-1}$	0
TC2	Quadratic temperature coefficient	$°\text{C}^{-2}$	0

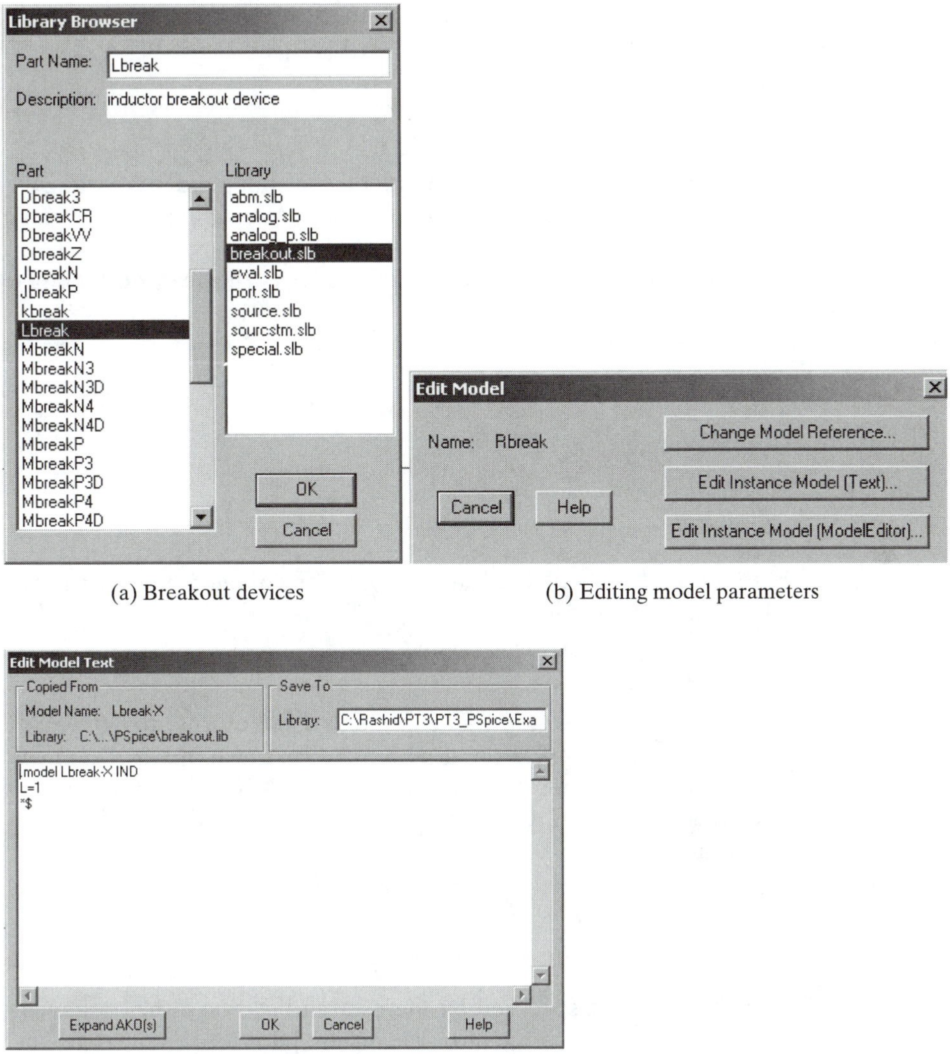

(a) Breakout devices

(b) Editing model parameters

(c) Model parameters for inductors

FIGURE 4.5

Breakout devices.

2. The initial condition (if any) applies only if the UIC (use initial condition) option is specified on the .TRAN command statement, which is described in Section 4.7.2.

Some Inductor Statements

```
L1       6     5      10MH
LLOAD    12    11     5UH       IC=0.2MA
LLINE    15    14     LMOD      5MH
```

```
LCHOKE   20   19    LMOD    2UH  IC=0.5A
.MODEL   LMOD  IND  (L=1   IL1=0.1   IL2=0.002   TC1=0.02   TC2=0.005)
```

4.3   MODELING OF TRANSIENT SOURCES

PSpice allows the generation of dependent (or independent) voltage and current sources. Independent sources can be time variant. A nonlinear source can also be simulated by a polynomial.

The independent voltage and current sources that can be modeled by PSpice follow:

Exponential
Pulse
Piecewise linear
Sinusoidal
Single-frequency frequency modulation.

Note. These sources are explained subsequently as voltages; however, the explanations are equally applicable to currents.

4.3.1   Exponential Source

The waveform and parameters of an exponential waveform are shown in Fig. 4.6 and Table 4.3. The schematic of Exponential Source is shown in Fig. 4.7(a), and the menu for setting the model parameters are shown in Fig. 4.7(b). In addition to the transient specifications, the dc (i.e., DC = 1 V) and ac (AC = 1 V) specifications can be assigned to the same source. TD1 is the rise-day time, TC1 is the time constant, TD2 is the fall-day time, and TD2 is their time constant. The symbol of exponential sources is EXP, and the general form is

```
EXP (V1 V2 TRD TRC TFD TFC)
```

where V_1 and V_2 *must* be specified by the user and can be either voltages or currents. (TSTEP is the incrementing time during transient [.TRAN] analysis.) In an EXP waveform, the voltage remains V_1 for the first TRD seconds. Then the voltage rises

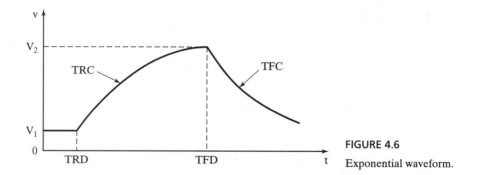

FIGURE 4.6
Exponential waveform.

TABLE 4.3 Model Parameters of EXP Sources

Name	Meaning	Units	Default
V1	Initial voltage	Volts	None
V2	Pulsed voltage	Volts	None
TRD	Rise delay time	Seconds	0
TRC	Rise-time constant	Seconds	TSTEP
TFD	Fall delay time	Seconds	TRD + TSTEP
TFC	Fall-time constant	Seconds	TSTEP

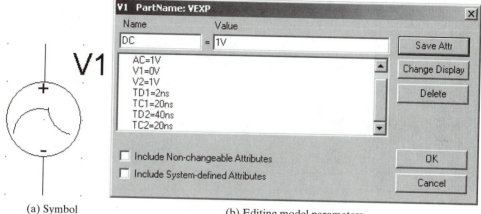

(a) Symbol   (b) Editing model parameters

FIGURE 4.7

Exponential source in PSpice schematic.

exponentially from V_1 to V_2, with a rise-time constant of TRC. After a time of TFD, the voltage falls exponentially from V_2 to V_1, with a fall-time constant of TFC.

Note. The values of an EXP waveform as well as the values of other time-dependent waveforms at intermediate time points are determined by PSpice by means of linear interpolation.

Typical model statements. For $V_1 = 0$, $V_2 = 1$ V, TRD = 2 ns, TRC = 20 ns, TFD = 60 ns, and TFC = 30 ns, the model statement is

```
EXP (0 1 2NS 20NS 60NS 30NS)
```

With TRD = 0, the statement becomes

```
EXP (0 1 0 20NS 60NS 30NS)
```

With $V_1 = -1$ V and $V_2 = 2$ V, it is

```
EXP (-1 2 2NS 20NS 60NS 30NS)
```

4.3.2 Pulse Source

The waveform and parameters of a pulse waveform are shown in Fig. 4.8 and Table 4.4. The schematic of Pulse Source is shown in Fig. 4.9(a), and the menu for setting the

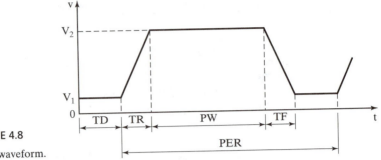

FIGURE 4.8

Pulse waveform.

TABLE 4.4 Model Parameters of Pulse Sources

Name	Meaning	Units	Default
V1	Initial voltage	Volts	None
V2	Pulsed voltage	Volts	None
TD	Delay time	Seconds	0
TR	Rise time	Seconds	TSTEP
TF	Fall time	Seconds	TSTEP
PW	Pulse width	Seconds	TSTOP
PER	Period	Seconds	TSTOP

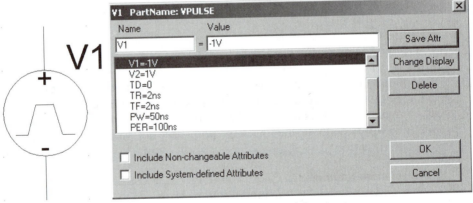

(a) Symbol   (b) Editing model parameters

FIGURE 4.9

Pulse source in PSpice schematic.

model parameters are shown in Fig. 4.9(b). In addition to the transient specifications, the dc (i.e., $DC = 10$ V) and ac ($AC = 1$ V) specifications can be assigned to the same source. The symbol of a pulse source is PULSE, and the general form is

```
PULSE (V1 V2 TD TR TF PW PER)
```

where V_1 and V_2 *must* be specified by the user and can be either voltages or currents. TSTEP and TSTOP are the incrementing time and stop time, respectively, during transient (.TRAN) analysis.

Typical statements. For $V_1 = -1$ V, $V_2 = 1$ V, TD $= 2$ ns, TR $= 2$ ns, TF $=$ 2 ns, PW $= 50$ ns, and PER $= 100$ ns, the model statement is

```
PULSE (-1 1 2NS 2NS 2NS 50NS 100NS)
```

With $V_1 = 0, V_2 = 1$ V, the model becomes

```
PULSE (0 1 2NS 2NS 2NS 50NS 100NS)
```

With $V_1 = 0, V_2 = -1$ V, the model becomes

```
PULSE (0 -1 2NS 2NS 2NS 50NS 100NS)
```

4.3.3 Piecewise Linear Source

A point in a waveform can be described by (T_i, V_i) or (T_i, I_i), and every pair of values (T_i, V_i) or (T_i, I_i) specifies the source value at time T_i. The voltage at times between the intermediate points is determined by PSpice by using linear interpolation. The schematic of Piecewise Linear Source is shown in Fig. 4.10(a) and the menu for setting the model parameters are shown in Fig. 4.10(b). Up to 10 points (time, volts/currents) can be specified. In addition to the transient specifications, the dc (i.e., DC $= 10$ V) and ac (AC $= 1$ V) specifications can be assigned to the same source. The symbol of a piecewise linear source is PWL, and the general form is

```
PWL (T1 V1 T2 V2 . . . TN VN)
```

The model parameters of a PWL waveform are given in Table 4.5.

Typical statement. The model statement for the typical waveform in Fig. 4.11 is

```
PWL (0 0 5 3 10US 3V 15US 6V 40US 6V 45US 2V 60US 2V 65US 0)
```

4.3.4 Single-Frequency Frequency Modulation

The schematic of Single-Frequency Frequency Modulation (SFFM) Source is shown in Fig. 4.12(a), and the menu for setting the model parameters are shown in Fig. 4.12(b).

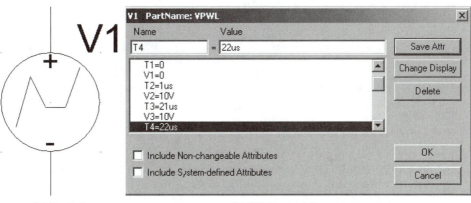

(a) Symbol   (b) Editing model parameters

FIGURE 4.10

Piecewise linear source in PSpice schematic.

TABLE 4.5	Model Parameters of PWL Sources		
Name	Meaning	Units	Default
T_i	Time at a point	Seconds	None
V_i	Voltage at a point	Volts	None

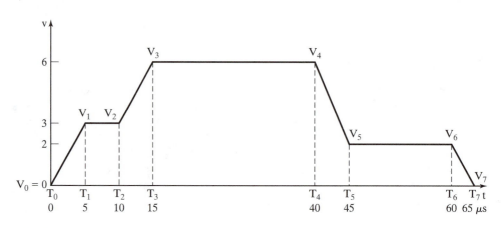

FIGURE 4.11

Piecewise linear waveform.

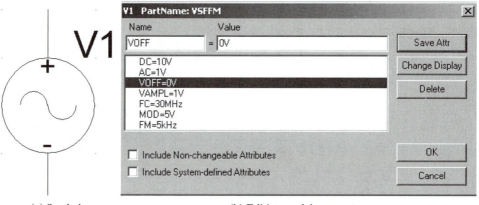

(a) Symbol (b) Editing model parameters

FIGURE 4.12

SFFM source in PSpice schematic.

In addition to the transient specifications, the dc (i.e., DC = 10 V) and ac (AC = 1 V) specifications can be assigned to the same source.

The symbol of a source with single-frequency frequency modulation is SFFM, and the general form is

```
SFFM (VO VA FC MOD FS)
```

TABLE 4.6	Model Parameters of SFFM Sources		
Name	Meaning	Units	Default
VO	Offset voltage	Volts	None
VA	Amplitude of voltage	Volts	None
FC	Carrier frequency	Hertz	1/TSTOP
MOD	Modulation index		0
FS	Signal frequency	Hertz	1/TSTOP

The model parameters of an SFFM waveform are given in Table 4.6.

V_O and V_A *must* be specified by the user and can be either voltages or currents. TSTOP is the stop time during transient (.TRAN) analysis. The waveform is of the form

$$V = V_O + V_A \sin[(2\pi F_C t) + M \sin(2\pi F_s t)]$$

Typical statements.   For $V_O = 0$, $V_A = 1$ V, $F_C = 30$ MHz, MOD $= 5$, and $F_S = 5$ kHz, the model statement is

```
SFFM (0 1V 30MEGHZ 5 5KHZ)
```

With $V_O = 1$ mV and $V_A = 2$ V, the model becomes

```
SFFM (1MV 2V 30MEGHZ 5 5KHZ)
```

4.3.5   Sinusoidal Source

The schematic of Sinusoidal Source is shown in Fig. 4.13(a), and the menu for setting the model parameters are shown in Fig. 4.13(b). In addition to the transient specifications, the dc (i.e., DC $= 10$ V) and ac (AC $= 1$ V) specifications can be assigned to the same source.

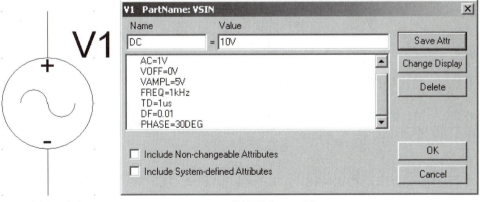

(a) Symbol   (b) Editing model parameters

FIGURE 4.13

Sinusoidal source in PSpice schematic.

TABLE 4.7   Model Parameters of SIN Sources

Name	Meaning	Units	Default
VO	Offset voltage	Volts	None
VA	Peak voltage	Volts	None
FREQ	Frequency	Hertz	1/TSTOP
TD	Delay time	Seconds	0
ALPHA	Damping factor	1/seconds	0
THETA	Phase delay	Degrees	0

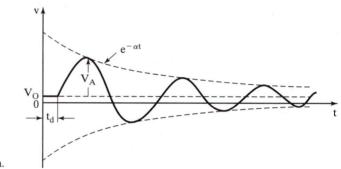

FIGURE 4.14

Damped sinusoidal waveform.

The symbol of a sinusoidal source is SIN, and the general form is

```
SIN (VO VA FREQ TD ALP THETA)
```

The model parameters of the SIN waveform are given in Table 4.7. V_O and V_A *must* be specified by the user and can be either voltages or currents. TSTOP is the stop time during transient (.TRAN) analysis. The waveform stays at 0 for a time of TD, and then the voltage becomes an exponentially damped sine wave. An exponentially damped sine wave is described by

$$V = V_O + V_A\, e^{-\alpha(t-t_d)} \sin[(2\pi f(t - t_d) - \theta]$$

and this is shown in Fig. 4.14.

Typical Statements

```
SIN   (0    1V    10KHZ    10US    1E5)
SIN   (1    5V    10KHZ    1E5     30DEG)
SIN   (0    2V    10KHZ    30DEG)
SIN   (0    2V    10KHZ)
```

4.4   TRANSIENT SOURCES

The transient sources are time variant and can be either independent or dependent. They can be voltages or currents, as shown in Fig. 4.15. The dependent sources are discussed in Section 3.6. The voltage/current sources in the PSpice schematic can assume the specifications of dc, ac, and transients, depending on the type of analysis invoked.

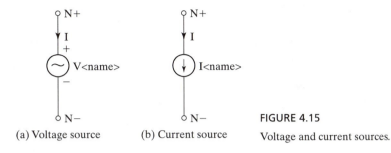

FIGURE 4.15

(a) Voltage source   (b) Current source   Voltage and current sources.

4.4.1 Independent Voltage Source

The symbol of an independent voltage source is V, and the general form for assigning dc and transient values is

```
V<name> N+ N- [DC <value>]
+       [(transient value)
+       [PULSE] [SIN] [EXP] [PWL] [SFFM] [source arguments]]
```

where $N+$ is the positive node and $N-$ is the negative node, as shown in Fig. 4.15(a). Positive current flows from node $N+$ through the voltage source to the negative node $N-$. The voltage source need not be grounded. For the dc and transient values, the default value is zero. None or all of the dc and transient values may be specified.

A source can be assigned either a dc value or a transient value. The source is set to the dc value in dc analysis. The time-dependent source (e.g., PULSE, EXP, or SIN) is assigned for transient analysis. A voltage source may be used as an **ammeter** in PSpice by inserting a zero-valued voltage source into the circuit for the purpose of measuring current.

Typical Statements

```
V1       15    0    6V
V2       15    0    DC   6V
VPULSE   10    0    PULSE (0 1 2NS 2NS 2NS 50NS 100NS)
VPULSE   12    0    PULSE (0 1 2NS 2NS 2NS 50NS 100NS)
VIN      12    3    DC   15V  SIN (0 2V 10KHZ)
```

Note. VIN assumes 15 V for dc analysis, and a sine wave of 2 V at 10 kHz for transient analysis. This allows source specifications for different analyses in the same statement.

4.4.2 Independent Current Source

The symbol of an independent current source is I, and the general form for assigning dc and transient values is

```
I<name> N+ N- [DC <value>]
+       [(transient value)
+       [PULSE] [SIN] [EXP] [PWL] [SFFM] [source arguments]]
```

Note. The first column with + (*plus*) signifies continuation of the PSpice statement. After the + sign, the statement can continue in any column.

$N+$ is the positive node and $N-$ is the negative node, as shown in Fig. 4.15(b). Positive current flows from node $N+$ through the current source to the negative node $N-$. The current source need not be grounded. The source specifications are similar to those for independent voltage sources.

Typical Statements

```
I1        15    0    2.5MA
I2        15    0    DC  2.5MA
IPULSE    10    0    PULSE (0 IV 2NS 2NS 2NS 50NS 100NS)
IIN       25    22   DC  2          SIN    (0 2V  10KHZ)
```

4.5   TRANSIENT OUTPUT VARIABLES

The output variables of transient analysis are similar to those of the dc sweep, discussed in Section 3.7.

Example 4.1:   Specifying the transient output voltages and currents of an *RLC* circuit

An RLC circuit with a step input is shown in Fig. 4.16(a). Write the various currents and voltages in forms that are allowed by PSpice. The step voltage in Example 4.1 is generated by a PWL waveform. The voltage at a node or the current flowing into a node can be marked so that *probe* displays the voltage or the current as the default at the end of the simulation. The markers menu is shown in Fig. 4.16(b).

Solution

Symbols	PSpice Variables	Meaning
i_R	I(R)	The current through resistor R
i_L	I(L)	The current through inductor L
i_C	I(C)	The current through capacitor C
i_{in}	I(VIN)	The current flowing into voltage source v_{in}
v_3	V(3)	Voltage of node 3 with respect to ground
$v_{2,3}$	V(2,3)	Voltage of node 2 with respect to node 3
$v_{1,2}$	V(1,2)	Voltage of node 1 with respect to node 2

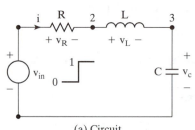

(a) Circuit   (b) Markers

FIGURE 4.16

An *RLC* circuit with a step input.

v_R	V(R)	Voltage of resistor R, where the first node (as defined in the circuit file) is positive with respect to the second node
v_L	V(L)	Voltage of inductor L, where the first node (as defined in the circuit file) is positive with respect to the second node
v_C	V(C)	Voltage of inductor L, where the first node (as defined in the circuit file) is positive with respect to the second node

Note. SPICE and older versions of PSpice do not permit measuring voltage across a resistor, an inductor, or a capacitor, for example, V(R1), V(L1), and V(C1)]. Such statements are applicable only to outputs by .PLOT and .PRINT commands.

4.6 TRANSIENT OUTPUT COMMANDS

The output commands are similar to those for the dc sweep. The .PRINT, .PLOT, and .PROBE statements for transient outputs are

```
.PRINT TRAN <output variables>
.PLOT TRAN <output variables>
        +[<lower limit) value>, <(upper limit) value>]
.PROBE
```

Some Statements

```
.PRINT TRAN V(5) V(4, 7) (0, 10V) IB(Q1) (0, 50MA) IC(Q1) (-50MA, 50MA)
.PLOT  TRAN V(5) V(4, 7) (0, 10V) IB(Q1) (0, 50MA) IC(Q1) (-50MA, 50MA)
```

Notes

1. Having two .PRINT statements for the same variables will not produce two tables. PSpice will ignore the first statement and produce output for the second statement.
2. The x-axis is the time, by default. In the last statement, the range for voltages V(5) and V(4,7) is 0 to 10 V, that for current IB(Q1) is 0 to 50 MA, and that for the current IC(Q1) is −50 to 50 MA.

4.7 TRANSIENT RESPONSE

A transient response determines the output in the time domain in response to an input signal in the time domain. Let us consider the RC circuit shown in Fig. 4.17(a), which is subjected to a step input voltage. Assuming that the switch S_1 is closed at time $t = 0$, the current i_c can be described by the equation

$$V_S = Ri_C + \frac{1}{C} \int i_C \, dt + V_C(t = 0) \tag{4.1}$$

where the initial condition is $V_C(t = 0) = 0$. Solving Eq. (4.1) gives the capacitor voltage $v_C(t)$ as

$$v_C(t) = V_S[1 - e^{-t/RC}] \tag{4.2}$$

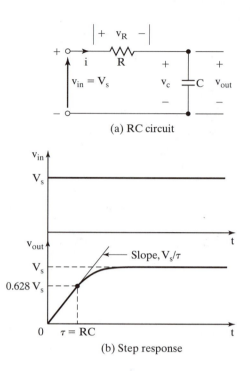

(a) RC circuit

(b) Step response

FIGURE 4.17

Transient response of an *RC* circuit.

The plot of $v_C(t)$ against time t gives the transient response of the capacitor voltage $v_C(t)$, shown in Fig. 4.17(b). The method for calculating the transient analysis bias point differs from that of the dc analysis bias point. The dc bias point is also known as the *regular bias point*. In the regular (dc) bias point, the initial values of the circuit nodes do not contribute to the operating point or to the linearized parameters. The capacitors and inductors are considered open circuited and short circuited, respectively, whereas in the transient bias point, the initial values of the circuit nodes are taken into account in calculating the bias point and the small-signal parameters of the nonlinear elements. The capacitors and inductors, which may have initial values, therefore remain as parts of the circuit.

The determination of the transient analysis requires statements involving the following:

.IC   Initial transient conditions
.TRAN   Transient analysis

4.7.1 .IC (Initial Transient Conditions)

The various nodes can be assigned to initial voltages during transient analysis, and the general form for assigning initial values is

```
.IC V(1)=A1 V(1)=A2 . . . V(N)=AN
```

where $A1, A2, A3, \ldots$ are the initial values for node voltages $V(1), V(2), V(3), \ldots$, respectively. These initial values are used by PSpice to calculate the transient analysis bias point and the linearized parameters of nonlinear devices for transient analysis. After the transient analysis bias point has been calculated, the transient analysis starts, and the nodes are released. It should be noted that these initial conditions do not affect the regular bias-point calculation during dc analysis or the dc sweep.

For the .IC statement to be effective, UIC (use initial conditions) *should not* be specified in the .TRAN command.

The .IC command is not necessary in the linear circuits that are illustrated by examples in this chapter. Students are encouraged to run circuits with an .IC statement.

Statement for Initial Transient Conditions

```
.IC V(1)=2.5 V(5)=1.7V V(7)=0.5
```

4.7.2   .TRAN (Transient Analysis)

Transient analysis can be performed by the .TRAN command, which has one of the general forms

```
.TRAN         TSTEP TSTOP [TSTART TMAX] [UIC]
.TRAN[/OP]    TSTEP TSTOP [TSTART TMAX] [UIC]
```

TSTEP is the printing increment, TSTOP is the final time (or stop time), and TMAX is the maximum size of internal time step. TMAX allows the user to control the internal time step. TMAX can be smaller or larger than the printing time, TSTEP. The default value of TMAX is TSTOP/50.

The transient analysis always starts at time = 0. However, it is possible to suppress the printing of the output for a time of TSTART. TSTART is the initial time at which the transient response is printed. In fact, PSpice analyzes the circuit from $t = 0$ to TSTART, but it does not print or store the output variables. Although PSpice computes the results with an internal time step, the results are generated by interpolation for a printing step of TSTEP. Figure 4.18 shows the relationships of TSTART, TSTOP, and TSTEP.

In transient analysis, only the node voltages of the transient analysis bias point are printed. However, the .TRAN command can control the output for the transient response bias point. An .OP command with a .TRAN command, namely, .TRAN/OP, will print the small-signal parameters during transient analysis.

If UIC is not specified as an option at the end of the .TRAN statement, PSpice calculates the transient analysis bias point before the beginning of transient analysis. PSpice uses the initial values specified with the .IC command.

If UIC (use initial conditions) is specified as an option at the end of the .TRAN statement, PSpice does not calculate the transient analysis bias point before the beginning of transient analysis. However, PSpice uses the initial values specified with the IC = initial conditions for capacitors and inductors, which are discussed in Section 4.2. Therefore, if UIC is specified, the initial values of the capacitors and inductors *must* be supplied. The .TRAN statement requires a .PRINT, .PLOT, or .PROBE statement to get the results of the transient analysis.

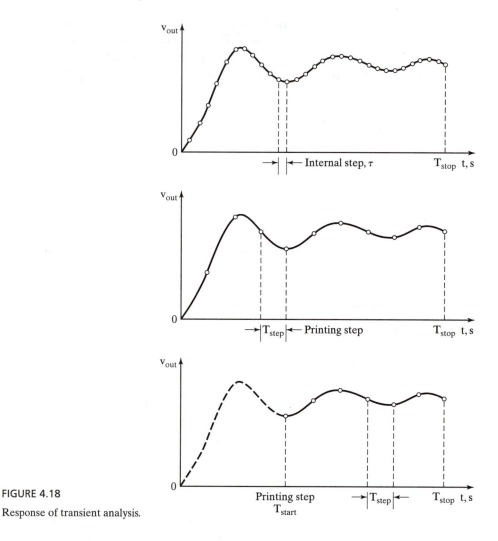

FIGURE 4.18

Response of transient analysis.

Statements for Transient Analysis

```
.TRAN       5US     1MS
.TRAN       5US     1MS 200US 0.1NS
.TRAN       5US     1MS 200US 0.1NS UIC
.TRAN/OP    5US     1MS 200US 0.1NS UIC
```

Example 4.2:   Transient pulse response of an *RLC* circuit

A pulse input, as shown in Fig. 4.19(b), is applied to the *RLC* circuit of Fig. 4.19(a). Use PSpice to calculate and plot the transient response from 0 to 400 µs with a time increment of 1 µs. The capacitor voltage $V(3)$ and the current through R_1 I(R1) are to be plotted. The circuit file's name is EX4-2.CIR, and the outputs are to be stored in file EX4-2.OUT. The results should also be made available for display and hard copy by the .PROBE command.

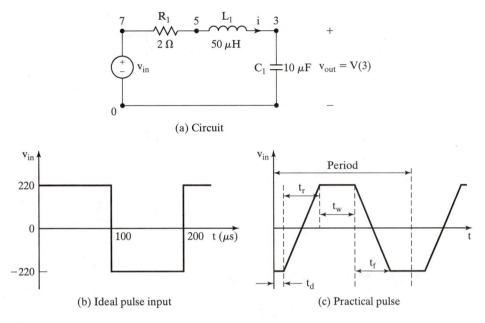

(a) Circuit

(b) Ideal pulse input

(c) Practical pulse

FIGURE 4.19

An *RLC* circuit with a pulse input.

Solution The PSpice schematic is shown in Fig. 4.20(a). The voltage and current markers display the output waveforms in probe at the end of the simulation. The input voltage is specified by a pulse source as shown in Fig. 4.20(b). The transient analysis is set at the analysis setup as shown in Fig. 4.20(c), and its specifications are set at the transient menu as shown in Fig. 4.20(d). The circuit file contains the following statements:

```
Example 4.2   Pulse response of an RLC circuit
▲ *   PULSE  (-VS +VS   TD  TR  TF  PW  PER)  ; Pulse input
    VIN   7   0   PULSE  (-220V  220V  0  1NS  1NS  100US  200US)
▲ ▲ R1   7   5   2
    L1  5   3   50UH
    C1  3   0   10UF
▲ ▲ ▲ *. TRAN   TSTEP  TSTOP            ; Command for transient analysis
    .TRAN      1US   400US
    *.PRINT  TRAN V(R1)    V(L1)  V(C1) ; Prints to the output file
    *.PLOT   TRAN  V(3)      I(R1)  ; Plots in the output file
    .PROBE                          ; Graphical waveform analyzer
.END                                ; END of circuit file
```

It should be noted that with the .PROBE command, there is no need for the .PLOT command; .PLOT generates the plot in the output file, while .PROBE sends graphical output to the monitor screen that can directly be dumped into a plotter or a printer. The .PRINT and .PLOT commands are made ineffective by placing an asterisk (*) in front of them. The results of the .PRINT and .PLOT statements can be obtained by printing the contents of the output file EX4-2.OUT.

The results of the transient response that are produced on the display by .PROBE command are shown in Fig. 4.21.

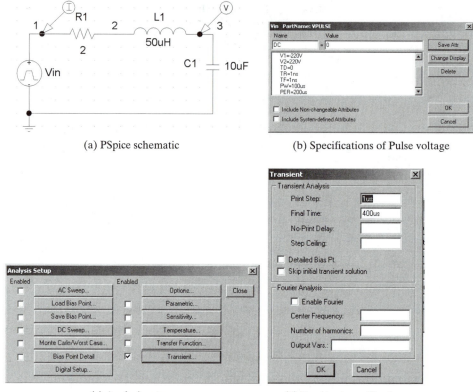

(a) PSpice schematic

(b) Specifications of Pulse voltage

(c) Analysis setup

(d) Transient specifications

FIGURE 4.20

PSpice schematic for Example 4.2.

Example 4.3: Effect of resistors on the transient pulse response of an *RLC* circuit

Three *RLC* circuits with $R = 2\ \Omega, 1\ \Omega$, and $8\ \Omega$ and are shown in Fig. 4.22(a). The inputs are identical step voltages, as shown in Fig. 4.22(b). Use PSpice to calculate and plot the transient response from 0 to 400 μs with an increment of 1 μs. The capacitor voltages are the outputs $V(3)$, $V(6)$, and $V(9)$, which are to be plotted. The circuit is to be stored in the file EX4-3. CIR, and the outputs are to be stored in the file EX4-3.OUT. The results should also be made available for display and hard copy by the .PROBE command.

Solution The description of the circuit file is similar to that of Example 4.2, except the input is a step voltage instead of a pulse voltage. The circuit may be regarded as three *RLC* circuits having three separate inputs. The step signal can be represented by piecewise linear source, which is described in general by

```
PWL (T1 V1 T2 V2 . . . TN VN)
```

where VN is the voltage at time TN.

Example 4.2   Pulse response of a RLC circuit

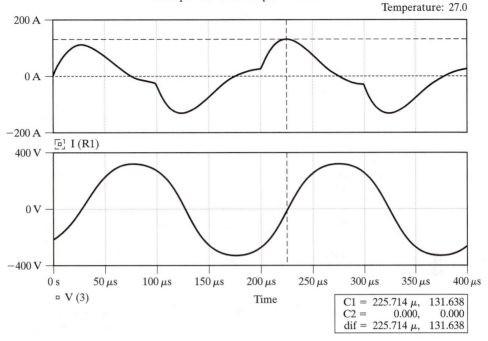

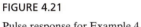

FIGURE 4.21

Pulse response for Example 4.2.

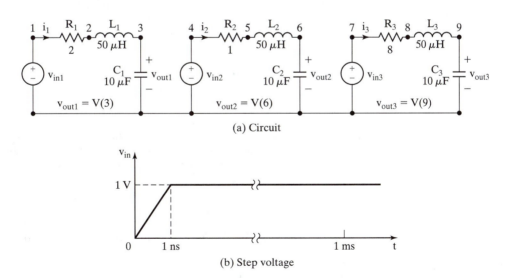

FIGURE 4.22

RLC circuits with step-pulse input voltages.

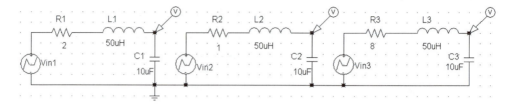

FIGURE 4.23

PSpice schematic for Example 4.3.

Assuming a rise time of 1 ns, the step voltage of Fig. 4.22(b) can be described by

```
PWL (0 0 1NS 1V 1MS 1V)
```

The PSpice schematic is shown in Fig. 4.23. The voltage and current markers display the output waveforms in probe at the end of the simulation. The input voltages are specified by three pulse sources, similar to Fig. 4.20(b). Transient analysis is set at the analysis setup as shown in Fig. 4.20(b), and its specifications are set at the transient menu as shown in Fig. 4.20(d).

The listing of the circuit file follows.

Example 4.3   Step response of series *RLC* circuits

```
▲      VI1  1   0    PWL  (0   0    1NS    1V    1MS    1V)  ; step of 1 V
       VI2  4   0    PWL  (0   0    1NS    1V    1MS    1V)  ; step of 1 V
       VI3  7   0    PWL  (0   0    1NS    1V    1MS    1V)  ; step of 1 V
▲ ▲    R1   1   2    2
       L1   2   3    50UH
       C1   3   0    10UF
       R2   4   5    1
       L2   5   6    50UH
       C2   6   0    10UF
       R3   7   8    8
       L3   8   9    50UH
       C3   9   0    10UF
▲ ▲ ▲  .TRAN  1US    400US                 ; Transient analysis
       *.PLOT   TRAN  V(3)   V(6)   V(9)    ; Plots in the output file
       .PROBE                              ; Graphical waveform analyzer
       *
.END                                       ; End of circuit file
```

With the .PROBE command, there is no need for .PLOT command. The .PLOT command is made ineffective by preceding it with an asterisk (*). The results of the .PLOT statement can be obtained by printing the contents of the output file EX4-3.OUT. The results of the transient analysis that are produced on the display by the .PROBE command are shown in Fig. 4.24.

Example 4.4:   Transient response of an *RLC* circuit with a sinusoidal input voltage

Repeat Example 4.2 if the input voltage is a sine-wave of $v_{in} = 10\sin(2\pi \times 5000t)$.

Solution   For a sinusoidal voltage $v_{in} = 10\sin(2\pi \times 5000t)$, the model is

```
SIN (0 10V 5KHZ)
```

Example 4.3   Step-response of series RLC circuits

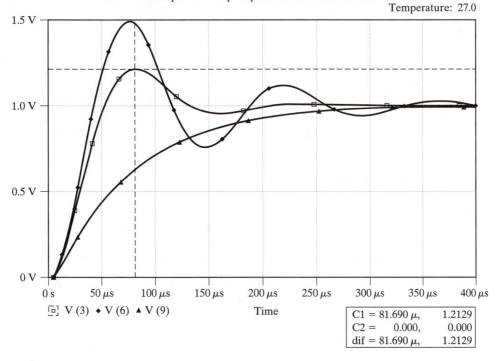

FIGURE 4.24

Step response of RLC circuits for Example 4.3.

The PSpice schematic is shown in Fig. 4.25(a). The voltage and current markers display the output waveforms in probe at the end of the simulation. The input voltage is specified by a sinusoidal source as shown in Fig. 4.25(b). The transient analysis is set at the analysis setup as shown in Fig. 4.20(c), and its specifications are set at the transient menu as shown in Fig. 4.20(d).

The circuit file contains the following statements:

Example 4.4   An *RLC* circuit with sinusoidal input voltage

```
▲ *     SIN   (VO    VA    FREQ)          ; Simple sinusoidal source
      VIN   7    0    SIN   (0  10V 5KHZ)   ; Sinusoidal input voltage
▲ ▲   R1   7    5    2
       L1   5    3    50UH
       C1   3    0    10UF
▲ ▲ ▲ .TRAN       1US    500US             ; Transient analysis
       .PLOT       TRAN   V(3)   V(7)       ; Plots in the output file
       .PROBE                               ; Graphical waveform analyzer
   .END                                     ; End of circuit file
```

The results of the transient response that are produced on the display by the .PROBE command are shown in Fig. 4.26. The results of the .PLOT statement can be obtained by printing the contents of the output file EX4-4.OUT.

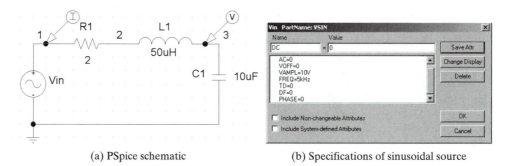

(a) PSpice schematic

(b) Specifications of sinusoidal source

FIGURE 4.25

PSpice schematic for Example 4.4.

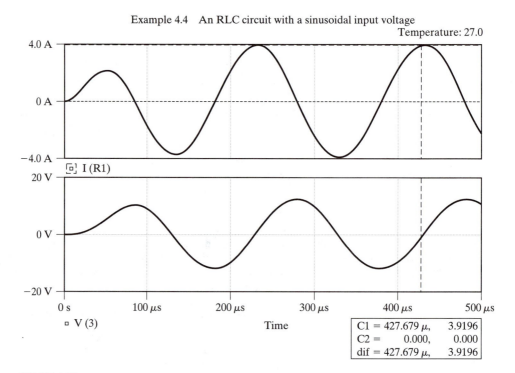

FIGURE 4.26

Transient response for Example 4.4.

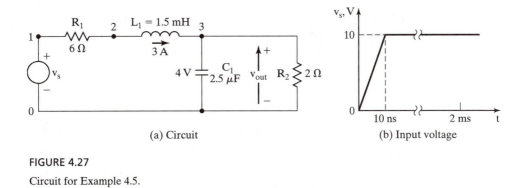

(a) Circuit (b) Input voltage

FIGURE 4.27

Circuit for Example 4.5.

Example 4.5: Transient step-response of an *RLC* circuit with an initial inductor current

For the circuit of Fig. 4.27(a), calculate and plot the transient response from 0 to 1 ms with a time increment of 5 μs. The output voltage is taken across resistor R_2, and the input voltage is shown in Fig. 4.27(b). The results should be made available for display and hard copy by using Probe. The model parameters for the resistor are $R = 1$, TC1 = 0.02, TC2 = 0.005; for the capacitor, $C = 1$, VC1 = 0.01, VC2 = 0.002, TC1 = 0.02, TC2 = 0.005; and for the inductor, $L = 1$, IL1 = 0.1, IL2 = 0.002, TC1 = 0.02, TC2 = 0.005. The operating temperature is 50°C.

Solution The PSpice schematic is shown in Fig. 4.28(a). The voltage markers are displays the input and output waveforms in probe at the end of the simulation. The step input voltage is specified by a PWL source as shown in Fig. 4.28(b). The breakout devices are used to specify the parameters of model RMD for R_1 and R_2 as shown in Fig. 4.28(c), of model LMOD for L_1 as shown

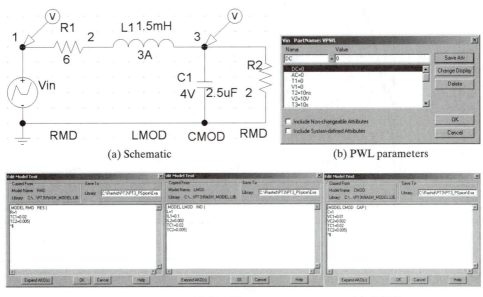

(a) Schematic (b) PWL parameters

(c) RMD parameters (d) LMOD parameters (e) CMOD parameters

FIGURE 4.28

PSpice schematic for Example 4.5.

in Fig. 4.28(d) and of model CMOD for C_1 as shown in Fig. 4.28(e). The transient analysis is set at the analysis setup as shown in Fig. 4.20(c), and its specifications are set at the transient menu as shown in Fig. 4.20(d). The circuit file contains the following statements:

Example 4.5   Transient response of an *RLC* circuit

```
▲   VS  1  0     PWL (0    0    10NS 10V 2MS 10V)   ; Step voltage of PWM waveform
▲ ▲ R1  1 2       RMOD       6OHM        ; Resistance with model RMOD
    L1  2 3       LMOD       1.5MH IC=3A ; Initial current of 3 A and model LMOD
    C1  3 0       CMOD       2.5UF IC=4V ; Initial voltage of 4 V and model name
                                           CMOD
    R2  3 0       RMOD       2OHM
    .TEMP 50                             ; Operating temperature of 50 degrees C
    *   Model statements for resistor, inductor, and capacitor
.MODEL  RMOD RES (R=1 TC1=0.02 TC2=0.005)
.MODEL  CMOD CAP (C=1 VC1=0.01 VC2=0.002 TC1=0.02 TC2=0.005)
.MODEL  LMOD IND (L=1 IL1=0.1  IL2=0.002 TC1=0.02 TC2=0.005)
▲ ▲ ▲ .TRAN  5US    1MS    UIC        ; Transient analysis with UIC
       *.PLOT TRAN  V(3)   V(1)       ; Not needed with .PROBE
       .PROBE                         ; Graphical waveform analyzer
.END                                  ; End of circuit file
```

The results of the simulation that are obtained by .PROBE are shown in Fig. 4.29. With .PROBE, the .PRINT command is redundant. The inductor has an initial current of 3 A, which is taken into consideration by UIC in the .TRAN command.

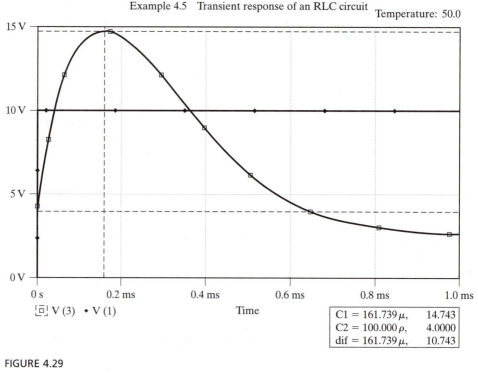

Example 4.5   Transient response of an RLC circuit   Temperature: 50.0

C1 = 161.739 μ,	14.743
C2 = 100.000 ρ,	4.0000
dif = 161.739 μ,	10.743

FIGURE 4.29

Transient response for Example 4.5.

Example 4.6: Transient step-response of an *RLC* circuit with an initial capacitor voltage

Repeat Example 4.5, assuming that the voltage across the capacitor is set by the .IC command instead of an IC condition, and UIC is *not* specified.

Solution In order to set the capacitor voltage by the .IC command, the following statement should be added:

```
.IC   V(3)=4V          ; Node 3 is set to 4 V
```

and the UIC is removed from the .TRAN statement as follows:

```
.TRAN   5US   1MS       ; Transient analysis without UIC
```

The PSpice schematic is shown in Fig. 4.30 in which the inductor L_1 has no initial current. The voltage marker displays the output waveform in probe at the end of the simulation. The step input voltage is specified by a PWL source as shown in Fig. 4.28(b). The breakout devices are used to specify the parameters of model RMD for R_1 and R_2 as shown in Fig. 4.28(c), of model LMOD for L_1 as shown in Fig. 4.28(d) and of model CMOD.

The result of the simulation is shown in Fig. 4.31. Notice that the response is completely different because the node voltage of the capacitor was set to 4 V.

4.8 SWITCHES

PSpice allows the simulation of a special kind of switch, shown in Fig. 4.32, whose resistance varies continuously depending on the voltage or current. When the switch is on, the resistance is R_{ON}, and when it is off, the resistance becomes R_{OFF}.

Two types of switches are permitted in PSpice:

Voltage-controlled switches
Current-controlled switches
Time-dependent switches.

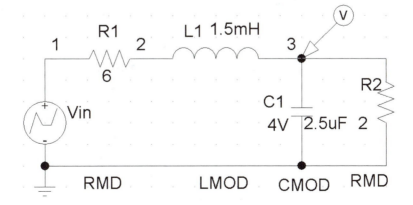

FIGURE 4.30

PSpice schematic for Example 4.6.

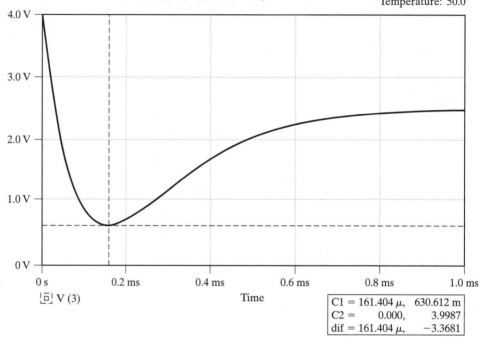

Example 4.6 Transient response of an RLC circuit

Temperature: 50.0

FIGURE 4.31

Transient response for Example 4.6.

Note. The voltage- and current-controlled switches are not available in SPICE2, but they are available in SPICE3.

4.8.1 Voltage-Controlled Switch

The symbol for a voltage-controlled switch is S. The name of this switch must start with S, and it takes the general form

```
S<name> N+ N- NC+ NC- SNAME
```

where $N+$ and $N-$ are the two nodes of the switch. The current is assumed to flow from $N+$ through the switch to node $N-$. NC+ and NC- are the positive and negative nodes

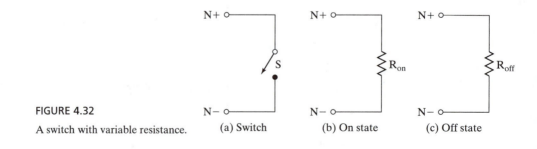

FIGURE 4.32

A switch with variable resistance.

(a) Switch

(b) On state

(c) Off state

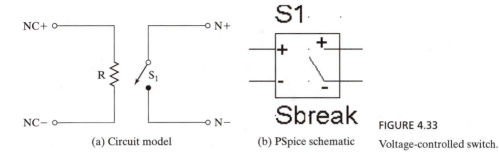

FIGURE 4.33

Voltage-controlled switch.

(a) Circuit model (b) PSpice schematic

TABLE 4.8 Model Parameters for Voltage-Controlled Switch

Name	Meaning	Units	Default
VON	Control voltage for on state	Volts	1.0
VOFF	Control voltage for off state	Volts	0
RON	On resistance	Ohms	1.0
ROFF	Off resistance	Ohms	10^6

of the controlling voltage source, as shown in Fig. 4.33(a). SNAME is the model name. The resistance of the switch varies depending on the voltage across the switch. The type name for a voltage-controlled switch is VSWITCH, and the model parameters are shown in Table 4.8. The PSpice schematic is shown in Fig. 4.33(b).

Voltage-controlled switch statement

```
S1      6  5  4   0   SMOD
.MODEL  SMOD  VSWITCH  (RON=0.5  ROFF=10E+6  VON=0.7  VOFF=0.0)
```

Notes

1. R_{ON} and R_{OFF} must be greater than zero and less than 1/GMIN. The value of GMIN can be defined as an option, as described in the .OPTIONS command in Section 6.10. The default value of conductance, GMIN, is $1E - 12$ mhos.
2. The ratio of R_{OFF} to R_{ON} should be less than $1E + 12$.
3. The difficulty due to high gain of an ideal switch can be minimized by choosing the value of R_{OFF} as high as permissible and that of R_{ON} as low as possible as compared to other circuit elements, within the limits of allowable accuracy.

Example 4.7: Transient step-response with a voltage-controlled switch

A circuit with a voltage-controlled switch is shown in Fig. 4.34. If the input voltage is $V_s = 200\sin(2000\pi t)$, plot the voltage at node 3 and the current through the load resistor R_L for a time duration of 0 to 1 ms with an increment of 5 µs. The model parameters of the switch are RON = 5M, ROFF = 10E + 9, VON = 25M, and VOFF = 0.0. The results should be available for display by Probe.

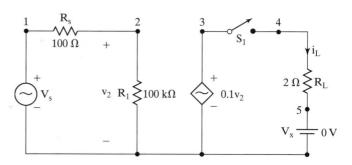

FIGURE 4.34

A circuit with a voltage-controlled switch.

Solution The PSpice schematic is shown in Fig. 4.35(a). The gain of the voltage-controlled voltage source E1 is set to 0.1. The voltage marker displays the output waveform in probe at the end of the simulation. The input voltage is specified by a sinusoidal source. The breakout voltage controlled switch (S1) is used to specify the switch model Sbreak as shown in Fig. 4.35(b), and its model parameters are RON = 5M, ROFF = 10E + 9, VON = 25M, and VOFF = 0. The voltage source $V_X = 0$ V is inserted to monitor the output current. The listing of the circuit file is as follows.

Example 4.7 A voltage-controlled switch

```
▲    VS    1    0    SIN  (0  200V  1KHZ) ; Sinusoidal voltage of 200 V peak
▲▲   RS    1    2    100OHM
     R1    2    0    100KOHM
     E1    3    0    2    0    0.1  ; V-controlled source with a gain of 0.1
     RL    4    5    2OHM
     VX    5    0    DC    OV        ; Measures the load current
     S1    3    4    3    0        SMOD ; V-controlled switch with model SMOD
     *    Switch model descriptions
     .MODEL   SMOD    VSWITCH (RON=5M ROFF=10E+9 VON=25M VOFF=0.0)
▲▲▲  .TRAN   5US    1MS        ; Transient analysis
     *.PLOT   TRAN    V(3)  I(VX)    ; Not needed with .PROBE
     .PROBE                  ; Graphical waveform analyzer
.END                        ; End of circuit file
```

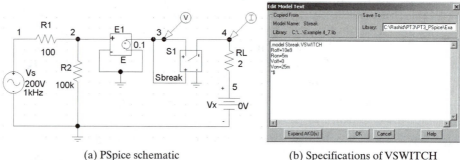

(a) PSpice schematic   (b) Specifications of VSWITCH

FIGURE 4.35

PSpice schematic for Example 4.7.

Example 4.7   A Voltage-Controlled Switch

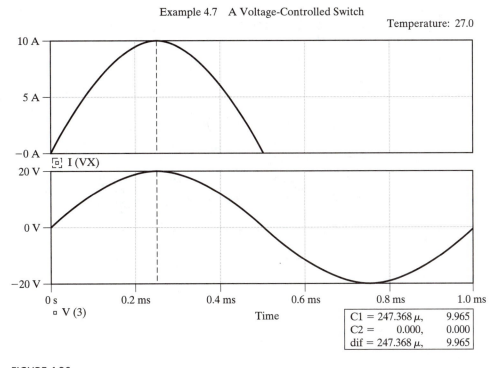

FIGURE 4.36

Transient response for Example 4.7.

The results of the simulation that are obtained by Probe are shown in Fig.4.36, which is the output of a diode rectifier. The switch S_1 behaves as a diode. The .PROBE statement has made .PLOT redundant.

Example 4.8:   DC Transients with a voltage-controlled switch

The *RLC* circuit of Fig. 4.37(a) is subjected to dc transients by opening and closing switch S_1, as shown in Fig. 4.37(b). Use PSpice to calculate and plot the inductor current i_L and the capacitor voltage v_C from 0 to 20 ms with a time increment of 5 μs. The model parameters of the switch are RON = 0.01, ROFF = 10E + 5, VON = 0.1V, and VOFF = 0V.

Solution   The PSpice schematic is shown in Fig. 4.38(a). The voltage marker displays the output voltage waveform in probe at the end of the simulation. The input voltage is dc and the controlling voltage for the switch is specified by a pulse source as shown in Fig. 4.38(a). The breakout voltage-controlled switch (S1) is used to specify the switch model Sbreak as shown in Fig. 4.38(b), and its model parameters are RON = 0.01, ROFF = 10E + 5, VON = 0.1, and VOFF = 0. The listing of the circuit file is as follows:

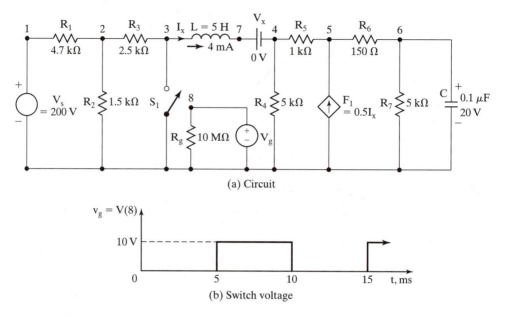

(a) Circuit

(b) Switch voltage

FIGURE 4.37

Dc circuit subjected to transients.

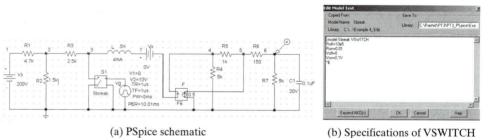

(a) PSpice schematic   (b) Specifications of VSWITCH

FIGURE 4.38

PSpice schematic for Example 4.8.

Example 4.8   Dc transients

```
▲   VS    1   0   DC      200V    ;  Voltage source of 100 V dc
    VG    8   0   PULSE  (0V   10V    5MS    1US    1US    5MS    10.01MS)
▲ ▲ RG    8   0   10MEG          ; High resistance for continuity
    R1    1   2   4.7K
    R2    2   0   1.5K
    R3    2   3   2.5K
    R4    4   0   5K
    R5    4   5   1K
    R6    5   6   150
    R7    6   0   5K
```

```
VX     7    4   DC    0V            ; Measures current through L₁
C1     6    0   0.1UF   IC=10V      ; Capacitor with initial voltage
L1     3    7   5H      IC=4MA      ; Inductor with initial current
F1     0    5   VX      0.5         ; Current-controlled current source
S1     3    0   8   0     SMOD      ; Voltage switch with model SMOD
.MODEL SMOD  VSWITCH (RON=0.01 ROFF=10E+5 VON=0.1V VOFF=0V)
.TRAN  5US  20MS   UIC             ; Transient analysis with UIC
.PROBE                             ; Graphical waveform analyzer
.END                               ; End of circuit file
```

The inductor current $I(L1)(=I(VX))$ and capacitor voltage $V(6)$, which are obtained by the .PROBE command, are shown in Fig. 4.39. Since the circuit has two energy-storage elements, it exhibits the characteristics of a second-order system.

4.8.2 Current-Controlled Switch

The symbol for a current-controlled switch is W. The name of the switch must start with W, and it takes the general form

```
W(name)  N+ N−  VN WNAME
```

where $N+$ and $N−$ are the two nodes of the switch. V_N is a voltage source through which the controlling current flows, as shown in Fig. 4.40(a). WNAME is the model name. The resistance of the switch depends on the current through the switch. The type name for a current-controlled switch is ISWITCH, and the model parameters are

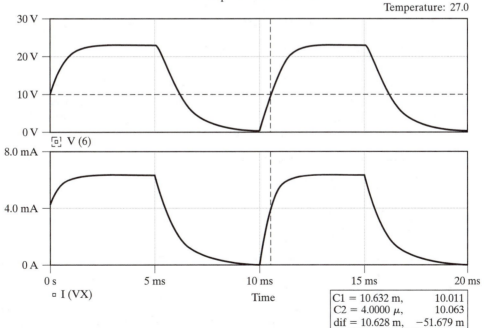

Example 4.8   DC Transients

Temperature: 27.0

C1 = 10.632 m,	10.011
C2 = 4.0000 µ,	10.063
dif = 10.628 m,	−51.679 m

FIGURE 4.39

Dc transient responses for Example 4.8.

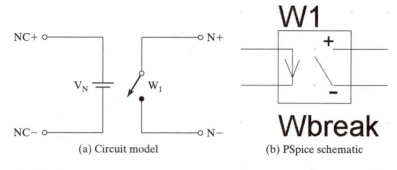

(a) Circuit model   (b) PSpice schematic

FIGURE 4.40

Current-controlled switch.

TABLE 4.9   Model Parameters for Current-Controlled Switch

Name	Meaning	Units	Default
ION	Control current for on state	Amps	1E − 3
IOFF	Control current for off state	Amps	0
RON	On resistance	Ohms	1.0
ROFF	Off resistance	Ohms	10^6

shown in Table 4.9. The PSpice schematic is shown in Fig. 4.40(b).

Current-controlled switch statement

```
W1      6  5  VN  RELAY
.MODEL RELAY ISWITCH (RON=0.5 ROFF=10E+6 ION=0.07 IOFF=0.0)
```

Note. The current through voltage source V_N controls the switch. The voltage source V_N must be an independent source, and it can have a zero or a finite value. The limitations of the parameters are similar to those for the voltage-controlled switch.

Example 4.9:   Transient response of an *LC* circuit with a current-controlled switch

A circuit with a current-controlled switch is shown in Fig. 4.41. Plot the capacitor voltage and the current through the inductor for a time duration of 0 to 160 μs with an increment of 1 μs. The model parameters of the switch are RON = 1E + 6 ROFF = 0.001, ION = 1MA, and IOFF = 0. The results should be available for display by Probe.

Solution   The PSpice schematic is shown in Fig. 4.42(a). The voltage and current markers display the output waveforms in probe at the end of the simulation. The initial voltage on the capacitor C1 specifies the input source. The breakout current-controlled switch (W1) is used to specify the parameters of the switch model Wbreak as shown in Fig. 4.42(b), and its model parameters are ION = 1MA, IOFF = 0, RON = 1E + 6, and ROFF = 0.01. The voltage source V_X = 0 V is inserted to monitor the controlling current. The listing of the circuit file is as follows:

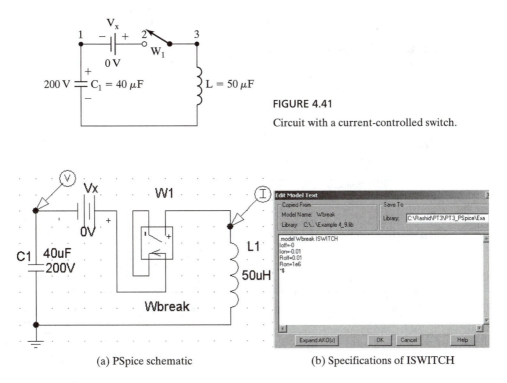

FIGURE 4.41

Circuit with a current-controlled switch.

(a) PSpice schematic

(b) Specifications of ISWITCH

FIGURE 4.42

PSpice schematic for Example 4.9.

Example 4.9   A current-controlled switch

```
▲ ▲   C1  1    0   40UF   IC=200        ; With an initial voltage of 200 V
      VX   2    1    DC   0V             ; Dummy voltage source of Vx=0
      W1   2    3    VX   SMOD           ; I-controlled switch with model SMOD
      .MODEL   SMOD   ISWITCH (RON=1E+6 ROFF=0.001 ION=1MA IOFF=0)
      L1   3    0    50UH
▲ ▲ ▲ .TRAN  1US  160US  UIC            ; Transient analysis with UIC
      .PLOT TRAN  V(1) I(VX)            ; Not needed with .PROBE
      .PROBE                            ; Graphical waveform analyzer
.END                                    ; End of circuit file
```

The results of the simulation that are obtained by Probe are shown in Fig. 4.43. The switch W_1 acts as diode and allows only positive current flow. The initial voltage on the capacitor is the driving source.

4.8.3   Time-Dependent Switches

The PSpice schematic supports two types of time-dependent switches:

Time-dependent close switch
Time-dependent open switch.

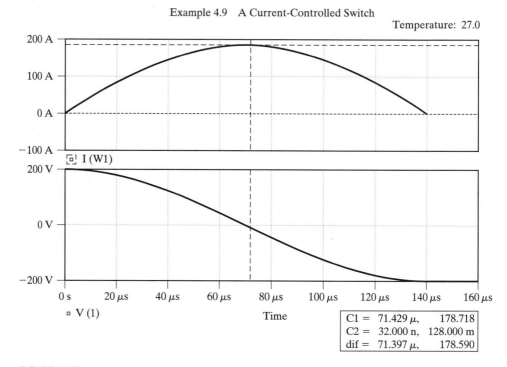

FIGURE 4.43

Transient response for Example 4.9.

The closing or the opening time and the transition time of the switch are specified by the switch parameters as shown in Table 4.10.

Time-Dependent Close Switch: This switch is normally open; setting the closing time closes it. Its schematic is shown in Fig. 4.44(a). The parameters are shown in Fig. 4.44(b).

   Figure 4.45(a) shows an RC circuit with a step input voltage, which is generated by a PWL source. The switch Sw_tClose normally remains open. At $t = 50$ μs, the switch closes, and the capacitor is changed exponentially with a time constant of R_1C_1 as shown in Fig. 4.45(b).

Time-Dependent Open Switch: This switch is normally closed; setting the opening time opens it. Its schematic is shown in Fig. 4.46(a). The parameters are shown in Fig. 4.46(b).

TABLE 4.10   Model Parameters for Close/Open Switch

Name	Meaning	Default
TClose/Topen	Time at which switch begins to close/open	0
ttran	Time required to switch states from off state to on state (must be realistic, not 0)	1 μs
Rclosed	Closed state resistance	10 mΩ
Ropen	Open state resistance (Ropen/Rclosed < 1E + 10)	1 MΩ

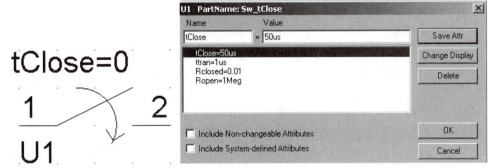

(a) Schematic   (b) Parameters of Sw_tClose

FIGURE 4.44

PSpice schematic for Sw_tClose switch.

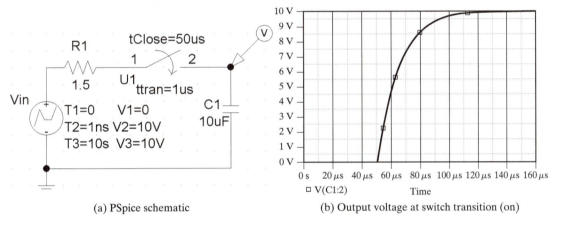

(a) PSpice schematic   (b) Output voltage at switch transition (on)

FIGURE 4.45

PSpice schematic for charging an *RC* circuit.

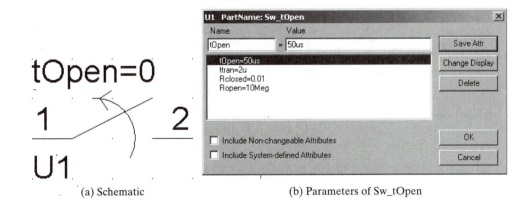

(a) Schematic   (b) Parameters of Sw_tOpen

FIGURE 4.46

PSpice schematic for Sw_tOpen switch.

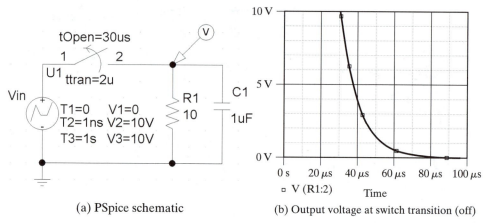

(a) PSpice schematic

(b) Output voltage at switch transition (off)

FIGURE 4.47

PSpice schematic for discharging an *RC* circuit.

Figure 4.47(a) shows an *RC* circuit with a step input voltage, which is generated by a PWL source. The switch Sw_tOpen remains normally close. At $t = 20$ μs, the switch opens and the capacitor is discharged exponentially with a time constant of $R_1 C_1$ as shown in Fig. 4.47(b).

SUMMARY

The symbols and statements for elements are

C   Capacitor

    `C<name> N+ N- CNAME CVALUE IC=VO`

L   Inductor

    `L<name> N+ N- LNAME LVALUE IC=IO`

R   Resistor

    `R<name> N+ N- RNAME RVALUE`

S   Voltage-controlled switch

    `S<name> N+ N- NC+ NC- SNAME`

W   Current-controlled switch

    `W<name> N+ N- VN WNAME`

The symbols and statements for transient sources are

EXP   Exponential source

    `EXP (V1 V2 TRD TRC TFD TFC)`

POLY   Polynomial source

    POLY(n) <(controlling) nodes> <(coefficients) values>

PULSE   Pulse source

    PULSE (V1 V2 TD TR TF PW PER)

PWL   Piecewise linear source

    PWL (T1 V1 T2 V2 . . . TN VN)

SFFM   Single-frequency frequency modulation

    SFFM (VO VA FC MOD FS)

SIN   Sinusoidal source

    SIN (VO VA FREQ TD ALP THETA)

E   Voltage-controlled voltage source

    E<name> N+ N− NC+ NC− <(voltage gain) value>

F   Current-controlled current source

    F<name> N+ N− VN <(current gain) value>

G   Voltage-controlled current source

    G<name> N+ N− NC+ NC− <(transconductance) value>

H   Current-controlled voltage source

    H<name> N+ N− VN <(transresistance) value>

I   Independent current source

    I<name> N+ N− [DC <value>]
    +   [(transient value)
    +   [PULSE] [SIN] [EXP] [PWL] [SFFM] [source arguments]]

V   Independent voltage source

    V<name> N+ N− [DC <value>]
    +   [(transient value)
    +   [PULSE] [SIN] [EXP] [PWL] [SFFM] [source arguments]]

The commands that are generally used for transient analysis follow:

.END	End of circuit
.IC	Initial transient conditions
.MODEL	Model
.OP	Operating point
.PLOT	Plot
.PRINT	Print
.PROBE	Probe
.TEMP	Temperature
.TRAN	Transient analysis

UIC   Use initial conditions
.WIDTH   Width

REFERENCES

[1]   Robert Boylestad, *Introductory Circuit Analysis*, 10th ed. Upper Saddle River, New Jersey: Prentice Hall, 2003.

[2]   R.C. Dorf and J.A. Svoboda, *Introduction to Electric Circuits*, 5th ed. New York: John Wiley & Sons, 2001.

[3]   William H. Hayt, Jr., Jack E. Kemmerly, and Steven M. Durbin, *Engineering Circuit Analysis*, 6th ed. Upper Saddle River, New Jersey: Prentice Hall, 2003.

[4]   J. David Irwin, *Basic Engineering Circuit Analysis*. New York: Macmillan, 1989.

[5]   J.D. Irwin, *Basic Engineering Circuit Analysis*, 5th ed. Upper Saddle River, New Jersey: Prentice Hall, 1996.

[6]   D. E. Johnson, J. Hilburn, and J. R. Johnson, *Basic Electric Circuit Analysis*. Englewood Cliffs, New Jersey: Prentice Hall, 1990.

[7]   J. W. Nilsson and S. A. Riedel, *Electric Circuits*, 6th ed. Upper Saddle River, New Jersey: Prentice Hall, 2000.

[8]   *PSpice Manual*, Irvine, California: MicroSim Corporation, 1992.

[9]   M.H. Rashid, *SPICE for Power Electronics and Electric Power*. Englewood Cliffs, New Jersey: Prentice Hall, 1993.

PROBLEMS

Write the PSpice statements for Problems 4.1 to 4.6, 4.19, and 4.20.

4.1 A capacitor C_1 is connected between nodes 5 and 6 and has a value of 10 pF and an initial voltage of -20 V.

4.2 A capacitor C_1 is connected between nodes 5 and 6 and has a nominal value of $C = 10$ pF. The operating temperature is $T = 55°$C. The capacitance, which is a function of its voltage and the operating temperature, is given by

```
C₁ = C * (1 + 0.01 * V + 0.002 * V²)
        * [1 + 0.03 * (T − T0) + 0.05 * (T − T0)²]
```

4.3 An inductor L_1 is connected between nodes 5 and 6; it has a value of 0.5 mH and carries an initial current of 0.04 mA.

4.4 An inductor L_1 that is connected between nodes 3 and 4 has a nominal value of $L = 1.5$ mH. The operating temperature is $T = 55°$C. The inductance, which is a function of its current and the operating temperature, is given by

```
L₁ = L * (1 + 0.01 * I + 0.002 * I²)
        * [1 + 0.03 * (T − T0) + 0.05 * (T − T0)²]
```

4.5 The various voltage or current waveforms that are connected between nodes 4 and 5 are shown in Fig. P4.5.

4.6 A voltage source that is connected between nodes 4 and 5 is given by

$$v = 2 \sin\{2\pi\ 50{,}000t + 5 \sin(2\pi\ 1000t\}$$

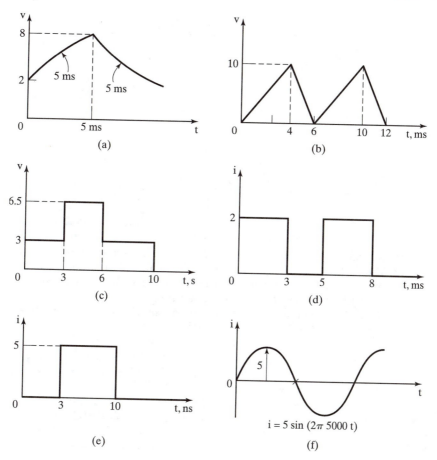

FIGURE P4.5

4.7 The *RLC* circuit of Fig. P4.7 is to be simulated to calculate and plot the transient response from 0 to 2 ms with an increment of 10 μs. The voltage across resistor *R* is the output. The input and output voltages are to be plotted in the output file. The results should also be made available for display and hard copy by the .PROBE command.

4.8 Repeat Problem 4.7 for the circuit of Fig. P4.8, where the output is taken across capacitor *C*.

4.9 Repeat Problem 4.7 for the circuit of Fig. P4.9, where the output is the current i_s through the circuit.

4.10 Repeat Problem 4.7 if the input is a step input, as shown in Fig. 4.19(b).

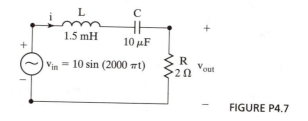

FIGURE P4.7

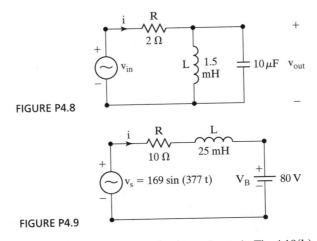

FIGURE P4.8

FIGURE P4.9

4.11 Repeat Problem 4.8 if the input is a step input, as shown in Fig. 4.19(b).

4.12 The *RLC* circuit of Fig. P4.12(a) is to be simulated to calculate and plot the transient response from 0 to 2 ms with an increment of 5 μs. The input is a step current, as shown in Fig. P4.12(b). The voltage across resistor *R* is the output. The input and output voltages are to be plotted in the output file. The results should also be made available for display and hard copy by the .PROBE command.

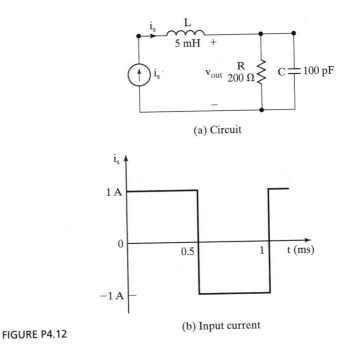

(a) Circuit

(b) Input current

FIGURE P4.12

4.13 Repeat Problem 4.12 for the circuit of Fig. P4.13, where the output is taken across capacitor *C*.

4.14 Plot the transient response of the circuit in Fig. P4.14 from 0 to 5 ms with a time increment of 25 μs. The output voltage is taken across the capacitor. Use Probe for graphical output.

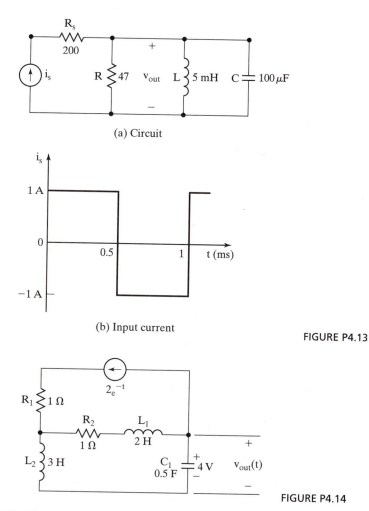

(a) Circuit

(b) Input current

FIGURE P4.13

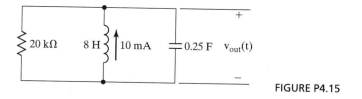

FIGURE P4.14

4.15 Repeat Problem 4.14 for the circuit in Fig. P4.15.

FIGURE P4.15

4.16 For the circuit in Fig. P4.16(a), calculate and plot the transient response of the output voltage from 0 to 2 ms with a time increment of 5 μs. The input voltage is shown in Fig. P4.16(b). The results should be available for display and hard copy by Probe.

4.17 Repeat Problem 4.16 for the input voltage shown in Fig. P4.17.

4.18 The parallel RLC circuit of Fig. P4.18(a) is supplied from a step input current as shown in Fig. P4.18(b). Use PSpice to calculate and plot the capacitor voltage v_C from 0 to 500 μs with a time increment of 1 μs for R = 200 Ω, 500 Ω, and 1 kΩ.

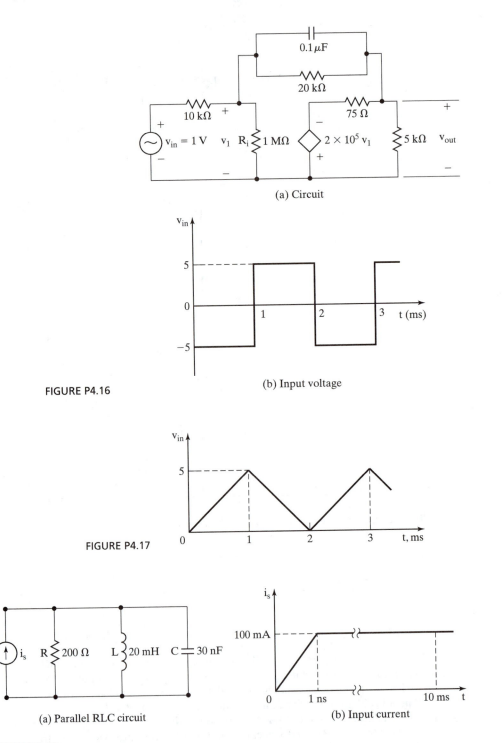

(a) Circuit

(b) Input voltage

FIGURE P4.16

FIGURE P4.17

(a) Parallel RLC circuit

(b) Input current

FIGURE P4.18

4.19 A switch that is connected between nodes 5 and 4 is controlled by a voltage source between nodes 3 and 0. The switch will conduct if the controlling voltage is 0.5 V. The on-state resistance is 0.5 Ω, and the off-state resistance is 2E + 6 Ω.

4.20 A switch that is connected between nodes 5 and 4 is controlled by a current. The voltage source V_1 through which the controlling current flows is connected between nodes 2 and 0. The switch will conduct if the controlling current is 0.55 mA. The on-state resistance is 0.5 Ω, and the off-state resistance is 2E + 6 Ω.

4.21 For the circuit in Fig. P4.21, plot the transient response of the load and source current for five cycles of the switching period, with a time increment of 25 µs. The model parameters of the voltage-controlled switches are RON = 0.025, ROFF = 1E + 8, VON = 0.05, and VOFF = 0. The output should also be available for display and hard copy by Probe.

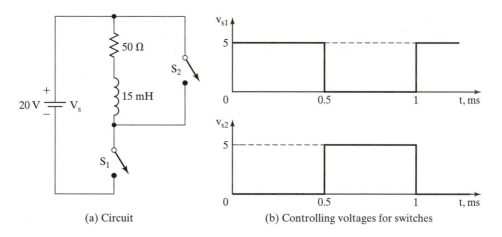

(a) Circuit   (b) Controlling voltages for switches

FIGURE P4.21

4.22 An *RLC* circuit is shown in Fig. P4.22. The voltage-controlled switch S_1 is closed to position *a* at $t = 0$ and then moved instantaneously to position *b* at $t = 10$ ms. Use PSpice to calculate and plot the capacitor voltage v_C from 0 to 20 ms with a time increment of 5 µs. Assume that the model parameters of the switch are RON = 0.01, ROFF = 10E + 5, VON = 0.1V, and VOFF = 0V.

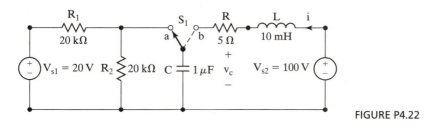

FIGURE P4.22

4.23 Use PSpice to plot the pulse response of the output voltage v_o for the *RC* circuit as shown in Fig. P4.23 for time constant $\tau = RC = 0.1$ ms, 1 ms, and 10 ms. The input is a pulse voltage v_S that switches between 1V to −1V at a frequency of 500 Hz.

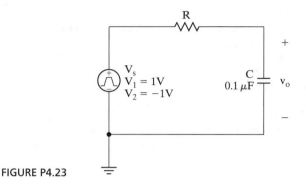

FIGURE P4.23

4.24 Use PSpice to plot the pulse response of the output voltage v_o for the CR circuit as shown in Fig. P4.24 for time constant $\tau = RC = 0.1$ ms, 1 ms, and 10 ms. The input is a pulse voltage v_S that switches between 1 V to -1 V at a frequency of 500 Hz.

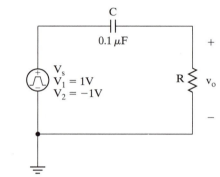

FIGURE P4.24

4.25 Use the parametric command in PSpice to vary the value of resistance R so that it is 1, 2, and 8 Ω for the circuit shown in Figure 4.23(a) for Example 4.3. Plot the capacitor voltage $v_C(t)$ against time.

C H A P T E R   5

AC Circuit Analysis

After completing this chapter, students should be able to

- Model ac (frequency response) voltage and current sources and specifying their parameters
- Understand the response of ac analysis and obtain their output variables
- Perform the ac analysis (frequency response) of a circuit and setting the ac parameters
- Perform multiple analyses of a circuit
- Print the magnitude and phase angles of output voltages and currents
- Understand the response of transient analysis and obtain their output variables
- Model linear and nonlinear magnetic elements and specify their parameters
- Model ideal and lossy transmission lines and specify their parameters
- Model transient voltage and current sources and specify their parameters.

5.1   INTRODUCTION

Sources in a circuit are time variant. They are alternating current, or ac, and have both magnitude and phase displacement. Practical sources are generally sinusoidal or near sinusoidal. The behavior of a circuit is normally evaluated with a sinusoidal source. The steady-state voltages and currents that are normally used to evaluate the performance of a circuit can be calculated by applying the circuit laws that are applicable to dc circuits. The transient behavior is also important under some conditions, but the analysis becomes more complex compared to steady-state analysis.

To simplify the steady-state analysis of ac circuits with sinusoidal inputs, the circuit elements (R, L, and C) are represented in complex numbers, and all voltages and currents are expressed in phasor qualities. For example, a sinusoidal voltage of V_m with a phase delay of ϕ is represented as

$$V = V_m \underline{/\phi}$$

Inductance (L) and capacitor (C) are expressed as impedances:

$$Z_L = j2\pi L = j\omega$$
$$Z_C = -j/(2\pi C) = -j/(\omega C)$$

SPICE is an ideal software tool for simulating a circuit and for studying the behavior of voltages, currents, and power flow under steady-state and transient conditions. We have discussed the SPICE representations of circuit elements and sources in Chapters 3 and 4. In this chapter, we will simulate the steady-state analysis of a circuit and include coupled and nonlinear magnetic elements. Students are encouraged to apply the basic circuit laws and to verify the SPICE results by hand calculations.

5.2   AC OUTPUT VARIABLES

In ac analysis, the output variables are sinusoidal quantities and are represented by complex numbers. An output variable can have magnitude in decibels, phase, group delay, real part, and imaginary part. The output variables listed in Sections 5.2.1 and 5.2.2 are augmented by adding a suffix as follows:

Suffix	Meaning
(none)	Peak magnitude
M	Peak magnitude
DB	Peak magnitude in decibels
P	Phase in radians
G	Group delay (ΔPHASE/ΔFREQUENCY)
R	Real part
I	Imaginary part

In PSpice schematic, the advanced markers can be selected from the markers menu to display the magnitude, phase, group delay, real part, and imaginary part as shown in Fig. 5.1.

5.2.1   Voltage Output

The statements for ac analysis are similar to those for the dc sweep and the transient analysis, provided the suffixes are added as follows:

Variables

VM(5)	Magnitude of voltage at node 5 with respect to ground
VM(4,2)	Magnitude of voltage at node 4 with respect to node 2
VDB(R1)	DB magnitude of voltage across resistor R_1, where the first node (as defined in the circuit file) is assumed to be positive with respect to second node
VP(D1)	Phase of anode voltage of diode D_1 with respect to cathode
VCM(Q3)	Magnitude of the collector voltage of transistor Q_3 with respect to ground
VDSP(M6)	Phase of the drain-source voltage of MOSFET M_6
VBP(T1)	Phase of voltage at port B of transmission line T_1

FIGURE 5.1

Advanced markers.

VR(2,3)   Real part of voltage at node 2 with respect to node 3
VI(2,3)   Imaginary part of voltage at node 2 with respect to node 3

5.2.2   Current Output

The statements for ac analysis are similar to those for the dc sweep and the transient responses. However, only the currents through the elements in Table 5.1 are available. For all other elements, a zero-valued voltage source must be placed in series with the device (or device terminal) of interest. Then a print or plot statement should be used for the current through this voltage source.

TABLE 5.1   Current through Elements for AC Analysis

First Letter	Element
C	Capacitor
I	Independent current source
L	Inductor
R	Resistor
T	Transmission line
V	Independent voltage source

Variables	Meaning
IM(R5)	Magnitude of current through resistor R_5
IR(R5)	Real part of current through resistor R_5
II(R5)	Imaginary part of current through resistor R_5
IM(VIN)	Magnitude of current through source v_{in}
IR(VIN)	Real part of current through source v_{in}
II(VIN)	Imaginary part of current through source v_{in}
IAG(T1)	Group delay of current at port A of transmission line T_1

Example 5.1:   Defining PSpice output variables

For the circuit in Fig. 5.2, write the various voltages and currents in forms that are allowed by PSpice. The dummy voltage source of 0 V is introduced to measure current, I_L.

Solution

PSpice	Variables	Meaning
V_2	VM(2)	The peak magnitude of voltage at node 2
$\angle V_2$	VP(2)	The phase angle of voltage at node 2
$V_{1,2}$	VM(1,2)	The peak magnitude of voltage between nodes 1 and 2
$\angle V_{1,2}$	VP(1,2)	The phase angle of voltage between nodes 1 and 2
I_R	IM(VX)	The magnitude of current through voltage source v_X
$\angle I_R$	IP(VX)	The phase angle of current through voltage source v_X
I_L	IM(L1)	The magnitude of current through inductor L_1
$\angle I_L$	IP(L1)	The phase angle of current through inductor L_1
I_C	IM(C1)	The magnitude of current through capacitor C_1
$\angle I_C$	IP(C1)	The phase angle of current through capacitor C_1

5.3   INDEPENDENT AC SOURCES

The schematic of the ac voltage source is shown in Fig. 5.3(a) and its menu for setting the model parameters are shown in Fig. 5.3(b). In addition to the ac specifications, the dc (i.e., DC = 10 V) specification can be assigned to the same source. ACMAG is the magnitude and ACPHASE is the phase delay angle in degrees.

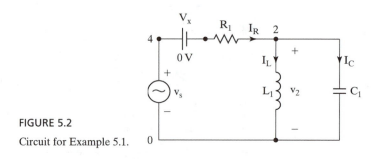

FIGURE 5.2

Circuit for Example 5.1.

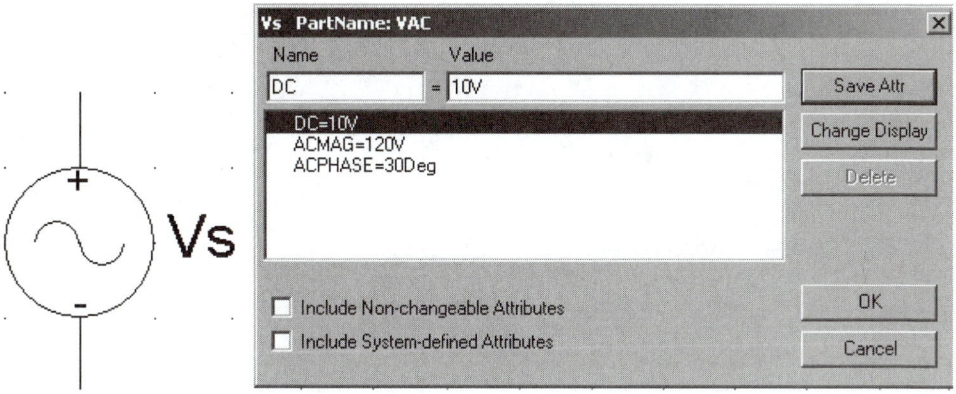

(a) Symbol   (b) Source parameters

FIGURE 5.3

Ac voltage source.

The statements for a voltage and current source have the following general forms, respectively:

```
V⟨name⟩  N+  N−   [AC⟨ (magnitude) value⟩ ⟨ (phase) value⟩]
I⟨name⟩  N+  N−   [AC⟨ (magnitude) value⟩ ⟨ (phase) value⟩]
```

The ⟨(magnitude) value⟩ is the peak value of sinusoidal voltage. The ⟨(phase) value⟩ is in degrees.

Some Typical Statements

```
VAC   5  6  AC  1V           ; Ac specification of 1V with 0 delay
VACP  5  6  AC  1V   45DEG   ; Ac specification of 1V with 45 delay
IAC   5  6  AC  1A           ; Ac specification of 1A with 0 delay
IACP  5  6  AC  1A   45DEG   ; Ac specification of 1V with 45 delay
```

5.4   AC ANALYSIS

The ac analysis calculates the frequency response of a circuit over a range of frequencies. Let us consider an *RC* circuit, as shown in Fig. 5.4(a). The output voltage is taken across the capacitor *C*. Using the voltage-divider rule, the voltage gain $G(j\omega)$, which is the ratio of the output voltage $V_{out}(j\omega)$ to the input voltage $V_{in}(j\omega)$ can be expressed as

$$G(j\omega) = \frac{V_{out}(j\omega)}{V_{in}(j\omega)} = \frac{-j/(\omega C)}{R - j/(\omega C)} = \frac{1}{1 + j\omega RC}$$

$$= \frac{1}{1 + j\omega\tau} \tag{5.1}$$

where $\tau = RC$.

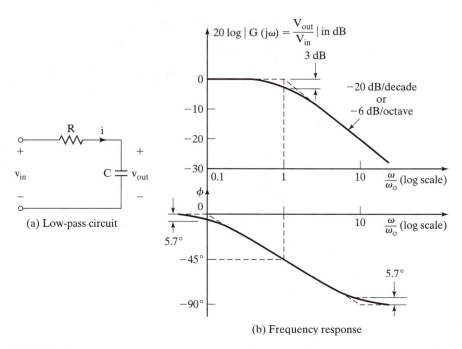

FIGURE 5.4

Frequency response of an RC circuit.

Thus, the magnitude $|G(j\omega)|$ can be found from

$$|G(j\omega)| = \frac{1}{[1 + (\omega\tau)^2]^{1/2}} = \frac{1}{[1 + (\omega/\omega_o)^2]^{1/2}} \qquad (5.2)$$

and the phase angle ϕ of $G(j\omega)$ is given by

$$\phi = -\tan^{-1}(\omega\tau) = -\tan^{-1}(\omega/\omega_o) \qquad (5.3)$$

where $\omega_o = 1/RC = 1/\tau$.

The frequency response is shown in Fig. 5.4(b). If the frequency is doubled, it is called an *octave* in frequency axis. If the frequency is increased by a factor of 10, it is called a *decade*. Thus, for a decade increase in frequency, the magnitude changes by -20 dB, and the magnitude plot is a straight line with a slope of -20 dB/decade, or -6 dB/octave. The magnitude curve is therefore defined by two straight-line asymptotes, which meet at the *corner frequency* (or *break frequency*), ω_o. The difference between the actual magnitude curve and the asymptotic curve is the largest at the break frequency. The error can be found by substituting $\omega = \omega_o$, $|G(j\omega)| = 1/\sqrt{2}$, and $20 \log_{10}(1/\sqrt{2}) = -3$ dB. This error is symmetrical with respect to the break frequency. The break frequency is also known as the *3-db frequency*.

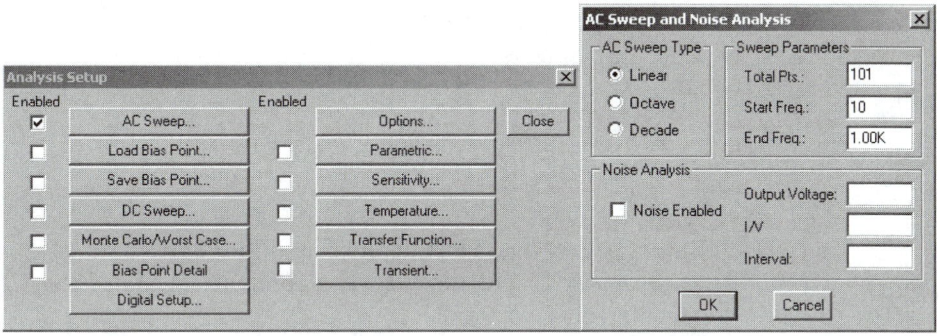

(a) Analysis setup   (b) Ac sweep specifications

FIGURE 5.5

Ac analysis setup.

If the circuit contains nonlinear devices or elements, it is necessary to obtain the small-signal parameters of the elements before calculating the frequency response. Prior to the frequency response (or ac analysis), PSpice determines the small-signal parameters of the elements. The method for calculating the bias point for ac analysis is identical to that for dc analysis. The details of the bias points can be printed by the .OP command.

The ac analysis is set at the analysis setup as shown in Fig. 5.5(a), and its specifications are set at the ac sweep menu as shown in Fig. 5.5(b). The noise analysis (discussed in Section 6.13) can also be enabled.

The command for performing frequency response takes one of the following general forms:

```
.AC  LIN  NP  FSTART  FSTOP
.AC  OCT  NP  FSTART  FSTOP
.AC  DEC  NP  FSTART  FSTOP
```

NP is the number of points in a frequency sweep. FSTART is the starting frequency, and FSTOP is the ending frequency. Only one of LIN, OCT, or DEC must be specified in the statement. LIN, OCT, or DEC specify the type of sweep, as follows:

LIN   *Linear sweep:* The frequency is swept linearly from the starting frequency to the ending frequency, and NP becomes the total number of points in the sweep. The next frequency is generated by adding a constant to the present value. LIN is used if the frequency range is narrow.

OCT   *Sweep by octave:* The frequency is swept logarithmically by octave, and NP becomes the number of points per octave. The next frequency is generated by multiplying the present value by a constant larger than unity. OCT is used if the frequency range is wide.

DEC   *Sweep by decade:* The frequency is swept logarithmically by decade, and NP becomes the number of points per decade. DEC is used if the frequency range is the widest.

PSpice does not print or plot any output by itself for ac analyses. The results of ac sweep are obtained by .PRINT, .PLOT, or .PROBE statements.

Some Statements for Ac Analysis

```
.AC  LIN  201  100HZ  300HZ
.AC  LIN  1    60HZ   120HZ
.AC  OCT  10   100HZ  10KHZ
.AC  DEC  100  1KHZ   1MEGHZ
```

Notes

1. FSTART must be less than FSTOP and cannot be zero.
2. NP = 1 is permissible, and the second statement calculates the frequency response at 60 Hz only.
3. Before performing the frequency-response analysis, PSpice automatically calculates the bias point to determine the linearized circuit parameters around the bias point.
4. All independent voltage and current sources that have ac values are inputs to the circuit. At least one source must have an ac value; otherwise, the analysis would not be meaningful.
5. If a group delay output is required by a suffix of G, as mentioned in Section 3.3, the frequency steps should be small, so that the output changes smoothly.

Example 5.2: Finding the frequency response of an *RLC* circuit

For the *RLC* circuit shown in Fig. 4.22(a), with $R = 2\ \Omega, 1\ \Omega$, and $8\ \Omega$, calculate and print the frequency response over the frequency range from 100 Hz to 100 kHz with a decade increment and 100 points per decade. The peak magnitude and phase angle of the voltage across the capacitors are to be plotted on the output file. The results should also be available for display and hard copy by the .PROBE command.

Solution The circuit file is similar to that of Example 4.3, except the statements for the type of analysis and output are different. The input voltage is a type, and the frequency is variable. We can consider a voltage source with a peak magnitude of 1 V.

The PSpice schematic is shown in Fig. 5.6. The voltage markers are displays of the output waveforms in probe at the end of the simulation. The input voltage is specified by an ac source as shown in Fig. 5.3(b). The ac analysis is set at the analysis setup as shown in Fig. 5.5(a), and its specifications are set at the ac sweep menu as shown in Fig. 5.5(b).

The frequency-response analysis is invoked by the .AC command. For NP = 100, FSTART = 100 Hz, and FSTOP = 100 kHz, the statement is

```
.AC  DEC  100  100  100KHZ
```

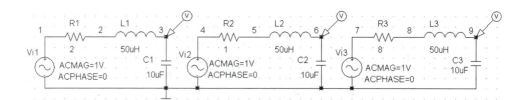

FIGURE 5.6

PSpice schematic for Example 5.2.

The magnitude and phase of voltage V(3) are specified as VM(3) and VP(3). The statement to plot is

```
.PLOT  AC  VM(3)  VP(3)
```

The circuit file contains the following statements:

Example 5.2   Frequency response of *RLC* circuit

```
▲   VI1   1   0   AC    1V     ; Ac voltage of 1 V
    VI2   4   0   AC    1V     ; Ac voltage of 1 V
    VI3   7   0   AC    1V     ; Ac voltage of 1 V
▲▲  R1    1   2   2
    L1    2   3   50UH
    C1    3   0   10UF
    R2    4   5   1
    L2    5   6   50UH
    C2    6   0   10UF
    R3    7   8   8
    L3    8   9   50UH
    C3    9   0   10UF
▲▲▲ *     .AC   DEC  NP    FSTART  FSTOP   ; Format for .AC command
          .AC   DEC  100   100HZ   100KHZ  ; Ac sweep
          .PLOT      AC    VM(3)   VP(3)       ; Plots on the output file
          .PROBE                         ; Graphical waveform analyzer
    .END                             ; End of circuit file
```

Example 5.2   Frequency response of RLC circuits

Temperature: 27.0

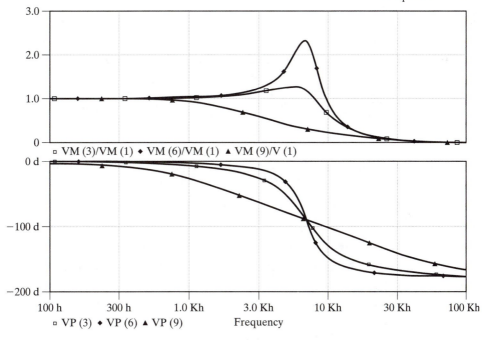

FIGURE 5.7

Frequency responses of the *RLC* circuit for Example 5.2.

The results of the frequency response that are obtained on the display by the .PROBE command are shown in Fig. 5.7. The results of the .PLOT statement can be obtained by printing the contents of the output file EX5-2.OUT. A circuit file can contain both .AC and .TRAN commands in order to perform two analyses.

Example 5.3: Finding the Thevenin's rms voltage and impedance of an ac circuit

An ac circuit is shown in Fig. 5.8. Use PSpice to calculate the rms input current I_1, Thévenin's rms voltage $V_{Th} = V_{ab}$ and impedance Z_{Th}.

Solution $V_s = 120$ V (rms), and peak voltage $V_m = \sqrt{2} \times 120 = 169.7$ V. Thévenin's voltage V_{Th} can be found by measuring open-circuit voltage V_{ab} between terminals a and b. The listing of the circuit file for determining V_{Th} and transient response follows.

```
Example 5.3   Ac circuit analysis
▲ VS  1  0  AC  120V ; Ac voltage of 120 V (peak)
  IX  5  0  AC  0A   ; Only for finding Thévenin's impedance
▲▲ R1 1  2   10
   R0 5  0   50
   L1 3  4   20MH
   C1 4  5   25UF
   C0 5  0   50UF
   VX 2  3   DC    0V ; Measures current through R₁
   F1 5  0   VX    4  ; Current-controlled current source
   E1 4  0   5  0 1M  ; Voltage-controlled voltage source
▲▲▲   .AC   LIN  1   60Hz  120HZ ; Only one ac sweep
      .PRINT    AC   IM(R1)  IP(R1)          ; Prints the input current
      .PRINT    AC   VR(5)  VI(5) VM(5) VP(5) ; Prints the output voltage
 .END                                         ; End of circuit file
```

The results of the ac sweep, which are obtained from the output file EX5-3.OUT, are as follows:

```
****     AC ANALYSIS                        TEMPERATURE =   27.000 DEG C
   FREQ       IM(R1)        IP(R1)       VM(5)       VP(5)
  6.000E+01  9.579E+00   -3.754E+01   1.107E+03   8.774E+01
```

$I_1 = 9.579 \; \underline{/-37.54°}$ A, and $V_{Th} = V_{ab} = 1107 \; \underline{/87.74°}$ V

Since the .TF command can be applied to ac circuits, we need to find a way to measure the output impedance Z_{Th}. The simplest way is to apply a test current I_x between terminals a and b with voltage source V_s short-circuited. The circuit file EX5-3.CIR can be modified by adding a test current I_x of 1 A between nodes 5 and 0, and by setting V_s to 0 V for ac analysis. These statements become

```
VS  1  0  AC  0V ; Ac voltage of 0 V
IX  5  0  AC  1A ; Only for finding Thévenin's resistance
```

The results of simulation for Thévenin's output impedance are as follows:

```
   VR(5)      VI(5)      VM(5)      VP(5)
 -1.646E+01  2.372E+01  2.887E+01  1.248E+02
```

$Z_{Th} = -16.46 + j\ 23.72 = 28.87 \; \underline{/124.8°} \; \Omega$

The PSpice schematic is shown in Fig. 5.9(a). The bubble points such as *A, B, C,* and *D* in the *port.slb* library connect terminals to other parts of the circuit and minimize wiring

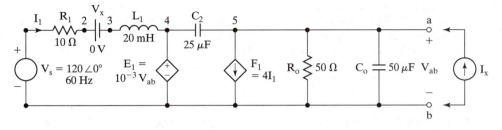

FIGURE 5.8

Ac circuit with controlled sources.

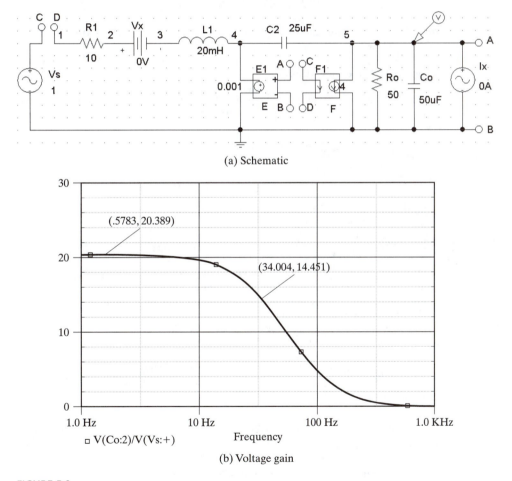

(a) Schematic

□ V(Co:2)/V(Vs:+)   Frequency

(b) Voltage gain

FIGURE 5.9

PSpice schematic for Example 5.3.

crossovers. The voltage markers are displays of the output waveforms in probe at the end of the simulation. The input voltage is specified by an ac source as shown in Fig. 5.3(b). The ac analysis is set at the analysis setup as shown in Fig. 5.5(a), and its specifications are set at the ac sweep menu as shown in Fig. 5.5(b). The PSpice plot of the voltage gain is shown in Fig. 5.9(b). The low-frequency gain is 20.389, and the break frequency is $f_b = 34$ at 14.451 Hz.

Example 5.4:   Finding the instantaneous, rms, and real parts of currents in a three-phase circuit

The ac circuit shown in Fig. 5.10 is supplied from a three-phase balanced supply. Use PSpice (a) to calculate and plot the instantaneous currents i_a, i_b, i_c, and i_n and total (average) input power P_{in} from 0 to 50 ms with a time increment of 5 μs; and (b) to calculate the rms magnitudes and phase angles of currents: I_a, I_b, I_c, and I_n.

Solution $V_{an} = 120$ V rms, and peak phase voltage $V_m = \sqrt{2} \times 120 = 169.7$ V. The listing of the circuit file follows:

Example 5.4   Three-phase circuit
```
▲ Van   1   0   AC   120V    0   SIN(0 169.7V 60HZ)
  Vbn   2   0   AC   120V  120   SIN(0 169.7V 60HZ 0 0 120DEG)
  Vcn   3   0   AC   120V  240   SIN(0 169.7V 60HZ 0 0 240DEG)
▲▲    RA   1    4   0.5
      RB   2    5   0.5
      RC   3    6   0.5
      RX   4    7    1
      RY   5    8    1
      RZ   6    9    1
      R1   7   10    5
      R2   8   11   10
      R3   9   12   10
      C1  10   12   150UF
      L2  11   12   120MH
      VX  12    0   DC   0V  ; Measures the neutral line current
▲▲▲   .TRAN   5US    50MS          ; Transient analysis
      .AC    LIN    1   60HZ 120HZ  ; Only one ac sweep
      *     Prints magnitude and phase of input currents and output voltages
      .PRINT   AC   IM(RA)  IP(RA)  VM(7,12) VP(7,12)
      .PRINT   AC   IM(RB)  IP(RB)  VM(8,12) VP(8,12)
      .PRINT   AC   IM(RC)  IP(RC)  VM(9,12) VP(9,12)
      .PRINT   AC   IM(VX)  IP(VX)   ; Neutral line current
      .PROBE                         ; Graphical waveform analyzer
.END                                 ; End of circuit file
```

(a) The currents I(RA), I(RB), I(RC), and I(VX), which are obtained by the .PROBE command, are shown in Fig. 5.11. The total (average) input power P_{in} is shown in Fig. 5.12. Due to unbalanced loads, the line currents are unbalanced, and a current flows through the neutral line. For a balanced supply with balanced loads, the neutral current will be zero.

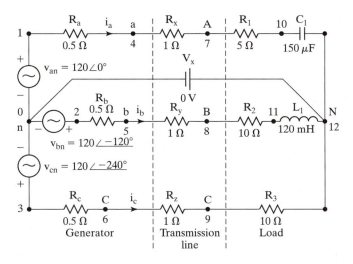

FIGURE 5.10

Three-phase circuit.

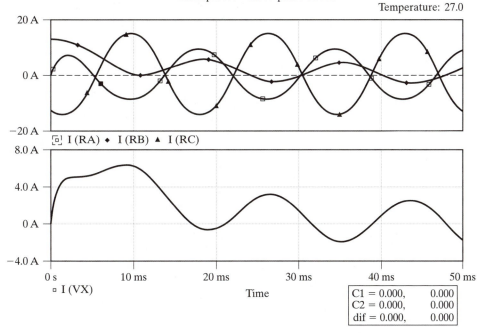

FIGURE 5.11

Instantaneous current for Example 5.4.

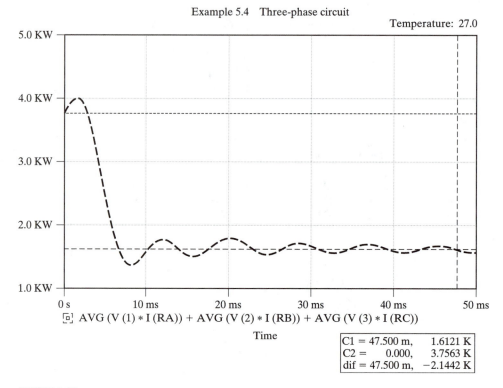

FIGURE 5.12

Instantaneous average input power for Example 5.4.

(b) The results of the ac sweep, which are obtained from the output file EX5-4.OUT, are as follows:

```
****      AC ANALYSIS                      TEMPERATURE =   27.000 DEG C
    IM(R1)       IP(R1)       VM(5)         VP(5)
  9.579E+00   -3.754E+01   1.107E+03    8.774E+01
    I₁ = 9.579 /-37.54° A,     and    V_Th = V_ab = 1107 /87.74° V
    IM(RA)       IP(RA)       VM(7,12)      VP(7,12)
  6.369E+00    6.982E+01    1.170E+02   -4.394E+00
    Iₐ = 6.369 /69.82° A,     and    V_AN = 117 /-4.394° V
    IM(RB)       IP(RB)       VM(8,12)      VP(8,12)
  2.571E+00    4.426E+01    1.191E+02    1.218E+02
    I_b = 2.571 /44.26° A,     and    V_BN = 119.1 /121.8° V
    IM(RC)       IP(RC)       VM(9,12)      VP(9,12)
  1.043E+01   -1.200E+02    1.043E+02   -1.200E+02
    I_c = 10.43 /-120° A,     and    V_CN = /-120° V
    IM (VX)      IP (VX)
  1.729E+00   -1.330E+02
    Iₙ = 1.729 /-133° A
```

Note: PSpice (that is, the netlist DOS version) is better for printing the magnitude and phase of voltages. Schematic is, however, better for drawing the schematic and plotting the voltages and currents.

5.5 MAGNETIC ELEMENTS

The magnetic elements are mutual inductors (transformers). The symbol for mutual coupling is K. The general form of coupled inductors is

```
K⟨name⟩   L⟨ (1st inductor) name⟩ L⟨ (2nd inductor) name⟩
+          ⟨ (coupling) value⟩
```

For couple inductors, K⟨name⟩ couples two or more inductors, and ⟨(coupling) value⟩ is the coefficient of coupling k. The value of the coefficient of coupling must be greater than 0 and less than or equal to 1, $0 < k \leq 1$.

The inductors can be coupled in either order. In terms of the *dot* convention as shown in Fig. 5.13(a), PSpice assumes a dot on the first node of each inductor. The mutual inductance is determined from

$$M = k \sqrt{L_1 L_2}$$

In the time domain, the voltages of coupled inductors are expressed as

$$v_1 = L_1 \frac{di_1}{dt} + M \frac{di_2}{dt}$$

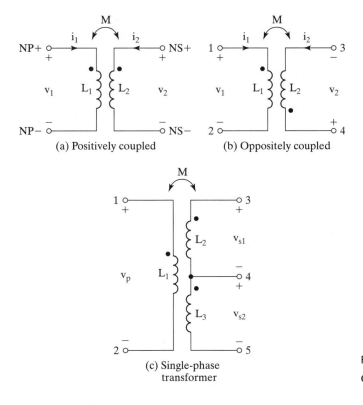

(a) Positively coupled

(b) Oppositely coupled

(c) Single-phase transformer

FIGURE 5.13

Coupled inductors.

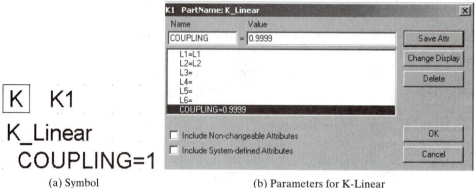

(a) Symbol   (b) Parameters for K-Linear

FIGURE 5.14

Coupled inductors, K-Linear.

and

$$v_2 = M\frac{di_1}{dt} + L_2\frac{di_2}{dt}$$

In the frequency domain, the voltages are expressed as

$$V_1 = j\omega L_1 I_1 + j\omega M I_2$$

and

$$V_2 = j\omega M I_1 + j\omega L_2 I_2$$

where ω is the frequency in radians per second.

The PSpice schematic is shown in Fig. 5.14(a). The K-Linear in the *analog.slb* library can couple up to eight (8) inductors as shown in Fig. 5.14(b) with two inductors L_1 and L_2.

Some Coupled Inductor Statements

```
KTR    LA  LB  0.9
KIND   L1  L2  0.98
```

The coupled inductors in Fig. 5.13(a) can be written as a single-phase transformer (with $k = 0.9999$):

```
*    PRIMARY
L1          1   2    0.5MH
*    SECONDARY
L2          3   4    0.5MH
*    MAGNETIC COUPLING
KXFRMER  L1  L2  0.9999
```

If the dot in the second coil is changed as shown in Fig. 5.13(b), the coupled inductors are written as

```
L1          1   2   0.5MH
L2          4   3   0.5MH
KXFRMER  L1  L2   0.9999
```

A transformer with a single primary coil and center-tapped secondary, as shown in Fig. 5.13(c), can be written as

```
*   PRIMARY
L1    1   2    0.5MH
*   SECONDARY
L2    3   4    0.5MH
L3    4   5    0.5MH
*   MAGNETIC COUPLING
K12  L1   L2   0.9999
K13  L1   L3   0.9999
K23  L2   L3   0.9999
```

These three statements can be written in PSpice as KALL L1 L2 L3 0.9999. For a non-linear inductor, the general form is

```
K⟨name⟩  L⟨ (inductor) name⟩  ⟨ (coupling) value⟩
+          ⟨ (model) name⟩    [(size) value]
```

Notes

1. The name Kxx need not be related to the names of the inductors it is coupling. However, it is a good practice, because it is convenient to identify the inductors involved in the coupling.

2. The polarity (or dot) is determined by the order of the nodes in the L ... statements and not by the order of the inductors in the K ... statement— for example (K12 L1 L2 0.9999) has the same result as (K12 L2 L1 0.9999).

For an iron-core transformer, k is very high and is greater than 0.9999. The model type name for a nonlinear magnetic inductor is CORE; the model parameters are shown in Table 5.2. The [(size) value] scales the magnetic cross section and defaults to 1. It represents the number of lamination layers, so that only one model statement can be used for a particular lamination type of core.

TABLE 5.2   Model Parameters for Nonlinear Magnetic Elements

Name	Meaning	Units	Default
AREA	Mean magnetic cross section	centimeters^2	0.1
PATH	Mean magnetic path length	centimeters	1.0
GAP	Effective air-gap length	centimeters	0
PACK	Pack (stacking)		1.0
MS	Magnetic saturation	A/meters	1E + 6
A	Shape parameter		1E + 3
C	Domain wall-flexing constant		0.2
K	Domain wall-pinning constant		500

If the ⟨(model) name⟩ is specified, then the mutual coupling inductor becomes a nonlinear magnetic core, and the inductors specify the number of turns instead of inductance. The list of the coupled inductors may be just one inductor. The magnetic core's $B-H$ characteristics are analyzed using the Jiles–Atherton model [9]. The procedures to adjust the model parameters to specified $B-H$ characteristics are described in Appendix C.

The statements for the coupled inductors in Fig. 5.13(a) are similar to the following:

```
*    Inductor L1 of 100 turns:
L1          1    2    100
*    Inductor L2 of 10 turns:
L2          3    4    10
*    Nonlinear coupled inductors with model CMOD
K12    L1   L2   0.9999   CMOD
*    Model for the nonlinear inductors
.MODEL CMOD CORE (AREA=2.0 PATH=62.8 GAP=0.1 PACK=0.98)
```

Note. A nonlinear magnetic model is not available in SPICE2.

Example 5.5:   Finding the rms currents and the gain-frequency plot of a coupled-inductor circuit

A circuit with two coupled inductors is shown in Fig. 5.15. If the input voltage is 120 V peak, calculate the magnitude and phase of the output current for frequencies from 60 to 120 Hz with a linear increment. The total number of points in the sweep is 2. The coefficient of coupling for the transformer is 0.999.

Solution   It is important to note that the primary and the secondary windings have a common node. Without this, PSpice will give an error message, because there is no dc path from the nodes of the secondary to the ground. The voltage source, $V_X = 0$ V, is connected to measure the output current I_L.

The PSpice schematic is shown in Fig. 5.16(a). K-Linear couples L1 and L2 with a coupling of 0.9999. The voltage marker displays the output waveform in probe at the end of the simulation. The PSpice plot of the voltage gain is shown in Fig. 5.16(b). The high-frequency gain is 1.0 and the break frequency is $f_b = 1.7$ kHz at a gain of 0.707. The circuit file contains the following statements:

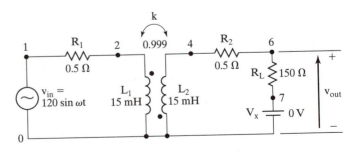

FIGURE 5.15

Circuit with two coupled inductors.

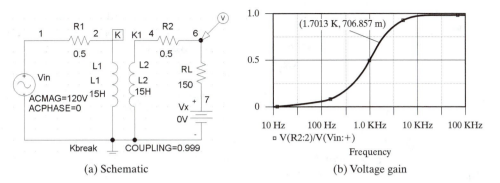

(a) Schematic

(b) Voltage gain

FIGURE 5.16

PSpice schematic for Example 5.5.

Example 5.5   Coupled inductors

```
▲  VIN  1   0   AC   120    ; Ac voltage is 120 V peak and 0 degree phase.
▲▲ R1  1   2   0.5
    *   The dot convention is followed in inductors L1 and L2.
    L1  2   0   15MH
    L2  0   4   15MH
    *    Magnetic coupling coefficient is 0.999.
    K12  L1   L2   0.999 ; Order of L1 and L2 is not significant.
    R2   4   6   0.5
    RL   6   7   150
    VX   7   0   DC  0V  ; Measure the load current.
▲▲▲  * AC linear sweep from 60 Hz to 120 Hz with 2 points
    .AC LIN  2   60HZ  120HZ
    *  Print the magnitude and phase of output current. Some versions of
    *  PSpice and SPICE do not permit reference to currents through
resistors.
    *    e.g.,   IM(RL)   IP(RL).
    .PRINT    AC  IM(VX)  IP(VX)  IM(RL)  IP(RL)
.END                   ; End of circuit file
```

The results of the simulation that are obtained from the output file EX5-5.OUT follow:

```
****      AC ANALYSIS                      TEMPERATURE =  27.000 DEG C
  FREQ          IM(VX)        IP(VX)      IM(RL)        IP(RL)
  6.000E+01    2.809E-01   -1.107E+02   2.809E-01   -1.107E+02
  1.200E+02    4.790E-01   -1.271E+02   4.790E-01   -1.271E+02
          JOB CONCLUDED
          TOTAL JOB TIME              8.18
```

Note: PSpice (that is, the netlist DOS version) is better for printing the magnitude and phase of voltages. Schematic is, however, better for drawing the schematic and plotting the voltages and currents.

5.6   TRANSMISSION LINES

The symbol for a lossless transmission line is *T*. A transmission line has two ports— input and output [2,7]. The general form of a transmission line is

```
T⟨name⟩  NA+  NA−  NB+  NB−  Z0=⟨value⟩  [TD=⟨value⟩]
+         [F=⟨value⟩   NL=⟨value⟩]
```

T⟨name⟩ is the name of the transmission line. NA+ and NA− are the nodes at the input port. NB+ and NB− are the nodes at the output port. NA+ and NB+ are defined as the positive nodes. NA− and NB− are defined as the negative nodes. The positive current flows from NA+ to NA− and from NB+ to NB−. Z_0 is the characteristic impedance.

The length of the line can be expressed in either of two forms: (1) The transmission delay TD may be specified, or (2) the frequency F may be specified together with NL, which is the normalized electrical length of the transmission line with respect to wavelength in the line at frequency F. If the frequency F is specified but NL is not, then the default value of NL is 0.25—that is, F has quarter-wave frequency. It should be noted that one of the options for expressing the length of the line must be specified— that is, TD or at least F must be specified. The block diagram of a transmission line is shown in Fig. 5.17(a).

A coaxial line, as shown in Fig. 5.17(b), can be represented by two propagating lines, where the first line (T_1) models the inner conductor with respect to the shield, and the second line (T_2) models the shield with respect to the outside. This is shown as follows:

```
T1   1   2   3   4   Z0=50    TD=1.5NS
T2   2   0   4   0   Z0=150   TD=1NS
```

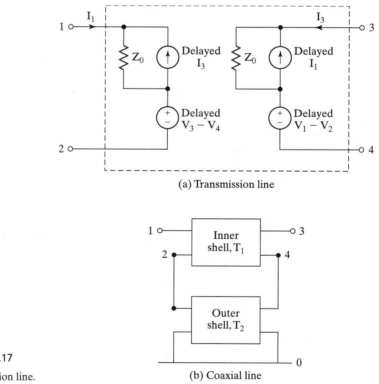

(a) Transmission line

FIGURE 5.17

Transmission line.

(b) Coaxial line

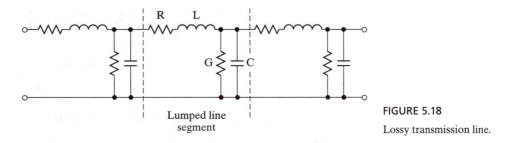

FIGURE 5.18

Lossy transmission line.

A lossy transmission line can be represented by "lumped line segments," as shown in Fig. 5.18. R, L, G, and C are the per-unit-length values of resistance, inductance, conductance, and capacitance, respectively.

The general form of a lossy transmission line is

```
T⟨name⟩   NA+   NA−   NB+   NB−
+      LEN=⟨value⟩  R=⟨value⟩  L=⟨value⟩  G=⟨value⟩  C=⟨value⟩
```

where LEN is the per-unit electrical length.

It should be noted that line parameters can be expressions; R and G can be Laplace's expressions, thereby permitting frequency-dependent effects, such as skin effect and dielectric loss. Expressions are discussed in Section 6.2.

Some Transmission Statements

```
T1     1    2    3    4    Z0=50   TD=10NS
T2     4    5    6    7    Z0=50   F=2MHZ
TTRM   9   10   11   12    Z0=50   F=2MHZ   NL=0.4
TR     2    3    7    9    LEN=1.1  R=0.45  L=450U  G=3.5U  C=105P
```

The PSpice schematic for lossless transmission line is shown in Fig. 5.19(a) and the line parameters are set from the menu as shown in Fig. 5.19(b). The PSpice schematic for

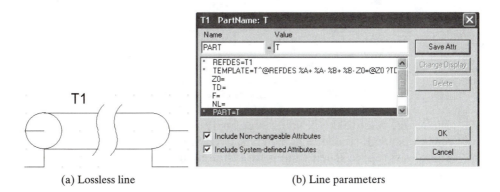

(a) Lossless line   (b) Line parameters

FIGURE 5.19

Lossless transmission line.

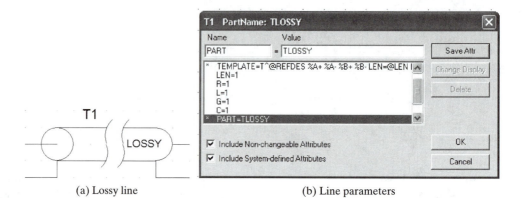

(a) Lossy line   (b) Line parameters

FIGURE 5.20

Lossy transmission line.

lossy transmission line is shown in Fig. 5.20(a) and its line parameters are set from the menu as shown in Fig. 5.20(b).

Note. During the transient (.TRAN) analysis, the internal time step of PSpice is limited to be no more than one-half of the smallest transmission delay. Thus, short transmission lines will cause long run times.

5.7   MULTIPLE ANALYSES

A source may be assigned the specifications of dc, ac, and transient analysis for different analyses, and the statements for voltage and current sources have the following general forms, respectively:

```
V⟨name⟩    N+      N−
+          [DC ⟨value⟩]
+          [AC ⟨(magnitude) value⟩ ⟨(phase) value⟩]
+          [⟨transient specifications⟩]

I⟨name⟩    N+      N−
+          [DC ⟨value⟩]
+          [AC ⟨(magnitude) value⟩ ⟨(phase) value⟩]
+          [⟨transient specifications⟩]
```

The source is set to the dc value in dc analysis. It is set to an ac value in ac analysis. The time-dependent source (e.g., PULSE, EXP, SIN, etc.) is assigned for transient analysis. This allows source specifications for different analyses in the same statement.

Typical Statements

```
VIN  25  22  DC  2V  AC  1V  30    SIN  (0   2V  10KHZ)
IIN  25  22  DC  2A  AC  1A  30DEG  SIN  (0   2A 10KHZ)
```

Notes

1. VIN assumes 2 V for dc analysis, 1 V with a delay angle of 30° for ac analysis, and a sine wave of 2 V at 10 kHz for transient analysis.
2. IIN assumes 2 A for dc analysis, 1 A with a delay angle of 30° for ac analysis, and a sine wave of 2 A at 10 kHz for transient analysis.

Example 5.6: Finding the frequency response and dc analysis of a series-parallel circuit

A series-parallel *RLC* circuit is shown in Fig. 5.21. The output voltage is taken across the capacitor. Write the circuit descriptions for the following:

(a) to calculate and plot the instantaneous output voltage v_{out} from 0 to 5 ms with a time increment of 10 μs

(b) to calculate and plot the dc transfer characteristics, V_{out} versus V_S, from 0 to 20 V with a 10 V increment

(c) to calculate and plot the output voltage V_{out} over the frequency range from 100 Hz to 100 kHz with a decade increment and 100 points per decade.

The peak magnitude and phase angle of the voltage across the capacitors are to be plotted on the output file. The results should also be available for display and hard copy by the .PROBE command.

Solution The step input voltage is described by a piecewise linear waveform

```
PWL (0 0 1NS 1V 10MS 1V)
```

The listing of the circuit file follows.

Example 5.6 Dc/ac/transient analysis
```
▲ VS   1   0   DC   20V   AC   10V   PWL   (0   0   1NS   1V   10MS   1V) ; Input voltages
▲▲  L    1   2      5MH
    R    2   3      5
    C    3   0      10UF
    RL   3   0      80
▲▲▲  .DC    VS    0V    20V   10V          ; Dc sweep
     .TRAN      10US      5MS              ; Transient analysis
     .AC     DEC   100     100HZ   10KHZ   ; Ac sweep
     .PROBE                                ; Graphical waveform analyzer
 .END                                      ; End of circuit file
```

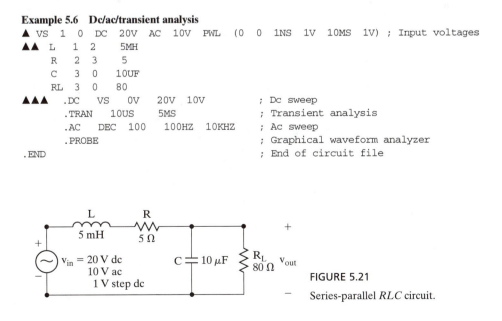

FIGURE 5.21

Series-parallel *RLC* circuit.

During the dc analysis, the capacitors are open-circuited, and inductors are short-circuited. Students are encouraged to plot the results of .TRAN, .DC, and .AC analysis and to verify the results with hand calculations. The values of *R, L,* or *C* can be varied by the .PARAM and .STEP commands discussed in Sections 6.11 and 6.15, respectively.

Note. Multiple analyses can be performed on a circuit. Depending on the analysis type invoked, the source can assume the specifications of the particular source type such as dc, ac, or transient (e.g., sinusoidal, pulse, or piecewise linear).

SUMMARY

```
V⟨name⟩    N+     N−
+          [DC ⟨value⟩]
+          [AC ⟨(magnitude) value⟩ ⟨ (phase) value⟩]
+          [(transient specifications)]

I⟨name⟩    N+     N−
+          [DC ⟨value⟩]
+          [AC ⟨ (magnitude) value⟩ ⟨(phase) value⟩]
+          [(transient specifications)]

K⟨name⟩    L⟨(1st inductor) name⟩   L⟨(2nd inductor) name⟩
+          ⟨(coupling) value⟩

K⟨name⟩    L⟨(inductor) name⟩    ⟨(coupling) value⟩
+          ⟨(model) name⟩ [(size) value]

T⟨name⟩    NA+    NA−    NB+    NB−    Z0=⟨value⟩    [TD=⟨value⟩]
+          [F=⟨value⟩    NL=⟨value⟩]

T⟨name⟩    NA+    NA−    NB+    NB−
+          LEN=⟨value⟩    R=⟨value⟩    L=⟨value⟩    G=⟨value⟩    C=⟨value⟩

T⟨name⟩    NA+    NA−    NB+    NB−    Z0=⟨value⟩    [TD=⟨value⟩]
+          [F=⟨value⟩ NL=⟨value⟩]
```

REFERENCES

[1]   R.C. Dorf and J.A. Svoboda, *Introduction to Electric Circuits*, 5th ed. New York: John Wiley & Sons, 2001.

[2]   Mohamed E. El-Hawary, *Electrical Power Systems—Design and Analysis*, Ch. 4. Reston, Virginia: Reston Publishing Co., 1983.

[3]   A. E. Fitzgerald, Charles Kingsley, and Stephen D. Umans, *Electric Machinery*, 6th ed. New York: McGraw-Hill, 2003.

[4]   Charles A. Gross, *Power System Analysis*, Ch. 4. New York: John Wiley & Sons, 1986.

[5]   J.D. Irwin, *Basic Engineering Circuit Analysis*, 5th ed. Upper Saddle River, New Jersey: Prentice Hall, 1996.

[6]   H. W. Jackson, *Introduction to Electric Circuits*. Englewood Cliffs, New Jersey: Prentice Hall, 1986.

[7]   J. W. Nilsson and S. A. Riedel, *Electric Circuits*, 6th ed. Upper Saddle River, New Jersey: Prentice Hall, 2000.

[8]   C. R. Paul, S. A. Nassar, and L. E. Unnewehr, *Introduction to Electrical Engineering*. New York: McGraw-Hill, 1992.

[9] *PSpice Manual.* Irvine, California: MicroSim Corporation, 1992.

[10] M. H. Rashid, *SPICE For Power Electronics and Electric Power.* Englewood Cliffs, New Jersey: Prentice Hall, 1993.

PROBLEMS

5.1 Use PSpice to calculate and print the frequency response of the circuit shown in Fig. P4.7. The results should be available for display and hard copy by the .PROBE command.

5.2 Use PSpice to calculate and print the frequency response of the circuit shown in Fig. P4.8. The results should be available for display and hard copy by the .PROBE command.

5.3 Use PSpice to calculate and print the frequency response of the circuit shown in Fig. P4.9. The results should be available for display and hard copy by the .PROBE command.

5.4 Use PSpice to calculate and print the frequency response of the circuit shown in Fig. P4.12. The results should be available for display and hard copy by the .PROBE command.

5.5 Use PSpice to calculate and print the frequency response of the circuit shown in Fig. P4.13. The results should be available for display and hard copy by the .PROBE command.

5.6 Plot the frequency response of the circuit in Fig. P5.6 from 10 Hz to 100 kHz with a decade increment and 10 points per decade. The output voltage is taken across the capacitor. Print and plot the magnitude and phase angle of the output voltage. Assume a source voltage of 1 V peak.

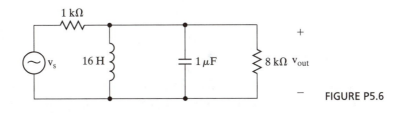

FIGURE P5.6

5.7 For the circuit in Fig. P5.7, the frequency response is to be calculated and printed over the frequency range from 1 Hz to 100 kHz with a decade increment and 10 points per decade. The peak magnitude and phase angle of the output voltage are to be plotted on the output file. The results should also be available for display and hard copy by the .PROBE command.

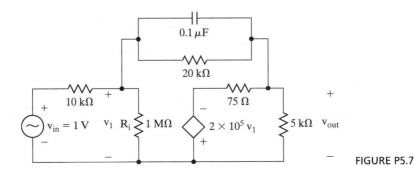

FIGURE P5.7

5.8 Repeat Problem 5.7 for the circuit in Fig. P5.8.

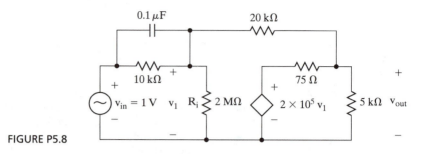

FIGURE P5.8

5.9 An ac circuit is shown in Fig. P5.9. Use PSpice **(a)** to calculate and plot the instantaneous output voltage v_{ab} from 0 to 33 ms with a time increment of 5 μs, and **(b)** to calculate the rms input current I_1, Thévenin rms voltage $V_{Th} = V_{ab}$, and Thévenin impedance Z_{Th}.

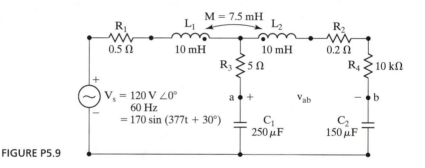

FIGURE P5.9

5.10 A feedback circuit is shown in Fig. P5.10. Use PSpice **(a)** to calculate and plot the voltage gain $|V_{out}/V_s|$ against the normalized frequency $u = f/f_o = 0$ to 5, where f_o is the resonant frequency, and **(b)** to calculate Thévenin rms voltage $V_{Th} = V_{ab}$ and Thévenin impedance Z_{Th}.

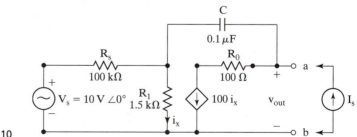

FIGURE P5.10

5.11 The ac circuit shown in Fig. P5.11 is supplied from a three-phase balanced supply. Use PSpice **(a)** to calculate and plot the instantaneous currents i_a, i_b, i_c, and i_n, and total (average) input power P_{in} from 0 to 50 ms with a time increment of 5 μs and **(b)** to calculate the rms magnitudes and phase angles of currents: $I_a, I_b, I_c,$ and I_n. Compare the results with those of Example 5.4.

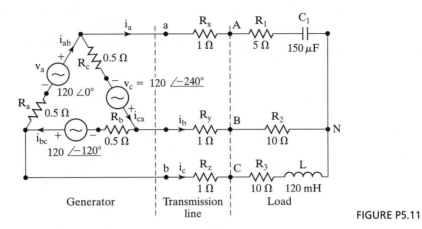

FIGURE P5.11

5.12 A single-phase transformer, as shown in Fig. P5.12, has a center-tapped primary, where $L_p = 1.5$ mH, $L_s = 1.3$ mH, and $K_{ps} = K_{sp} = 0.999$. Write PSpice statements.

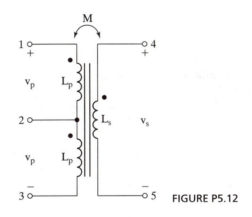

FIGURE P5.12

5.13 A three-phase transformer, which is shown in Fig. P5.13, has $L_1 = L_2 = L_3 = 1.2$ mH, and $L_4 = L_5 = L_6 = 0.5$ mH. The coupling coefficients between the primary and secondary windings of each phase are $K_{14} = K_{41} = K_{25} = K_{52} = K_{36} = K_{63} = 0.9999$. There is no cross-coupling with other phases. Write PSpice statements.

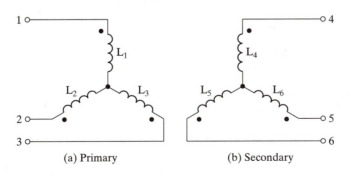

(a) Primary   (b) Secondary

FIGURE P5.13

5.14 Use PSpice to plot the frequency response of the low-pass *RC* circuit in Fig. P5.14. Assume that $V_s = 1$ V (peak ac), $R = 1$ kΩ, and $C = 0.1$ μF. The frequency *f* is varied from 1 Hz to 100 kHz. Plot the magnitude and phase angle of the voltage gain V_o/V_s against the frequency.

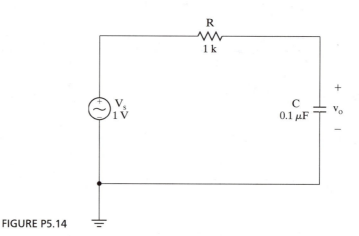

FIGURE P5.14

5.15 Use PSpice to plot the frequency response of the high-pass *CR*-circuit in Fig. P5.15. Assume that $V_s = 1$ V (peak ac), $R = 1$ kΩ, and $C = 0.1$ μF. The frequency *f* is varied from 1 Hz to 100 kHz. Plot the magnitude and phase angle of the voltage gain V_o/V_s against the frequency.

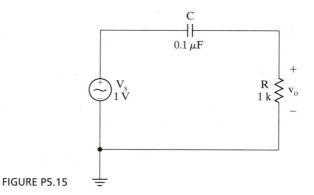

FIGURE P5.15

5.16 Use a parametric command in PSpice to vary the value of the resistance *R* to 1, 2, and 8 Ω for the circuit shown in Figure 5.6 for Example 5.2. Plot the frequency response of the capacitor voltage V_c.

Advanced SPICE Commands and Analysis

After completing this chapter, students should be able to

- Perform behavioral modeling of voltage and current-controlled sources in the form of value, table, Laplace, frequency, or a mathematical function
- Define and call a subcircuit
- Define a mathematical function and assign a global node or parameter
- Set a node at a specific voltage and include a library file
- Understand the option parameters and set their values to avoid conference problems, and control the output
- Perform parametric and step variations on the analysis
- Assign tolerances of components and model parameters
- Perform advanced type of analyses such as Fourier series of output voltages and currents, noise analysis, worst-case analysis, and Monte Carlo analysis.

6.1   INTRODUCTION

In addition to the commands described in Chapters 3, 4, and 5, PSpice has other commands that can enhance and simplify the analysis of electrical and electronics circuits. PSpice allows one (1) to model an element based on its parameters, (2) to model a small circuit that is repeated a number of times in the main circuit, (3) to use a model that is defined in another file, (4) to create a user-defined function, (5) to use parameters instead of number values, and (6) to use parameter variations. The following commands and features are available:

Behavioral modeling
.SUBCKT Subcircuit
.ENDS End of subcircuit

.FUNC Function
.GLOBAL Global
.INC Include file
.LIB Library file
.NODESET Nodeset
.OPTIONS Options
.PARAM Parameter
.FOUR Fourier analysis
.NOISE Noise analysis
.SENS Sensitivity analysis
.STEP Parametric analysis
.DC Dc parametric sweep
.MC Monte Carlo analysis
.DEV/LOT Device tolerances
.WCASE Worst-case analysis.

6.2   BEHAVIORAL MODELING

PSpice allows modeling of a device in terms of the relation between its input and output. The relationship is specified as an extension to a voltage-controlled voltage source (E) or a voltage-controlled current source (G) and takes one of the following forms:

VALUE
TABLE
LAPLACE
FREQ

This type of modeling is known as *behavioral modeling* and is only available with the Analog Behavioral Modeling Option of PSpice. However, the student's version of PSpice allows it.

The PSpice schematic supports many behavioral models such as DIFF, DIFFER, INTG, MULTI, SUM, SQRT, etc. These models can be selected by choosing *ab.slb library* from the library menu as shown in Fig. 6.1.

6.2.1   Value

The VALUE extension allows a relation to be described by a mathematical expression and can be specified by the following general forms:

```
E⟨name⟩   N+   N−    VALUE = { ⟨expression⟩ }
G⟨name⟩   N+   N−    VALUE = { ⟨expression⟩ }
```

The keyword VALUE indicates that the relation is described in an ⟨expression⟩. The ⟨expression⟩ itself must be enclosed by braces { }. It can have the functions shown in

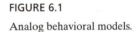

FIGURE 6.1

Analog behavioral models.

TABLE 6.1 Functions

Function	Meaning	
ABS(x)	$\lvert x \rvert$	(absolute value)
SQRT(x)	$\sqrt{x}$	
EXP(x)	e^x	
LOG(x)	$\ln(x)$	(log of base e)
LOG10(x)	$\log(x)$	(log of base 10)
PWR(x, y)	$\lvert x \rvert^y$	
PWRS(x, y)	$+\lvert x \rvert^y$	(if $x > 0$)
	$-\lvert x \rvert^y$	(if $x < 0$)
SIN(x)	$\sin(x)$	(x in radians)
COS(x)	$\cos(x)$	(x in radians)
TAN(x)	$\tan(x)$	(x in radians)
ARCTAN(x)	$\tan^{-1}(x)$	(result in radians)

Table 6.1, and can contain the arithmetical operators $(+, -, *,$ and $/)$ along with parentheses.

Notes

1. VALUE *should* be followed by a space.
2. The ⟨expression⟩ *must* fit on one line.

Typical Statements

```
ESQROOT  2  3  VALUE = {2V*SQRT(V(5)*V(2))}      ; square roots
EPWR     1  2  VALUE = {V(4,3)*I(VX)}            ; product of v and i
ELOG     3  0  VALUE = {10V*LOG(I (VS)/10mA)}   ; log of current ratio
GVCO     4  5  VALUE = {15MA*SIN(6.28*10kHz*TIME*(10V*V(7,3)))}
GRATIO   3  6  VALUE = {V(8,1)/V(7,3)}           ; voltage ratio
```

The value can be used to simulate linear and nonlinear resistances (or conductances). A resistance can be regarded as a current-controlled voltage source. For example, the statement

```
ERES  2  3  VALUE = {I(VX)*5K}
```

is a linear resistance with a value of 5 kΩ. VX is connected in series with ERES, and it measures the current through ERES.

Similarly, a conductance can be regarded as a voltage-controlled current source. For example, the statement

```
GCOND  2  3  VALUE = {V(2,3)*0.01}
```

is a linear conductance with a value of 0.01 mho. It should be noted that the controlling nodes are the same as the output nodes.

6.2.2 Table

The TABLE extension allows a relation to be described by a table and can be specified by the following general forms:

```
E⟨name⟩   N+   N−   TABLE  { ⟨expression⟩ } =
+                          ⟨ ⟨(input) value⟩, ⟨(output) value⟩ ⟩ *
G⟨name⟩   N+   N−   TABLE  { ⟨expression⟩ } =
+                          ⟨ ⟨(input) value⟩, ⟨(output) value⟩ ⟩ *
```

The keyword TABLE indicates that the relation is described by a table of data. The table consists of pairs of values. The first value in a pair is the input, and the second value is the corresponding output. The ⟨expression⟩ is the input value and is used to find the corresponding output from the look-up table. If an input value falls between two entries, the output is found by linear interpolation. If the input falls outside the table's range, the output remains a constant that corresponds to the smallest (or largest) input.

Notes

1. TABLE *must* be followed by a space.
2. The input to the table is ⟨expression⟩, which *must* fit in one line.
3. The TABLE's input *must* be in order from lowest to highest.

Typical Statements

The TABLE can represent the current-voltage characteristics of a diode,

```
EDIODE   5   6    TABLE { I(VDIODE) } =
+        (0.0,    0.5)    (10E−3, 0.870)   (20E−3, 0.98)   (30E−3, 1.058)
+     (40E−3, 1.115)   (50E−3, 1.173)   (60E−3, 1.212)   (70E−3, 1.250)
```

where I(VDIODE) is the diode current and produces an output voltage EDIODE. The TABLE can be used to represent a constant power load $P = 2\,kW$, with a voltage-controlled current source,

```
GCONST   2   3   TABLE   {2K/V(2,3)} = (−2K, −2K) (2K, 2K)
```

where GCONST will try to dissipate 2 kW of power regardless of the voltage across it. If the voltage V(2,3) is very small, the formula 2K/V(2,3) can give a high value of current. The TABLE limits the current between −2 kA and 2 kA.

6.2.3   Laplace

The LAPLACE extension allows a relation to be described by a Laplace transform function and can be specified by the following general forms:

```
E⟨name⟩   N+   N−   LAPLACE   { ⟨expression⟩ } = { ⟨transform⟩ }
G⟨name⟩   N+   N−   LAPLACE   { ⟨expression⟩ } = { ⟨transform⟩ }
```

The keyword LAPLACE indicates that the relation is described in Laplace's domain of s. The ⟨expression⟩ is the input to the Laplace's transform, and it follows the same rules as in Section 6.2.1. The output, which is defined by the ⟨transform⟩ relation, gives the value of E⟨name⟩ or G⟨name⟩.

Typical Statements
The output voltage of a lossy integrator with a time constant of 1 ms and an input voltage V(2,3) can be described by

```
EINTR   4   0   LAPLACE {V(2, 3)} = {1/(1 + 0.001*s)}
```

The output voltage of a lossless integrator with an input voltage V(2,3) can be described by

```
EINTR   4   0   LAPLACE {V(2, 3)} = {1/s}
```

A frequency-dependent impedance (with R and L) can be written as

```
GRL   2   3   LAPLACE {V(2, 3)} = {1/(1 + 0.001*s)}
```

The frequency-dependent impedance of a capacitor can be written as

```
GCAP   5   4   LAPLACE {V(5, 4)} = {s}
```

Notes

1. LAPLACE *must* be followed by a space.
2. ⟨expression⟩ and ⟨transform⟩ *must* each fit on one line.
3. Voltages, currents, and TIME must not appear in a Laplace transform.
4. The LAPLACE device uses much more computer memory than does the built-in capacitor (C) device and should be avoided, if possible.

6.2.4   FREQ

The FREQ extension allows a relation to be described by a frequency-response table and can be specified by the following general forms:

```
E⟨name⟩ N+ N− FREQ { ⟨expression⟩ } =
+ ⟨ ⟨(frequency) value⟩, ⟨(magnitude in dB) value⟩, ⟨(phase) value⟩ ⟩ *
G⟨name⟩ N+ N− FREQ { ⟨expression⟩ } =
+ ⟨ ⟨(frequency) value⟩, ⟨(magnitude in dB) value⟩, ⟨(phase) value⟩ ⟩ *
```

The keyword FREQ indicates that the relation is described by a table of frequency responses. The table contains the magnitude (in dB) and the phase (in degrees) of the response for each frequency. The ⟨expression⟩ is the input to the table at a specified frequency and follows the same rules as in Section 6.2.1. The output, which is defined by the tabular relation at the input frequency, gives the value of E⟨name⟩ or G⟨name⟩.

   If the frequency falls between two entries, the magnitude and phase of the output are found by linear interpolation. Phase is interpolated linearly, and the magnitude is interpolated logarithmically with frequencies. If the input frequency falls outside the table's range, the magnitude and the phase remain constants corresponding to the smallest (or largest) frequency.

Typical Statements

The output voltage of a low-pass filter with input voltage V(2) can be expressed by

```
ELOWPASS    2   0   FREQ {V(2)} = (0,     0,     0)
+                  (5 kHz,     0,   −57.6)   (6kHz0,   40,   −69.2)
```

Notes

1. FREQ *should* be followed by a space.
2. ⟨expression⟩ *must* fit on one line.
3. The FREQ's frequencies *must* be in order from lowest to highest.

6.3   .SUBCKT (SUBCIRCUIT)

PSpice allows one to define a small circuit as a *subcircuit*, which then can be called upon in several places in the main circuit. The general form for a subcircuit definition (or description) is

```
.SUBCKT   SUBNAME   [⟨(two or more) nodes⟩]
```

Whenever there is a subcircuit definition, there is normally a call statement in order to use the subcircuit during the simulation. The symbol for a subcircuit call is X, and the general form of a call statement is

```
X⟨name⟩   [⟨(two or more) nodes⟩]   SUBNAME
```

SUBNAME is the name of the subcircuit definition, and ⟨(two or more) nodes⟩ are the nodes of the subcircuit. X⟨name⟩ causes the referenced subcircuit to be inserted into the circuit, so that the given nodes in the subcircuit call replace the argument nodes in the definition. The subcircuit name SUBNAME may be considered as

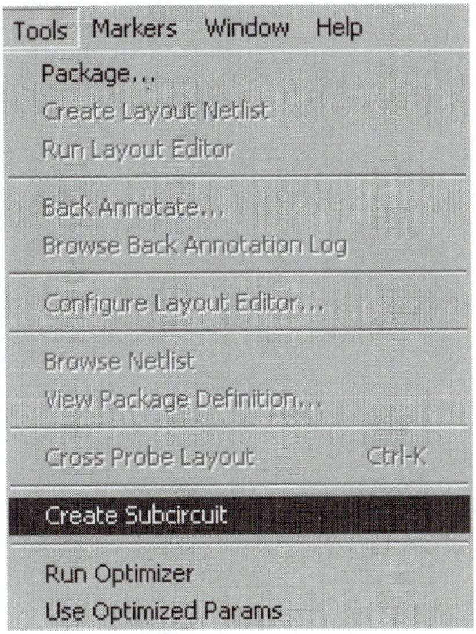

FIGURE 6.2

Analog behavioral models.

equivalent to a subroutine name in FORTRAN programming, where X⟨name⟩ is the call statement and ⟨(two or more) nodes⟩ are the variables or arguments of the subroutine.

Subcircuits can be nested; subcircuit A can call other subcircuits, but the nesting cannot be circular. That is, if subcircuit A contains a call to subcircuit B, then subcircuit B must not contain a call to subcircuit A.

The number of nodes in the subcircuit calling statement must be the same as that of its definition. The subcircuit definition can have only element statements and .MODEL statements, but must not contain other dot statements.

Note: The netlist for a subcircuit can be created using the command *Create Subcircuit* from the *Tools* menu as shown in Fig. 6.2.

6.4   .ENDS (END OF SUBCIRCUIT)

A subcircuit must end with an .ENDS statement. The end of a subcircuit definition has the general form

```
.ENDS   SUBNAME
```

SUBNAME is the name of the subcircuit, and it indicates which subcircuit description is to be terminated. If the .ENDS statement is missing, all subcircuit descriptions are terminated.

End of Subcircuit Statements

```
.ENDS   OPAMP
```
```
.ENDS
```

Note. The name of the subcircuit can be omitted; however, it is advisable to identify the name of the subcircuit to be terminated, especially if there is more than one subcircuit.

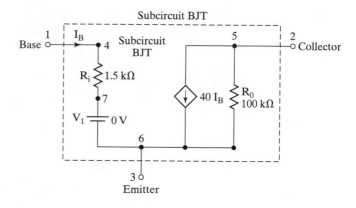

FIGURE 6.3

Bipolar junction transistor (BJT) subcircuit.

Example 6.1: Describing a subcircuit

The equivalent circuit of a bipolar junction transistor (BJT) is shown in Fig. 6.3. Write the subcircuit call and subcircuit description.

Solution The list of statements for the subcircuit call and description follows.

```
*   The call statement X1 to be connected to input nodes 1, 2, and 3. The subcircuit
*   name is BJT. Nodes 1, 2, and 3 are referred to the main circuit file;
*   they do not interact with the nodes of the subcircuit.
X1    1       2          3            BJT
*    base    emitter   collector   model name
*   The subcircuit definition for BJT.  Nodes 4, 5, 6, and 7 are referred
*   to the subcircuit; they do not interact with the nodes of the
*   main circuit.
.SUBCKT      BJT     4      6       5
*      model   name  base  emitter  collector
RI    4    7      1.5K
RO    5    6      100K
*  A dummy voltage source to measure the controlling current IB
VX    7    6    DC     0V
F1    5    6    VX     40     ; Current-controlled current source
.ENDS    BJT                  ; End of subcircuit definition BJT
```

6.5   .FUNC (FUNCTION)

PSpice allows one to generate user-defined functions. If there are several similar expressions in a single circuit file, a user-defined function can simplify the circuit description and can be accessed by an .INC statement (Section 6.7) near the beginning of the circuit file. The general form of a function statement is

.FUNC   FNAME (arg)   ⟨function⟩

The ⟨function⟩ can have the functions shown in Table 6.1 and may refer to other previously defined functions. FNAME (arg) is the name of the function with argument (arg). FNAME must not be the same as the built-in functions in Table 6.1, for example, sin or cos. The number of arguments (arg) must be the same as the number of variables in the ⟨function⟩. FNAME can be without arguments, but the parentheses are still required; that is, FNAME ( ) is acceptable.

Some Function Statements

```
.FUNC    E(x)          exp(x)
.FUNC    Sinh(x)       (E(x)+E(-x))/2
.FUNC    MIN(C,D)      (C+D-ABS(C-D))/2
.FUNC    MAX(C,D)      (C+D+ABS(C+D))/2
.FUNC    IND(I(VX))    (A0+A1*I(VX)+A2*I(VX)*I(VX))  ; Polynomial form
```

Notes

1. The definition of the ⟨function⟩ *must* fit on one line.
2. In-line comments *must not* be used after the ⟨function⟩ definition.
3. The last statement illustrates a current-dependent nonlinear inductor of polynomial form.

6.6   .GLOBAL (GLOBAL)

PSpice allows one to define global nodes, which are accessible by all subcircuits. The general statement is

```
.GLOBAL   N
```

where *N* is the node number. For example,

```
.GLOBAL   4
```

makes node 4 global to the circuit file and subcircuit(s).

6.7   .INC (INCLUDE FILE)

The statement for including the contents of another file in the circuit file has the following form:

```
.INC   NFILE
```

NFILE is the name of the file to be included and can be any character string that is a legal file name for computer systems. NFILE may contain any statements except a title line and can have a comment line(s). It can contain a .END statement, which simply indicates the end of the NFILE. Up to four levels of "including" are allowed in an .INC statement. It should be noted that the .INC statement simply brings everything from the included file into the circuit file and takes up space in main memory (RAM).

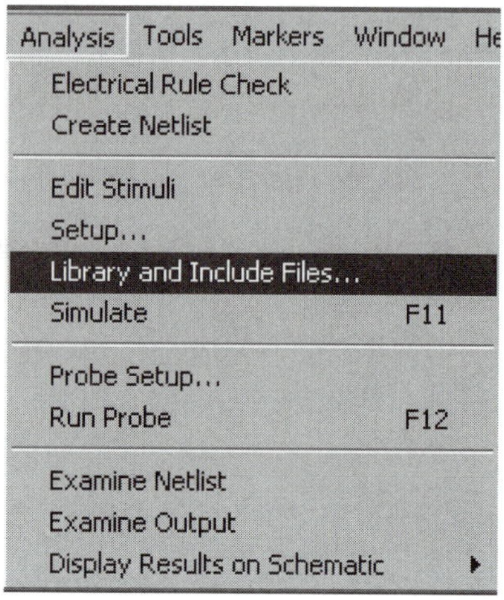

FIGURE 6.4

Command for library and include files.

Some Include File Statements

```
.INC  OPAMP.CIR
.INC  a:INVERTER.CIR
.INC  c:\LIB\NOR.CIR
```

In PSpice schematic, the library and include files are selected using the command Library and Include Files from the analysis menu as shown in Fig. 6.4.

6.8   .LIB (LIBRARY FILE)

A library file may be referenced in the circuit file by using the following statement:

```
.LIB  FNAME
```

FNAME is the name of the library file to be called. A library file may contain comments, .MODEL statements, subcircuit definitions, .LIB statements, and .END statements. No other statements are permitted. If FNAME is omitted, PSpice looks for the default file, EVAL.LIB, that comes with PSpice programs. The library file FNAME may call for another library file.

When a .LIB command calls for a file, it does not bring the whole text of the library file into the circuit file. It simply reads those models or subcircuits that are called by the main circuit file. As a result, only those models or subcircuit descriptions that are needed by the main circuit file take up the main memory (RAM) space.

Note. Check the PSpice files for the default library file name.

Some Library File Statements

```
.LIB                ; (Default file is EVAL.LIB.)
.LIB EVAL.LIB       ; (Library file EVAL.LIB is on the default
                        drive.)
.LIB C:\LIB\EVAL.LIB ; (Library file EVAL.LIB is in directory file
                        LIB on drive C.)
.LIB D:\LIB\EVAL.LIB ; (Library file EVAL.LIB is in directory file
                        LIB on drive D.)
```

6.9 .NODESET (NODESET)

In calculating the operating bias point, some or all of the nodes of the circuit may be assigned initial guesses to help dc convergence by the statement as

```
.NODESET  V(1)=A1 V(2)=A2 ... V(N)=AN
```

$V(1), V(2), \ldots$ are the node voltages, and $A1, A2, \ldots$ are the respective values of the initial guesses. Once the operating point is found, the .NODESET command has no effect during the dc sweep or transient analysis. This command can be used to avoid a convergence problem, for example, on flip-flop circuits to break the tie-in condition. This command is not normally used, and it should not be confused with the .IC command, which sets the initial conditions of the circuits during the operating-point calculations for transient analysis.

Statement for Nodeset

```
.NODESET V(4)=1.5V V(6)=0 V(25)=1.5V
```

6.10 OPTIONS

PSpice allows various options to control and to limit parameters for the various analysis. Figure 6.5 shows the options menu in the analysis setup. The general form is

```
.OPTIONS [⟨⟨options⟩ name⟩] [⟨⟨options⟩ name⟩=⟨value⟩]
```

The options can be listed in any order. There are two types of options: (1) those without values and (2) those with values. The options without values are used as flags

FIGURE 6.5

Options menu.

of various kinds, and only the option name is mentioned. Table 6.2 shows the options without values.

The options with values are used to specify certain optional parameters. The option names and their values are specified. Table 6.3 shows the options with values. The commonly used options are NOPAGE, NOECHO, NOMOD, TNOM, CPTIME, NUMDGT, GMIN, and LIMPTS.

Options Statements

```
.OPTIONS  NOPAGE NOECHO NOMOD DEFL=20U DEFW=15U DEFAD=50P DEFAS=50P
.OPTIONS  ACCT LIST RELTOL=.005
```

TABLE 6.2   List of Options Without Values

Option	Effects
NOPAGE	Suppresses paging and printing of a banner for each major section of output
NOECHO	Suppresses listing of the input file
NODE	Causes output of netlist (node table)
NOMOD	Suppresses listing of model parameters
LIST	Causes summary of all circuit elements (or devices) to be output
OPTS	Causes values for all options to be output
ACCT	Summary and accounting information is output at the end of all the analysis.
WIDTH	Same as .WIDTH OUT = statement

TABLE 6.3   List of Options with Values

Option	Effects	Units	Default
DEFL	MOSFET channel length (L)	Meter	100u
DEFW	MOSFET channel width (W)	Meter	100u
DEFAD	MOSFET drain diffusion area (AD)	$Meter^{-2}$	0
DEFAS	MOSFET source diffusion area (AS)	$Meter^{-2}$	0
TNOM	Default temperature (also the temperature at which model parameters are assumed to have been measured)	Degrees Celsius	27
NUMDGT	Number of digits output in print tables		4
CPTIME	CPU time allowed for a run	Second	1E6
LIMPTS	Maximum points allowed for any print table or plot		201
ITL1	Dc and bias-point "blind" iteration limit		40
ITL2	Dc and bias-point "educated guess" iteration limit		20
ITL4	Iteration limit at any point in transient analysis		10
ITL5	Total iteration limit for all points in transient analysis (ITL5 = 0 means ITL5 = infinite)		5000
RELTOL	Relative accuracy of voltages and currents		0.001
TRTOL	Transient analysis accuracy adjustment		7.0
ABSTOL	Best accuracy of currents	Amp	1pA
CHGTOL	Best accuracy of charges	Coulomb	0.01pC
VNTOL	Best accuracy of voltages	Volt	1uV
PIVREL	Relative magnitude required for pivot in matrix solution		1E–13
GMIN	Minimum conductance used for any branch	Ohm^{-1}	1E–12

Job statistics summary. If the option ACCT is specified on the .OPTIONS statement, PSpice will print various statistics about the run at the end. This option is not normally required for most circuit simulations. The list follows the format of the output:

Item	Meaning
NUNODS	Number of distinct circuit nodes before subcircuit expansion.
NCNODS	Number of distinct circuit nodes after subcircuit expansion. If there are no subcircuits, NCNODS = NUNODS.
NUMNOD	Total number of distinct nodes in circuit. This is NCNODS plus the internal nodes generated by parasitic resistances. If no device has parasitic resistances, NUMNOD = NCNODS.
NUMEL	Total number of devices (or elements) in circuit after subcircuit expansion. This includes all statements that do not begin with . or X.
DIODES	Number of diodes after subcircuit expansion.
BJTS	Number of bipolar transistors after subcircuit expansion.
JFETS	Number of junction FETs after subcircuit expansion.
MFETS	Number of MOSFETs after subcircuit expansion.
GASFETS	Number of GaAs MESFETs after subcircuit expansion.
NUMTEM	Number of different temperatures.
ICVFLG	Number of steps of dc sweep.
JTRFLG	Number of print steps of transient analysis.
JACFLG	Number of steps of ac analysis.
INOISE	1 or 0: noise analysis was/was not performed.
NOGO	1 or 0: run did/did not have an error.
NSTOP	The circuit matrix is conceptually (not physically) of dimension NSTOP $\times$ NSTOP.
NTTAR	Actual number of entries in circuit matrix at beginning of run.
NTTBR	Actual number of entries in circuit matrix at end of run.
NTTOV	Number of terms in circuit matrix that come from more than one device.
IFILL	Difference between NTTAR and NTTBR.
IOPS	Number of floating-point operations needed to complete one solution of circuit matrix.
PERSPA	Percent sparsity of circuit matrix.
NUMTTP	Number of internal time steps in transient analysis.
NUMRTP	Number of times in transient analysis that a time step was too large and had to be cut back.

NUMNIT	Total number of iterations for transient analysis.
MEMUSE/MAXMEM	Amount of circuit memory used/available in bytes. There are two memory pools. Exceeding either one will abort the run.
COPYKNT	Number of bytes that were copied in the course of doing memory management for this run.
READIN	Time spent reading and error checking the input file.
SETUP	Time spent setting up the circuit-matrix pointer structure.
DCSWEEP	Time spent and iteration count for calculating dc sweep.
BIASPNT	Time spent and iteration count for calculating bias point and bias point for transient analysis.
MATSOL	Time spent solving circuit matrix. (This time is also included in the total for each analysis.) The iteration count is the number of times the rows or columns were swapped in the course of solving it.
ACAN	Time spent and iteration count for ac analysis.
TRANAN	Time spent and iteration count for transient analysis.
OUTPUT	Time spent preparing .PRINT tables and .PLOT plots.
LOAD	Time spent evaluating device equations. (This time is also included in the total for each analysis.)
OVERHEAD	Other time spent during run.
TOTAL JOB TIME	Total run time, excluding the time to load the program files PSPICE1.EXE and PSPICE2.EXE into memory.

6.11   .PARAM (PARAMETER)

PSpice allows one to use a parameter instead of a numerical value. This parameter can be integrated into arithmetic expressions. The parameter definition is one of the following forms:

    .PARAM ⟨ PNAME = ⟨value⟩ or { ⟨ expression ⟩ } ⟩*

The keyword .PARAM is followed by a list of names with values or expressions. The ⟨value⟩ must be a constant and does not need to be enclosed in braces. However, the ⟨expression⟩ does need braces, { }, and must use only previously defined parameters.

PNAME is the parameter name, and it cannot be a predefined parameter, such as TIME (time), TEMP (temperature), VT (thermal voltage), or GMIN (shunt conductance for semiconductor p–n junctions).

Figure 6.6 shows the symbol and the parameters of PARAM.

Some .PARAM Statements

```
.PARAM VSUPPLY = 22V
.PARAM VCC = 15 V, VEE = −15V
.PARAM BANDWIDTH = {50kHz/20}
.PARAM PI = 3.14159, TWO_PI = {2*3.14159}
```

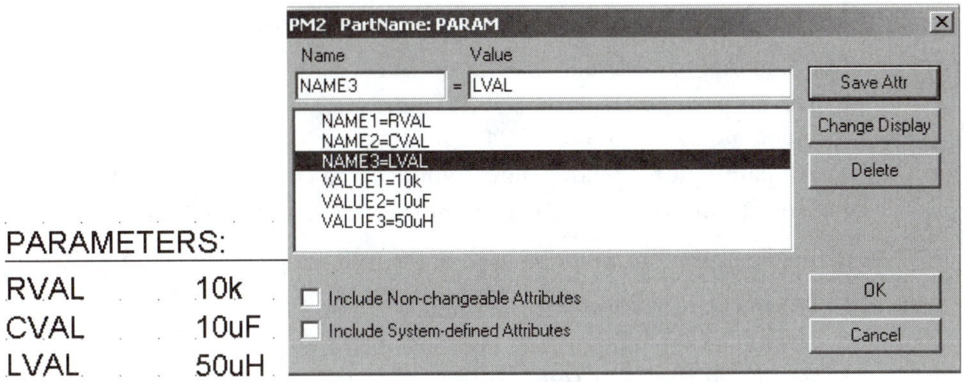

PARAMETERS:

RVAL	10k
CVAL	10uF
LVAL	50uH

(a) Symbol

PM2 PartName: PARAM

Name	Value	
NAME3	= LVAL	Save Attr

NAME1=RVAL
NAME2=CVAL
NAME3=LVAL
VALUE1=10k
VALUE2=10uF
VALUE3=50uH

Change Display

Delete

☐ Include Non-changeable Attributes
☐ Include System-defined Attributes

OK

Cancel

(b) Parameters

FIGURE 6.6

Parametric definitions.

Once a parameter is defined, it can be used in place of a numerical value; for example, both of the statements

```
VCC   1   0    {SUPPLY}
VEE   0   5    {SUPPLY}
```

will change the value of SUPPLY. As an additional example, each of the statements

```
.FUNC    IND(I(VX))   (A0+A1*I(VX)+A2*I(VX)*I(VX)) ; Polynomial form
.PARAM   INDUCTOR = IND(I(VX))
L1    1    3     {INDUCTOR}
```

will change the value of INDUCTOR in a polynomial form.

It should be noted that the parameters defined by the .PARAM statement are global and can be used anywhere in the circuit, including inside subcircuits. However, parameters can be made local to subcircuits by having parameters as arguments to subcircuits. For example,

```
.SUBCKT FILTER 1  2 PARAMS: CENTER=100kHz, BANDWIDTH=10kHz
```

is a subcircuit definition for a bandpass filter with nodes 1 and 2, and with parameters CENTER (center frequency), and BANDWIDTH (bandwidth). The keyword PARAMS separates the nodes list from the parameter list. The parameters can be given new values by changing the values of the arguments while calling the subcircuit FILTER. For example,

```
X1   4   6   FILTER   PARAMS: CENTER=200kHz
```

will override the default value of 100 kHz with a CENTER value of 200 kHz.

A defined parameter can be used in the following cases:

1. All model parameters
2. Device parameters, such as AREA, L, NRD, Z0, and IC-values on capacitors and inductors and TC1 and TC2 for resistors
3. All parameters of independent voltage and current sources (V and I devices)
4. Values in a .NODESET statement (Section 6.9) and .IC statement (Section 4.7.1).

A defined parameter *cannot* be used in the following cases:

1. For the transmission-line parameters NL and F
2. For the in-line temperature coefficients for resistors
3. For the $E, F, G,$ and H device polynomial coefficient values
4. For replacing node numbers
5. For values in analysis statements (.TRAN, .AC, .DC, etc.).

6.12 FOURIER ANALYSIS

The results of a transient analysis are in discrete forms. These sampled data can be used to calculate the coefficients of a Fourier series. A periodic waveform can be expressed in a Fourier series as

$$v(\theta) = C_0 + \sum_{n=1}^{\infty} C_n \sin(n\theta + \phi_n)$$

where $\theta = 2\pi ft$
  f = frequency in hertz
  C_0 = dc component
  C_n = nth harmonic component

PSpice uses the results of the transient analysis to perform the Fourier analysis up to nth harmonics. The statement takes one of the following general forms:

```
.FOUR   FREQ N V1   V2   V3 ... VN
.FOUR   FREQ N I1   I2   I3 ... IN
```

FREQ is the fundamental frequency. V1, V2, ... (or I1, I2, ...) are the output voltages (or currents) for which the Fourier analysis is desired. A .FOUR statement must be used with a .TRAN statement. The output voltages (or currents) must have the same forms as in the .TRAN statement for transient analysis. If the number of harmonics N is not specified, the dc component, fundamental, and second through ninth harmonics (or 10 coefficients) are calculated by default.

Fourier analysis is performed over the interval (TSTOP-PERIOD) to TSTOP, where TSTOP is the final (or stop) time for the transient analysis, and PERIOD is one period of the fundamental frequency. Therefore, the duration of the transient analysis must be at least one period long, PERIOD. At the end of the analysis, PSpice determines the dc component and the amplitudes up to the nth harmonic.

PSpice does print a table, automatically, showing the results of Fourier analysis and does not require the .PRINT, .PLOT, or .PROBE statement.

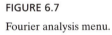

FIGURE 6.7

Fourier analysis menu.

In the PSpice schematic, Fourier analysis is enabled in the transient menu of the analysis setup as shown in Fig. 6.7.

Statement for Fourier Analysis

```
.FOUR  100KHZ  V(2, 3), V(3), I(R1), I(VIN)
```

Example 6.2: Finding the Fourier coefficients

For Example 4.7, calculate the coefficients of the Fourier series for the load current if the fundamental frequency is 1 kHz (up to the eleventh harmonic is desired).

Solution The PSpice schematic is shown in Fig. 6.8(a). Fourier analysis, which is done after the transient analysis, is selected from the transient analysis menu of the analysis setup as shown in Fig. 6.8(b). The listing of the circuit file follows:

Example 6.2 Fourier analysis

```
▲   VS  1   0  SIN (0 200V 1KHZ)  ; Sinusoidal voltage of 200 V peak
▲▲  RS  1   2    100OHM
    R1  2   0    100KOHM
    E1  3   0    2   0   0.1 ; V-controlled source with a gain of 0.1
    RL  4   5    2OHM
    VX  5   0    DC  0V      ; Measures the load current
    S1  3   4    3   0    SMOD  ; V-controlled switch with model SMOD
    * Switch model descriptions
    .MODEL SMOD   VSWITCH (RON=5M ROFF=10E+9 VON=25M VOFF=0.0)
▲▲▲    TRAN   5US 1MS             ; Transient analysis
```

```
*    Fourier analysis of load current at a fundamental frequency of 1 kHz
.FOUR  1KHZ  11   I(VX)      ; Fourier analysis
.PLOT  TRAN  V(3) I(VX)      ; Not needed with .PROBE
.PROBE                       ; Graphical waveform analyzer
.END                         ; End of circuit file
```

The results of the Fourier analysis, which are obtained from the output file EX6-2.OUT, follow:

```
****      FOURIER ANALYSIS                          TEMPERATURE = 27.000 DEG C
FOURIER COMPONENTS OF TRANSIENT RESPONSE I(VX)
DC COMPONENT = 3.171399E+00
```

HARMONIC NO	FREQUENCY (HZ)	FOURIER COMPONENT	NORMALIZED COMPONENT	PHASE (DEG)	NORMALIZED PHASE (DEG)
1	1.000E+03	4.982E+00	1.000E+00	3.647E−05	0.000E+00
2	2.000E+03	2.115E+00	4.245E−01	−9.000E+01	−9.000E+01
3	3.000E+03	3.896E−07	7.819E−08	−8.424E+01	−8.424E+01
4	4.000E+03	4.234E−01	8.499E−02	−9.000E+01	−9.000E+01
5	5.000E+03	1.318E−07	2.646E−08	−7.547E+01	−7.547E+01
6	6.000E+03	1.818E−01	3.648E−02	−9.000E+01	−9.000E+01
7	7.000E+03	4.353E−08	8.737E−09	2.692E+01	2.692E+01
8	8.000E+03	1.012E−01	2.032E−02	−9.000E+01	−9.000E+01
9	9.000E+03	4.223E−08	8.476E−09	2.707E+01	2.707E+01
10	1.000E+04	6.460E−02	1.297E−02	−9.000E+01	−9.000E+01
11	1.000E+04	7.012E−08	1.407E−08	1.778E+01	1.778E+01

```
TOTAL HARMONIC DISTORTION = 4.351438E+01 PERCENT
```

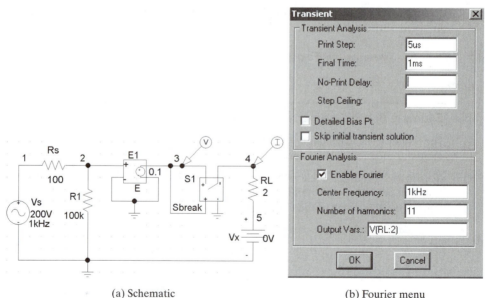

(a) Schematic   (b) Fourier menu

FIGURE 6.8

PSpice schematic for Example 6.2.

Notes

1. If the order of highest harmonics N is not specified in the .FOUR statement, PSpice calculates up to the ninth harmonic, or 10 coefficients. For example, the statement

```
.FOUR 1kHZ I(VX)
```

   will print up to the ninth harmonic in the output file.
2. One could add the statement .FOUR 1KHZ I(VX) in the circuit file of Example 4.7.

6.13   NOISE ANALYSIS

Resistors and semiconductor devices generate noise. The level of the noise depends on the frequency. The various types of noise that are generated by resistors and semi-conductor devices are discussed in Appendix B. Noise analysis is performed in conjunction with ac analysis and requires an .AC command. For each frequency of the ac analysis, the noise level of each generator in a circuit (e.g., resistors and transistors) is calculated, and their contributions to the output nodes are computed by summing the rms noise values. The gain from the input source to the output voltage is calculated. From this gain, the equivalent input noise level at the specified source is calculated by PSpice.

The statement for performing noise analysis has the form

```
.NOISE  V(N+, N-)  SOURCE M
```

where $V(N+, N-)$ is the output voltage across nodes $N+$ and $N-$. The output could be at a node N, such as $V(N)$.

SOURCE is the name of an independent voltage or current source at which the equivalent input noise will be generated. It should be noted that SOURCE is not a noise generator; rather it is a place at which to compute the equivalent noise input. For a voltage source, the equivalent input is in $V/\sqrt{Hz}$; and for a current source, it is in $A/\sqrt{Hz}$.

M is the print interval that permits PSpice to print a table for the individual contributions of all generators to the output nodes for every mth frequency. There is no need for a .PRINT or .PLOT command for printing a table of all contributions. If the value of M is not specified, then PSpice does not print a table of individual contributions. The output noise and equivalent noise can also be printed by the .PRINT or .PLOT command.

In the PSpice schematic, the noise analysis is enabled in the ac sweep menu of the analysis setup as shown in Fig. 6.9.

Statements for Noise Analysis
```
.NOISE   V(4,5)    VIN
.NOISE   V(6)      IIN
.NOISE   V(10)     V1  10
```

Note: The .PROBE command cannot be used for noise analysis.

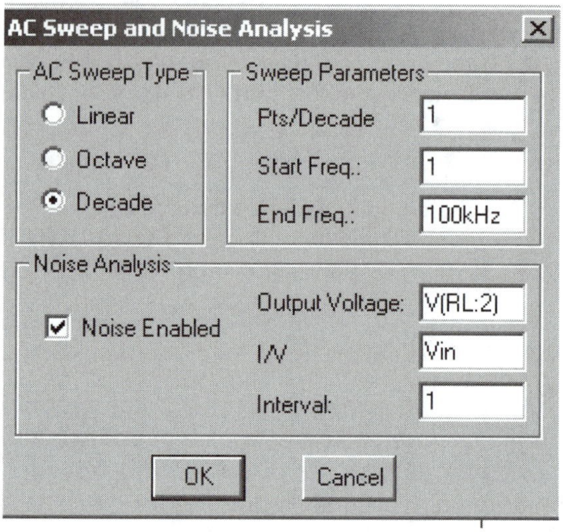

FIGURE 6.9

Noise analysis menu.

Example 6.3: Finding the equivalent input and output noise

For the circuit in Fig. 6.10(a), calculate and print the equivalent input and output noise if the frequency of the source is varied from 1 Hz to 100 kHz. The frequency should be increased by a decade with 1 point per decade. The amplifier is represented by the subcircuit of Fig. 6.10(b). The operating temperature is 40°C.

Solution The PSpice schematic is shown in Fig. 6.11(a). Noise analysis, which follows the ac analysis, is selected from the ac Sweep menu of the analysis setup as shown in Fig. 6 11(b). The input source is of ac type. The listing of the circuit file follows.

Example 6.3 Noise analysis

```
▲    VIN   1   0   AC    1V        ; Ac input voltage of 1 V
▲▲   R1    1   2   1K
     R2    2   0   20K
     RE    3   0   250
     RL    4   5   2K
     VX    5   0   DC     0V          ; Measures the load current
     X1    2   3   4     BJT         ; Subcircuit call for subcircuit BJT
     *     The subcircuit definition for BJT. Nodes 4, 5, 6, and 7 are
     *     referred to the subcircuit; they do not interact with
     *     the nodes of the main circuit
     .SUBCKT    BJT        4       6         5
     *         model name  base  emitter  collector
     RI    4   7    1.5K
     RO    5   6    100K
     VX    7   6   DC   0V    ; Measures the controlling current IB
     F1    5   6   VX   40    ; Current-controlled current source
     .ENDS    BJT                 ; End of subcircuit definition BJT
     .TEMP  40                    ; Operating temperature is 40 degrees
     .OPTIONS  NOPAGE  NOECHO    ; Options
▲▲▲  * Ac sweep from 1 Hz to 100 kHz with a decade increment and 1 point
     * per decade
     .AC    DEC    1    1HZ    100KHz
     * Noise analysis without printing details of individual contributions
```

```
     .NOISE    V(4)    VIN
     * PSpice prints the details of equivalent input and output noise.
     .PRINT   NOISE    ONOISE    INOISE
     .PROBE                            ; Graphical waveform analyzer
.END                                  ; End of circuit file
```

The results of noise analysis, which are stored in the output file EX6-3.OUT due to the .PRINT statement, follow (the node voltages are printed automatically by PSpice):

**** AC ANALYSIS			TEMPERATURE = 40.000 DEG C
FREQ	ONOISE	INOISE	
1.000E+00	4.325E−08	7.245E−09	
1.000E+01	4.325E−08	7.245E−09	
1.000E+02	4.325E−08	7.245E−09	
1.000E+03	4.325E−08	7.245E−09	
1.000E+04	4.325E−08	7.245E−09	
1.000E+05	4.325E−08	7.245E−09	

```
                JOB  CONCLUDED
                TOTAL  JOB  TIME                    1.53
```

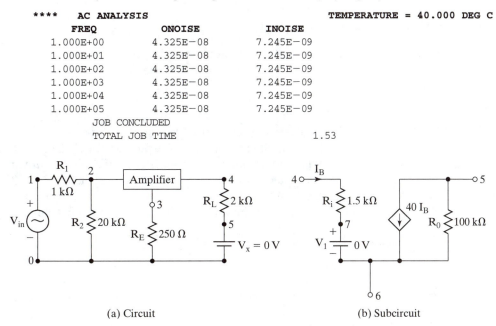

(a) Circuit   (b) Subcircuit

FIGURE 6.10

Circuit for Example 6.3.

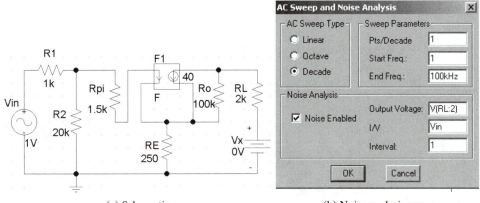

(a) Schematic   (b) Noise analysis menu

FIGURE 6.11

PSpice schematic for Example 6.3.

Note: The individual contributions of each element can be obtained by adding the parameter *M* in the NOISE statement. For instance, to print the individual contributions at every second frequency, the NOISE statement will read

```
.NOISE  V(4)    VIN 2
```

Students are encouraged to run the circuit file EX6-3.CIR with this statement and to look at the output file EX6-3.OUT. The effects of including *M*(=2) will be obvious.

6.14 .SENS (SENSITIVITY ANALYSIS)

The sensitivity of output voltages or currents with respect to every circuit and device parameter can be calculated by the .SENS statement, which has the following general form:

```
.SENS  〈(one or more output) variables)〉
```

The .SENS statement calculates the bias point and the linearized parameters around the bias point. In this analysis, the inductors are assumed to be short circuits, and capacitors are assumed to be open circuits. If the output variable is a current, then that current must be through a voltage source. The sensitivity of each output variable with respect to all the device values and model parameters is calculated, and the .SENS statement prints the results automatically. Therefore, it should be noted that a .SENS statement may generate a huge amount of data if many output variables are specified.

   Statement for Sensitivity Analysis
```
.SENS  V(5)   V(2.3)   I(V2)   I(V5)
```

Example 6.4: Finding the sensitivity of the output voltage with respect to each circuit element

For the circuit in Fig. 6.10(a), calculate and print the sensitivity of output voltage *V*(4) with respect to each circuit element. The amplifier is represented by the subcircuit in Fig. 6.10(b). The operating temperature is 40°C.

Solution The PSpice schematic is shown in Fig. 6.12(a). The output is specified in the sensitivity analysis menu of the analysis setup as shown in Fig. 6.12(b). The listing of the circuit file follows.

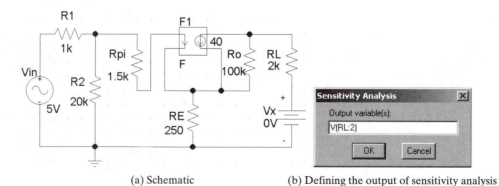

(a) Schematic   (b) Defining the output of sensitivity analysis

FIGURE 6.12

PSpice schematic for Example 6.4.

Example 6.4   Dc sensitivity analysis

```
▲    VIN   1   0   DC    5V       ; Dc input voltage of 5 V
▲▲   R1    1   2   1K
     R2    2   0   20K
     RE    3   0   250
     RL    4   5   2K
     VX    5   0   DC    0V        ; Measures the load current
     X1    2   3   4     BJT       ; Subcircuit call for subcircuit BJT
     *     The subcircuit definition for BJT. Nodes 4, 5, 6, and 7 are
     *     referred to the subcircuit; they do not interact with
     *     the nodes of the main circuit.
     .SUBCKT      BJT        4        6          5
     *            model name base     emitter    collector
     RI    4   7   1.5K
     RO    5   6   100K
     VX    7   6   DC    0V        ; Measures the controlling current IB
     F1    5   6   VX    40        ; Current-controlled current source
     .ENDS   BJT                   ; End of subcircuit definition BJT
     .TEMP  40                     ; Operating temperature is 40 degrees.
     .OPTIONS  NOPAGE  NOECHO      ; Options
▲▲▲  * Sensitivity analysis calculates the dc bias point and prints
     * the current through the input source, I(VIN), before computing
     * the sensitivity.
     * It calculates and prints the sensitivity analysis of output
     * voltage V(4) with respect to all elements in the circuit.
     .SENS  V(4)                   ; Sensitivity analysis
.END                              ; End of circuit file
```

The results of the sensitivity analysis follow (the node voltages are also printed automatically):

```
****    SMALL SIGNAL BIAS SOLUTION            TEMPERATURE = 27.000 DEG C
NODE    VOLTAGE    NODE    VOLTAGE    NODE    VOLTAGE    NODE    VOLTAGE
(   1)  5.0000   (    2)  4.3986   (   3)   3.8263   (   4)   −29.8470
(   5)  0.0000   ( X1.7)  3.8263
        VOLTAGE SOURCE CURRENTS
        NAME          CURRENT
        VIN          −6.014E−04
        VX           −1.492E−02
        X1.V1         3.815E−04
        TOTAL  POWER  DISSIPATION   3.01E−03  WATTS
```

```
****        DC SENSITIVITY ANALYSIS             TEMPERATURE = 40.000 DEG C
DC SENSITIVITIES OF OUTPUT V(4)
```

ELEMENT NAME	ELEMENT VALUE	ELEMENT SENSITIVITY (VOLTS/UNIT)	NORMALIZED SENSITIVITY (VOLTS/PERCENT)
R1	1.000E+03	3.590E−03	3.590E−02
R2	2.000E+04	−6.564E−05	−1.313E−02
RE	2.500E+02	9.600E−02	2.400E−01
RL	2.000E+03	−1.486E−02	−2.972E−01
X1.RI	1.500E+03	2.391E−03	3.587E−02
X1.R5	1.000E+05	−1.426E−06	−1.426E−03
VIN	5.000E+00	−5.969E+00	−2.985E−01
VX	0.000E+00	9.958E−01	0.000E+00
X1.V1	0.000E+00	6.268E+00	0.000E+00

```
JOB CONCLUDED
TOTAL JOB TIME          1.2
```

Notes

1. V(4) is most sensitive to changes in R_E and R_L. An increase in R_E causes V(4) to increase, whereas an increase in R_L causes V(4) to decrease.
2. One could combine the .AC, .NOISE, and .SEN V(4) commands in the circuit file of Example 6.4 by modifying VIN 1 0 AC 1V DC 5V.

6.15   .STEP (PARAMETRIC ANALYSIS)

The .STEP command can be used to evaluate the effects of parameter variations. It has one of the following general forms:

```
.STEP   LIN      SWNAME   SSTART    SEND    SINC
.STEP   OCT      SWNAME   SSTART    SEND    NP
.STEP   DEC      SWNAME   SSTART    SEND    NP
.STEP   SWNAME   LIST  ⟨value⟩*
```

SWNAME is the sweep variable name. SSTART, SEND, and SINC are the start value, the end value, and the increment value of the sweep variables, respectively. NP is the number of steps. LIN, OCT, or DEC specifies the type of sweep, as follows:

LIN   *Linear sweep:* SWNAME is swept linearly from SSTART to SEND. SINC is the step size. LIN is used if the variable range is narrow.

OCT   *Sweep by octave:* SWNAME is swept logarithmically by octave, and NP becomes the number of steps per octave. The next variable is generated by multiplying the present value by a constant larger than unity. OCT is used if the variable range is wide.

DEC   *Sweep by decade:* SWNAME is swept logarithmically by decade, and NP becomes the number of steps per decade. The next variable is generated by multiplying the present value by a constant larger than unity. DEC is used if the variable range is very wide.

LIST   *List of values:* There are no start and end values. The values of the sweep variables are listed after the keyword LIST.

The SWNAME can be one of the following types:

Source.   A name of an independent voltage or current source. During the sweep, the source's voltage or current is set to the sweep value.

Model Parameter.   A model type name and then model name followed by a model parameter name in parenthesis. The parameter in the model is set to the sweep value. The model parameters L and W for MOS devices and any temperature parameters, such as TC1 and TC2 for the resistor, *cannot* be swept.

Temperature.   The keyword TEMP followed by the keyword LIST. The temperature is set to the sweep value. For each value of sweep, the model parameters of all circuit components are updated to that temperature.

Global Parameter. The keyword PARAM followed by a parameter name. The parameter is set to sweep. During the sweep, the global parameter's value is set to the sweep value, and all expressions are affected by the sweep parameter value.

Some Step Statements

```
.STEP   VCE   -5V   10V   5V
```

sweeps the voltage VCE linearly.

```
.STEP   LIN   IIN   -10mA   5mA   0.1mA
```

sweeps the current IIN linearly.

```
.STEP   RES   RMOD(R)   0.9   1.1   0.001
```

linearly sweeps the model parameter R of the resistor model RMOD.

```
.STEP   DEC   NPN   QM(IS)   1E-18   1E-14   10
```

sweeps with a decade increment the parameter IS of the NPN transistor.

```
.STEP   TEMP   LIST   0 50   80   100   150
```

sweeps the temperature TEMP as listed.

```
.STEP   PARAM   Frequency   8.5kHz   10.5kHZ   50Hz
```

linearly sweeps the parameter PARAM frequency.

Notes

1. The sweep start value SSTART may be greater than or less than the sweep end value SEND.
2. The sweep increment SINC must be greater than zero.
3. The number of points NP must be greater than zero.

4. If the .STEP command is included in a circuit file, all specified analyses (.DC, .AC, .TRAN, etc.) are completed for each step.

In the PSpice schematic, the parameters are specified in the parametric menu of the analysis setup as shown in Fig. 6.13.

Example 6.5:   Finding the parametric effect on the transient response of an *RLC* circuit

The *RLC* circuit of Fig. 6.14(a) is supplied from a step input voltage, as shown in Fig. 6.14(c). Use PSpice to calculate and plot the capacitor voltage v_c from 0 to 400 μs with a time increment of 1 μs for $R_1 = 1\ \Omega, 2\ \Omega$, and 8 Ω. The capacitor voltage V(3) is the output that it is to be plotted. The results should also be available for display and hard copy by the .PROBE command.

Solution In Example 4.3, we found the transient responses for three values of R_1 by considering three separate *RLC* circuits. This is inefficient, because PSpice allows one to vary the values of a parameter. Thus, we will define R as a parameter and assign its values. The PSpice schematic

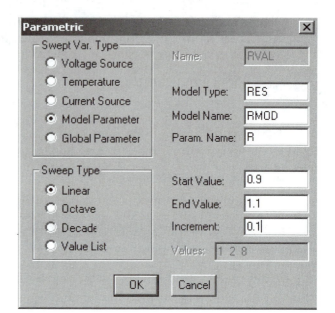

FIGURE 6.13

Defining sweep parameters.

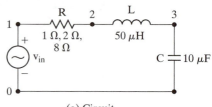

(a) Circuit

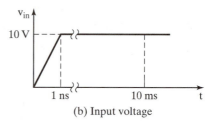

FIGURE 6.14

An *RLC* circuit with a step input voltage.

(b) Input voltage

is shown in Fig. 6.15(a). The sweep parameters for the variable resistor *R* are selected from the parametric menu of the analysis setup as shown in Fig. 6.15(b). The listing of the circuit file follows.

Example 6.5   Step response of an *RLC* circuit by parametric analysis

```
▲   VIN   1   0    PWL (0   0   1NS   1V   1MS   1V)    ; Step input of 1 V
▲▲  .PARAM    VAL = 1              ; Defining parameter VAL
    R    1    2    {VAL}    ; Resistance with variable values
    L    2    3    50UH
    C    3    0    10UF
▲▲▲    .STEP    PARAM   VAL   LIST   1   2   8    ; Assigning STEP values
    .TRAN    1US    400US              ; Transient analysis
    .PROBE                             ; Graphical waveform analyzer
.END                                   ; End of circuit file
```

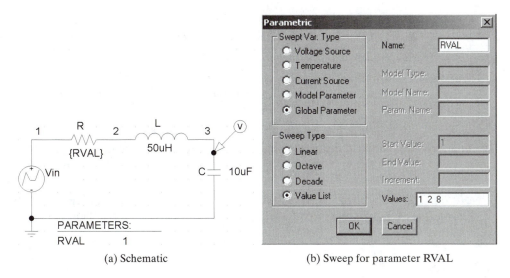

(a) Schematic

(b) Sweep for parameter RVAL

FIGURE 6.15

PSpice schematic for Example 6.5.

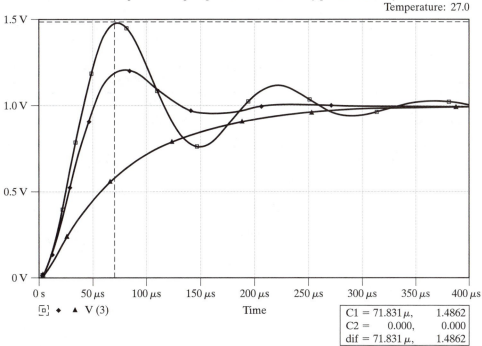

FIGURE 6.16

Step responses of an *RLC* circuit for Example 6.5.

The capacitor voltage V(3), which is obtained by the .PROBE command, is shown in Fig. 6.16 for three values of R. The responses depict the effects of damping on a second-order system.

Example 6.6:   Finding the parametric effect on the frequency response of an *RLC* circuit

An *RLC* circuit is shown in Fig. 6.17. Plot the frequency response of the current through the circuit and the magnitude of the input impedance. The frequency of the source is varied from 100 Hz to 100 kHz with a decade increment and 10 points per decade. The values of the inductor L are 5 mH, 15 mH, and 25 mH.

Solution   The PSpice schematic is shown in Fig. 6.18(a). The sweep parameters for the variable inductor L are selected from the parametric menu of the analysis setup as shown in Fig. 6.18(b). The listing of the circuit file follows.

Example 6.6   Input impedance characteristics by parametric analysis

```
▲    VIN    1   0    AC    1V          ; Ac voltage of 1 V
▲▲   R      2   3    50
       .PARAM  VAL = 1MH               ; Defining parameter VAL
       L      3   4    {VAL}           ; Inductance with variable values
       C      4   0    1UF
       VX     1   2    DC    0V        ; Measures the current
▲▲▲    .STEP  PARAM  VAL  5MH  25MH  10MH  ; Step variations for VAL
       .AC    DEC 100 100HZ 100KHZ        ; Ac sweep from 10 Hz to 100 kHz
       .PLOT  AC  I(VX)                   ; Plots in the output file
       .PROBE                            ; Graphical waveform analyzer
  .END                                    ; End of circuit file
```

The frequency response of the current through the circuit and the magnitude of the input impedance are shown in Fig. 6.19. As the inductance is increased, the resonant frequency decreases.

6.16   .DC (DC PARAMETRIC SWEEP)

PSpice allows one to evaluate the effects of parameter variations in dc analysis. These parameters could be elements and their model parameters. The general statement for the dc parametric sweep is

```
.DC  SWNAME   LIST⟨value⟩*
```

where SWNAME is the sweep variable name and can be one of the following types:

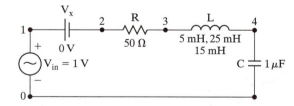

FIGURE 6.17

An *RLC* circuit for Example 6.6.

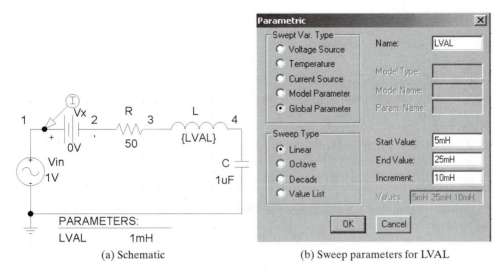

(a) Schematic

(b) Sweep parameters for LVAL

FIGURE 6.18

PSpice schematic for Example 6.6.

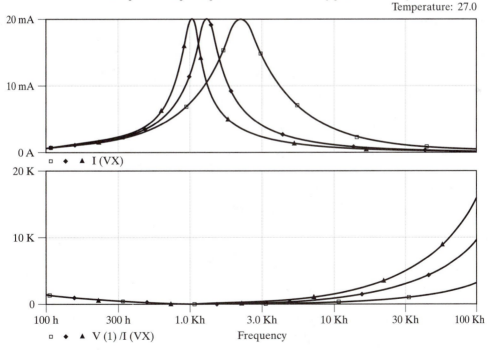

FIGURE 6.19

Frequency response for Example 6.6.

Source. A name of an independent voltage or current source. During the sweep, the source's voltage or current is set to the sweep value.

Model Parameter. A model type name and then a model name followed by a model parameter name in parenthesis. The parameter in the model is set to the sweep value. The model parameters L and W for MOS devices and any temperature parameters, such as TC1 and TC2 for the resistor, *cannot* be swept.

Temperature. The keyword TEMP followed by the keyword LIST. The temperature is set to the sweep value. For each value of sweep, the model parameters of all circuit components are updated to that temperature.

Global Parameter. The keyword PARAM followed by a parameter name. The parameter is set to sweep. During the sweep, the global parameter's value is set to the sweep value, and all expressions are evaluated.

> *Statements for dc Parametric Sweep*
>
> ```
> .DC   RES   RMOD (R)   0.9   1.1   0.001
> ```

linearly sweeps the model parameter R of the resistor model RMOD.

> ```
> .DC   DEC   NPN   QM (IS)   1E−18   1E−14   10
> ```

sweeps with a decade increment the parameter IS of the NPN transistor.

> ```
> .DC   TEMP   LIST   0 50   80   100   150
> ```

sweeps the temperature TEMP as listed values.

> ```
> .DC   PARAM   Vsupply   −15V   15V   0.5V
> ```

linearly sweeps the parameter PARAM Vsupply.

Example 6.7: Finding the parametric effect on the frequency response of an *RLC* circuit

For the amplifier circuit of Fig. 6.2, calculate and plot the dc transfer characteristic, V_{out} versus V_{in}. The input voltage V_{in} is varied from 0 to 100 mV with an increment of 2 mV. The load resistance is varied from 10 kΩ to 30 kΩ with a 10-kΩ increment.

Solution The PSpice schematic is shown in Fig. 6.20(a). The sweep parameters for the variable load resistor R_L are selected from the parametric menu of the analysis setup as shown in Fig. 6.20(b). The listing of the circuit file follows.

> **Example 6.7 Dc parametric sweep**
>
> ```
> ▲   VIN  1  0   DC   100MV        ; Dc input voltage of 100 mV
> ▲▲   R1   1   2    1K
>       R2   2   0    20K
>       RE   3   0    250
>      .PARAM   VAL = 10K                  ; Defining parameter VAL
> ```

```
RL      4    5      {VAL}          ; Load resistance with variable values
VX      5    0      DC    0V       ; Measures the load current
X1      2    3      4     BJT      ; Subcircuit call for subcircuit BJT
*  The subcircuit definition for BJT
.SUBCKT     BJT       4       6         5
*      model name    base   emitter   collector
RI      4    7      1.5K
RO      5    6      100K
VX      7    6      DC    0V       ; Measures the controlling current IB
F1      5    6      VX    40       ; Current-controlled current source
.ENDS   BJT                        ; End of subcircuit definition BJT
.STEP  PARAM  VAL  10K 30K 10K     ; Step variations for VAL
.DC     VIN       100MV    2MV     ; Dc sweep from 0 to 100 mV
.PLOT   DC  V(4)                   ; Plots in the output file
.PROBE                             ; Graphical waveform analyzer
.END                               ; End of circuit file
```

The transfer characteristic is shown in Fig. 6.21.

6.17 MONTE CARLO ANALYSIS

PSpice allows one to perform the Monte Carlo (statistical) analysis of a circuit for variations of model parameters. The variations are specified by DEV and LOT tolerances in .MODEL statements. The general form of the statement for Monte Carlo analysis is

```
.MC   ⟨(# runs) value)    ⟨(analysis)⟩    ⟨(output variable)⟩
+     ⟨(function)⟩         [(option)]*
```

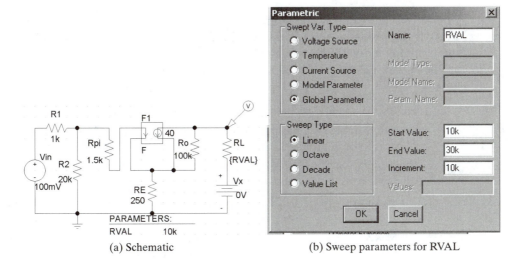

(a) Schematic   (b) Sweep parameters for RVAL

FIGURE 6.20

PSpice schematic for Example 6.7.

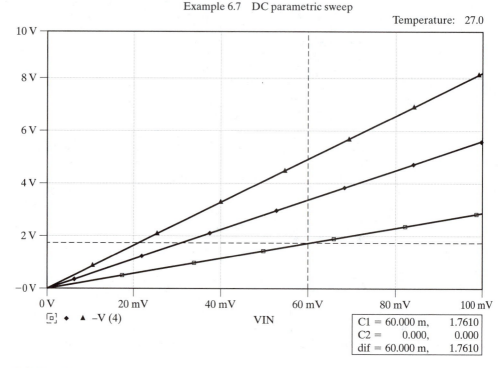

FIGURE 6.21

Transfer characteristic for Example 6.7.

where ⟨(# runs) value⟩ is the number of runs to be evaluated and has an upper limit of 2,000. If the results are to be viewed with Probe, then the upper limit is 100. The first run is completed with the nominal values of all components, but the subsequent runs are done with values generated randomly by PSpice. Thus, ⟨(# runs) value⟩ should be greater than 1.

⟨(analysis)⟩ specifies type of analysis and must be dc, ac, or transient:

⟨(output variable)⟩ is the output variable for which the Monte Carlo (statistical) analysis is to be completed and has the same format as the .PRINT output variable:

⟨(function)⟩ specifies the desired information of the ⟨(output variable)⟩ by one of the following keywords:

YMAX	Finds the greatest difference in each waveform from the nominal run.
MAX	Finds the maximum value of each waveform.
MIN	Finds the minimum value of each waveform.
RISE_EDGE (⟨value⟩)	Finds the first occurrence of the waveform crossing above the threshold ⟨value⟩.

FALL_EDGE (⟨value⟩)　　Finds the first occurrence of the waveform crossing below the threshold ⟨value⟩.

[(option)]* allows one to obtain information about the details of the parameters used in each run and to control the amount of output by one of the following keywords:

LIST　　Prints out the model parameter values used for each component during each run.

OUTPUT　　(output type) specifies the output after the nominal (first) run. If OUTPUT is omitted, the output is only from the nominal run. The (output type) is indicated by one of the following:

ALL　　Generates output from all runs.

FIRST ⟨value⟩　　Generates output only during first *n* runs.

EVERY ⟨value⟩　　Generates output every *n*th run.

RUNS ⟨value⟩*　　Generates output only for the listed runs, up to 25.

RANGE　　(⟨(low) value, ⟨(high) value⟩) limits the range for the ⟨(function)⟩. An asterisk (*) can be used to specify all ⟨values⟩. For example,

```
YMAX RANGE (*, 0.5)
MAX RANGE (-1, *)
```

Some Statements for Monte Carlo Analysis

```
.MC    5    TRAN   V(3)     YMAX
.MC    6    DC     V(5)     YMAX    LIST
.MC    10   DC     V(9)     YMAX    LIST    OUTPUT    ALL
.MC    15   AC     V(2,7)   YMAX    YMAX    RANGE     (*, 1.5)
```

In the PSpice schematic, the Monte Carlo or worst-case menu is selected from the analysis setup and the various options are shown in Fig. 6.22.

6.18　DEV/LOT DEVICE AND LOT TOLERANCES

The device tolerances are used for variations in component values. Each variation is independent; that is, the variation of one component does not affect that of another component. A device tolerance can be specified by the keyword DEV as a percentage value

```
DEV = 15%
```

or as an absolute value

```
DEV = 0.1
```

The lot tolerances assume the same variations for all components having the same .MODEL statement; that is, devices that have the same .MODEL statement will

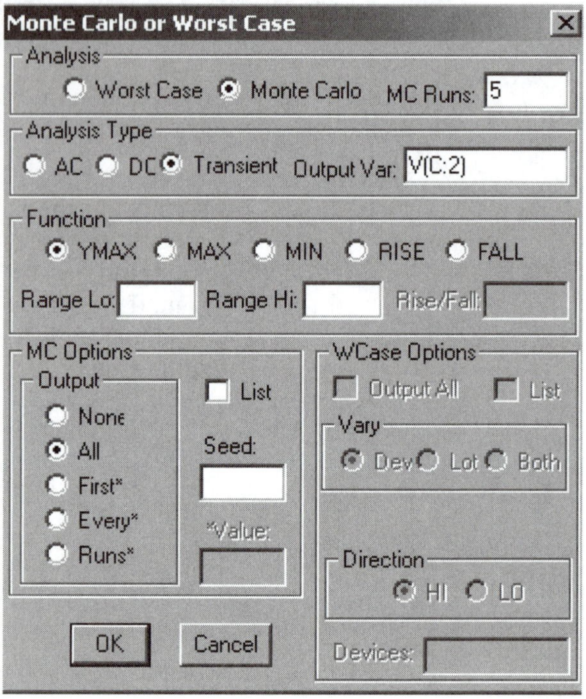

FIGURE 6.22

Monte-Carlo or worst-case menu.

vary together by the same amount. Thus, the variation is not independent. A lot tolerance can be specified by the keyword LOT as a percentage value

```
LOT = 5%
```

or as an absolute value

```
LOT = 0.01
```

The distribution of the deviation can be specified as uniformly distributed by including the keyword UNIFORM,

```
DEV/UNIFORM = 15%
```

Similarly, the distribution of the deviation can be specified as Gaussian distribution by the keyword GAUSS

```
DEV/GAUSS = 15%
```

It should be noted that if the distribution is not specified, it is, by default, UNIFORM.

Some Statements for Tolerances

```
R    2    7    RMOD    2K                          ; Resistance of 2 kΩ
.MODEL  RMOD   RES     (R=1  DEV = 15%)            ; Uniform by default
.MODEL  RMOD   RES     (R=1  DEV/UNIFORM = 15%)    ; Uniform distribution
.MODEL  RMOD   RES     (R=1  DEV/GAUSS = 15%)      ; Gaussian distribution
```

Example 6.8: Monte Carlo analysis of an *RLC* circuit

The *RLC* circuit of Fig. 6.14(a) is supplied from a step input voltage as shown in Fig. 6.14(b). The circuit parameters are as follows:

```
R = 2 Ω ± 15% (uniform deviation);
L = 50 μH ± 20% (uniform deviation);
C = 10 μH ± 10% (Gaussian deviation).
```

The model parameters are as follows: for the resistor, $R = 1$; for the capacitor, $C = 1$; and for the inductor, $L = 1$.

Use PSpice to perform Monte Carlo analysis of the capacitor voltage v_C from 0 to 400 μs with a time increment of 1 μs. Remember the following:

(a) The greatest difference from the nominal run is to be printed for five runs.

(b) The first occurrence of the capacitor voltage crossing below 1 V is to be printed.

Solution The PSpice schematic is shown in Fig. 6.23(a). The options for the maximum deviation (YMAX) are selected from the Monte Carlo menu of the analysis setup as shown in Fig. 6.23(b). The options for the time of the fall-edge are shown in Fig. 6.23(c). The listing of the circuit file follows.

Example   6.8   Monte Carlo analysis for transient response

```
▲   VIN   1   0   PWL   (0   0   1NS   1V   10MS   1V) ; Step input of 1 V
▲▲   R    1   2   RMOD   2            ; Resistance with model RMOD
     L    2   3   LMOD   50UH         ; Inductance with model RMOD
     C    3   0   CMOD   10UF         ; Capacitance with model CMOD
        .MODEL   RMOD   RES   (R=1 DEV=15%)
        .MODEL   LMOD   IND   (L=1 DEV/UNIFORM=20%)
        .MODEL   CMOD   CAP   (C=1 DEV/GAUSS=10%)
▲▲▲   .TRAN   1US   400US             ; Transient analysis
       .MC   5   TRAN   V(3) YMAX             ; Monte Carlo Analysis
      *.MC   5   TRAN   V(3) FALL_EDGE (1V) ; Monte Carlo Analysis
       .PROBE                         ; Graphical waveform analyzer
    .END                              ; End of circuit file
```

(a) The results of Monte Carlo analysis, which are obtained from the output file EX6-8.OUT, are as follows:

```
****    SORTED DEVIATIONS OF V(3)            TEMPERATURE = 27.000 DEG C
                           MONTE CARLO SUMMARY
Mean Deviation =            −.0229
Sigma          =             .0567
        RUN                          MAX DEVIATION FROM NOMINAL
Pass   2       .0732 (1.29 sigma) higher at T = 50.9930E−06
               (108.62% of Nominal)
Pass   4       .0653 (1.15 sigma) lower at T = 90.9940E−06
               (94.611% of Nominal)
Pass   3       .0635 (1.12 sigma) lower at T = 98.9940E−06
               (94.656% of Nominal)
Pass   5       .0362 (.64 sigma) lower at T = 50.9930E−06
               (95.74% of Nominal)
```

(b) To find the first occurrence of the capacitor voltage crossing below 1 V, the .MC statement is changed as follows:

```
.MC   5   TRAN   V(3)   FALL_EDGE (1V)       ; Monte Carlo analysis
```

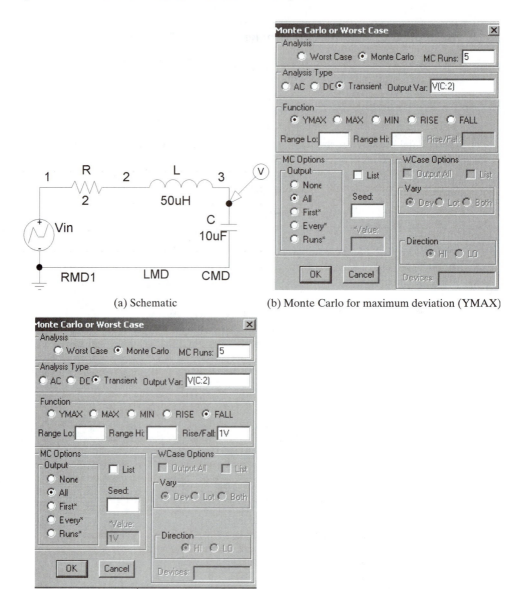

(a) Schematic

(b) Monte Carlo for maximum deviation (YMAX)

(c) Monte Carlo for finding the time of fall-edge

FIGURE 6.23

PSpice schematic for Example 6.8.

Following are the results of the Monte Carlo analysis:

```
                          MONTE CARLO SUMMARY
         RUN              FIRST FALLING-EDGE VALUE THRU 1
    Pass       5          135.0600E-06
                          ( 104.03% of Nominal)
    NOMINAL              129.8300E-06
```

```
Pass      4             129.6100E-06
                        ( 99.827% of Nominal)
Pass      3             124.6000E-06
                        ( 95.973% of Nominal)
Pass      2             119.4800E-06
                        ( 92.026% of Nominal)
```

6.19   SENSITIVITY/WORST-CASE ANALYSIS

PSpice allows one to perform the sensitivity and worst-case analysis of a circuit for variations of model parameters. The variations are specified by DEV and LOT tolerances on .MODEL statements. The general form of the sensitivity/worst-case analysis is

.WCASE ⟨(analysis)⟩ ⟨(output variable)⟩ ⟨(function)⟩ [(option)]*

This command is similar to the .MC command, except that it does not include ⟨(# runs) value⟩. With a .WCASE command, PSpice first calculates the sensitivity of the ⟨(output variable)⟩ to each model parameter. Once all the sensitivities are evaluated, one final run is done with all model parameters, and PSpice produces the worst case ⟨(output variable)⟩.

Note. Either the .MC or .WCASE command can be used, but *not* both in the same circuit.

Some Statements for Sensitivity/Worst-Case Analysis

```
.WCASE    TRAN    V(3)      YMAX
.WCASE    DC      V(5)      YMAX    LIST
.WCASE    DC      V(9)      YMAX    LIST    OUTPUT    ALL
.WCASE    AC      V(2,7)    YMAX    YMAX    RANGE     (*, 1.5)
```

Example 6.9:   Worst-case analysis of an *RLC* circuit

Run the sensitivity/worst-case analysis for Example 6.8.

Solution   The PSpice schematic is the same as that in Fig. 6.24(a). The options for the maximum deviation (YMAX) are selected from the worst-case menu of the analysis setup as shown in Fig. 6.24(a). The options for the time of the fall edge are shown in Fig. 6.24(b). Complete the following steps to obtain the solution:

(a) To run the sensitivity/worst-case analysis for finding the greatest difference from the nominal run, the .MC statement in Example 6.8 is changed to a .WCASE statement, as follows:

```
.WCASE   TRAN   V(3)   YMAX            ; Sensitivity/worst-case analysis
```

The plots of the greatest difference and the nominal run are shown in Fig. 6.25. The results of the worst-case analysis, which are obtained from the output file EX6-9.OUT, are as follows:

```
****    SORTED DEVIATIONS OF V(3)                 TEMPERATURE = 27.000 DEG C
                   WORST-CASE SUMMARY
RUN                MAX DEVIATION FROM NOMINAL
ALL DEVICES         .246 higher at T = 58.9930E-06
                   (124.59% of Nominal)
```

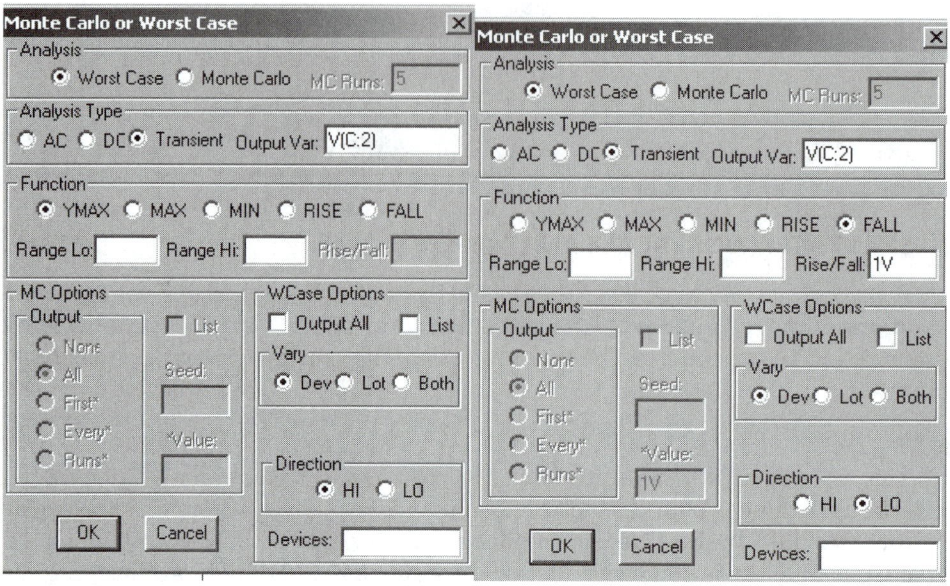

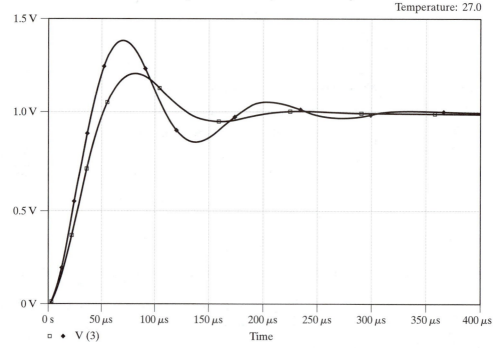

(a) Maximum deviation (YMAX)   (b) Finding the time of fall-edge

FIGURE 6.24

PSpice schematic for Example 6.9.

Example 6.9 Sensitivity/worst-case analysis

FIGURE 6.25

Plots of the greatest difference and the nominal run for Example 6.9.

Example 6.9   Sensitivity/worst-case analysis

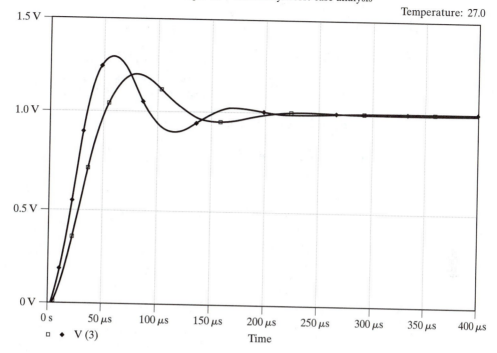

FIGURE 6.26

Plots of the first crossing below 1 V and the nominal run for Example 6.9.

(b) To find the first occurrence of the capacitor voltage crossing below 1 V, the .WCASE statement is changed as follows:

```
.WCASE  TRAN  V(3)  FALL_EDGE (1V) ; Sensitivity/worst-case analysis
```

The plots of the first crossing of the capacitor voltage below 1 V and the nominal run are shown in Fig. 6.26. The following are the results of the worst-case analysis:

WORST-CASE SUMMARY

RUN	FIRST FALLING EDGE VALUE THRU 1
NOMINAL	129.8300E−06
ALL DEVICES	91.9470E−06
	(70.82% of Nominal)

SUMMARY

The statements for behavioral modeling are as follows:

```
E⟨name⟩    N+  N−  VALUE = { ⟨expression⟩ }
E⟨name⟩    N+  N−  TABLE { ⟨expression⟩ } =
+                  ⟨ ⟨(input) value⟩, ⟨(output) value⟩ ⟩*
```

```
E⟨name⟩    N+  N−   LAPLACE { ⟨expression⟩ } = { ⟨transform⟩ }
E⟨name⟩    N+  N−   FREQ { ⟨expression⟩ } =
+ ⟨⟨(frequency) value⟩, ⟨(magnitude in dB) value⟩, ⟨(phase) value⟩⟩*

G⟨name⟩    N+  N−      VALUE = { ⟨expression⟩ }
G⟨name⟩    N+  N−      TABLE { ⟨expression⟩ } =
+                      ⟨ ⟨(input) value⟩, ⟨(output) value⟩ ⟩*
G⟨name⟩    N+  N−      LAPLACE { ⟨expression⟩ } = { ⟨transform⟩ }
G⟨name⟩    N+  N−      FREQ { ⟨expression⟩ } =
+ ⟨⟨(frequency) value⟩, ⟨(magnitude in dB) value⟩, ⟨(phase) value⟩⟩*
```

The following are advanced commands and features:

.SUBCKT	Subcircuit
.ENDS	End of subcircuit
.FUNC	Function
.GLOBAL	Global
.INC	Include file
.LIB	Library file
.NODESET	Nodeset
.OPTIONS	Options
.PARAM	Parameter
.FOUR	Fourier analysis
.NOISE	Noise analysis
.SENS	Sensitivity analysis
.STEP	Parametric analysis
.DC	Dc parametric sweep
.MC	Monte Carlo analysis
DEV/LOT	Device tolerances
.WCASE	Worst-case analysis

REFERENCES

[1] P. Antognetti and G. Massobri, *Semiconductor Device Modeling with SPICE*. New York: McGraw-Hill Book Co., 1988.

[2] *Books on PSpice*. http://www.pspice.com/publications/books.asp

[3] *OrCAD 9.2 Demo*. San Jose: Cadence Design Systems, Inc., 2001. http://www.cadencepcb.com/products/downloads/orcaddemo/default.asp

[4] *PSpice 9.1 Student Version*. San Jose: Cadence Design Systems, Inc., 2001. http://www.cadencepcb.com/products/downloads/PSpicestudent/default.asp

[5] *PSpice Design Community*. San Jose: Cadence Design Systems, Inc., 2001. http://www.PSpice.com

[6] *PSpice Models from Vendors*. http://www.pspice.com/models/links.asp

[7] M. H. Rashid, *SPICE for Power Electronics and Electronic Power*. Englewood Cliffs, New Jersey: Prentice Hall, 1993.

PROBLEMS

6.1 For the circuit in Fig. P6.1, calculate and print the sensitivity of output voltage V_{out} with respect to each circuit element. The operating temperature is 50°C.

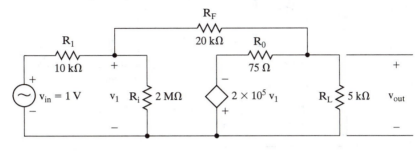

FIGURE P6.1

6.2 For the circuit in Fig. P6.1, calculate and print (a) the voltage gain, $A_v = V_{out}/V_{in}$; (b) the input resistance, R_{in}; and (c) the output resistance, R_{out}.

6.3 For the circuit in Fig. P6.1, calculate and plot the dc transfer characteristic, V_{out} versus V_{in}. The input voltage is varied from 0 to 10 V with an increment of 0.5 V for $R_F = 1\,k\Omega$, 20 $k\Omega$, and 50 $k\Omega$.

6.4 For the circuit in Fig. P6.1, calculate and print the equivalent input and output noise if the frequency of the source is varied from 10 Hz to 1 MHz. The frequency should be increased by a decade with two points per decade.

6.5 For the circuit in Fig. P6.5, the frequency response is to be calculated and printed over the frequency range from 1 Hz to 100 kHz with a decade increment and 10 points per decade. The peak magnitude and phase angle of the output voltage are to be plotted from the output file. The results should also be available for display and hard copy by the .PROBE command for $R_F = 1\,k\Omega$, 20 $k\Omega$, and 50 $k\Omega$.

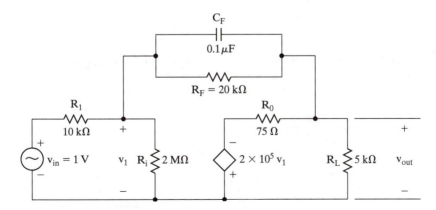

FIGURE P6.5

6.6 Repeat Problem 6.5 for the circuit in Fig. P6.6, for $R_F = 1\,\text{k}\Omega$, $20\,\text{k}\Omega$, and $50\,\text{k}\Omega$.

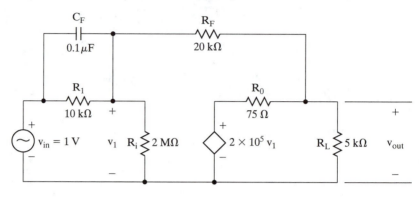

FIGURE P6.6

6.7 For the circuit in Fig. P6.5, calculate and plot the transient response of the output voltage from 0 to 2 ms with a time increment of 5 μs. The input voltage is shown in Fig. P6.7. The results should be available for display and hard copy by Probe for $R_F = 10\,\text{k}\Omega$, $20\,\text{k}\Omega$, and $30\,\text{k}\Omega$.

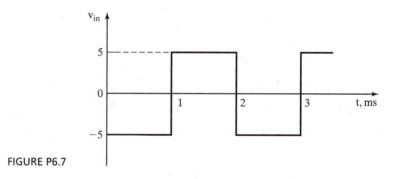

FIGURE P6.7

6.8 Repeat Problem 6.7 for the input voltage shown in Fig. P6.8.

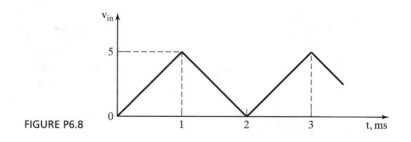

FIGURE P6.8

6.9 For the circuit in Fig. P6.6, calculate and plot the transient response of the output voltage from 0 to 2 ms with a time increment of 5 μs. The input voltage is shown in Fig. P6.7. The results should be available for display and hard copy by Probe for $R_F = 10\,\text{k}\Omega, 20\,\text{k}\Omega$, and 30 kΩ.

6.10 Repeat Problem 6.9 for the input voltage shown in Fig. P6.8.

6.11 Use PSpice to perform a Monte Carlo analysis for seven runs and for the dc response of Problem 6.3. The model parameter for resistors is $R = 1$. The circuit parameters having uniform deviations are the following:

$$R_F = 20\,\text{k}\Omega \pm 20\%$$
$$R_i = 2\,\text{M}\Omega \pm 5\%$$
$$R_L = 5\,\text{k}\Omega \pm 10\%$$
$$R_o = 75\,\Omega \pm 5\%$$
$$R_1 = 10\,\text{k}\Omega \pm 15\%$$

(a) The greatest difference from the nominal run is to be printed.
(b) The maximum value of the output voltage is to be printed.
(c) The minimum value of the output voltage is to be printed.
(d) The first occurrence of the output voltage crossing below 1 V is to be printed.

6.12 Use PSpice to perform the worst-case analysis for Problem 6.11.

6.13 Use PSpice to perform a Monte Carlo analysis for six runs and for the frequency response of Problem 6.5. The model parameters are $R = 1$ for resistors and $C = 1$ for the capacitor. The circuit parameters having uniform deviations are the following:

$$R_F = 20\,\text{k}\Omega \pm 20\%$$
$$R_i = 2\,\text{M}\Omega \pm 5\%$$
$$R_L = 5\,\text{k}\Omega \pm 10\%$$
$$R_o = 75\,\Omega \pm 5\%$$
$$R_1 = 10\,\text{k}\Omega \pm 15\%$$
$$C_F = 0.1\,\mu\text{F} \pm 10\%$$

(a) The greatest difference from the nominal run is to be printed.
(b) The maximum value of the output voltage is to be printed.
(c) The minimum value of the output voltage is to be printed.
(d) The first occurrence of the output voltage crossing below 1 V is to be printed.

6.14 Repeat Problem 6.13 for the frequency response of Problem 6.6.

6.15 Use PSpice to perform the worst-case analysis for Problem 6.13.

6.16 Use PSpice to perform the worst-case analysis for Problem 6.14.

6.17 Use PSpice to perform a Monte Carlo analysis for 10 runs and for the transient response of Problem 6.7. The model parameters are $R = 1$ for resistors and $C = 1$ for the capacitor. The circuit parameters having uniform deviations are the following:

$$R_F = 20\,\text{k}\Omega \pm 20\%$$
$$R_i = 2\,\text{M}\Omega \pm 5\%$$
$$R_L = 5\,\text{k}\Omega \pm 10\%$$

$$R_o = 75\ \Omega \pm 5\%$$
$$R_1 = 10\ k\Omega \pm 15\%$$
$$C_F = 0.1\mu F \pm 10\%$$

(a) The greatest difference from the nominal run is to be printed.
(b) The maximum value of the output voltage is to be printed.
(c) The minimum value of the output voltage is to be printed.
(d) The first occurrence of the output voltage crossing below 1 V is to be printed.

6.18 Repeat Problem 6.17 for the transient response of Problem 6.8.

6.19 Repeat Problem 6.17 for the transient response of Problem 6.9.

6.20 Use PSpice to perform the worst-case analysis for Problem 6.17.

6.21 Use PSpice to perform the worst-case analysis for Problem 6.18.

6.22 Use PSpice to perform the worst-case analysis for Problem 6.19.

6.23 Use PSpice to find the worst-case minimum and maximum voltages $V_{ab(\max)}$ and $V_{ab(\min)}$ across nodes a and b for the circuit as shown in Fig. P6.23. Assume uniform tolerances of $\pm 10\%$ for all resistances and an operating temperature of 25°C.

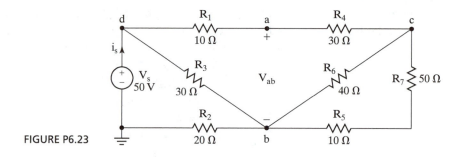

FIGURE P6.23

6.24 Use PSpice to find the worst-case minimum and maximum load resistances $R_{L(\max)}$ and $R_{L(\min)}$ for the circuit in Fig. P6.24 so that the maximum power is delivered to the load. Assume uniform tolerances of $\pm 10\%$ for all resistances except R_L and an operating temperature of 25°C.

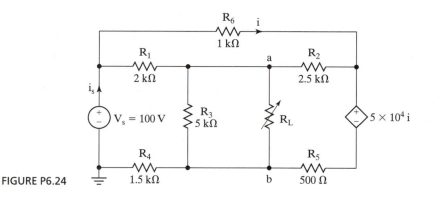

FIGURE P6.24

CHAPTER 7

Semiconductor Diodes

After completing this chapter, students should be able to

- Understand the $v-i$ characteristics of diodes, the dc operating/biasing point and small-signal model of a diode
- Model a diode in SPICE and specify its mode parameters
- Plot the diode characteristics in SPICE
- Perform dc, ac, and transient analysis of diode circuits
- Perform worst-case analysis of diode circuits for parametric variations of model parameters and tolerances.

7.1   INTRODUCTION

A semiconductor diode may be specified in PSpice by a diode statement in conjunction with a model statement. The diode statement specifies the diode name, the nodes to which the diode is connected, and its model name. The model incorporates an extensive range of diode characteristics (e.g., dc and small-signal behavior, temperature dependency, and noise generation). The model parameters take into account temperature effects, various capacitances, and physical properties of semiconductors.

7.2   DIODE CHARACTERISTICS

The typical $v-i$ characteristic of a diode is shown in Fig. 7.1 and can be expressed by an equation known as the *Shockley diode equation*, or [3]

$$I_D = I_S(e^{V_D/\eta V_T} - 1) \tag{7.1}$$

where I_D = forward current through the diode, A
V_D = forward voltage with anode positive with respect to cathode, V
I_S = leakage (or reverse saturation) current, typically in the range of 10^{-6} A to 10^{-16} A
η = an empirical constant known as the *emission coefficient* (or *ideality factor*), whose value varies from 1 to 2.

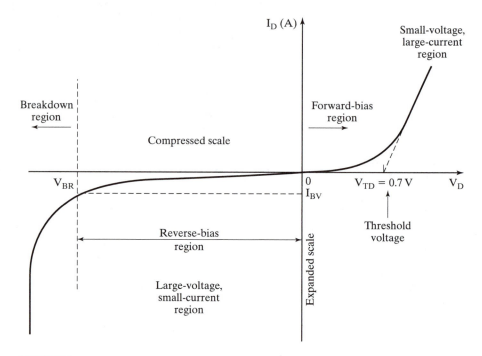

FIGURE 7.1

Diode characteristics.

The emission constant η depends upon the material and the physical construction of diodes. Its value ranges from 1 to 2, but it should, ideally, be 1.

The value V_T in Eq. (7.1) is a constant called *thermal voltage*, given as

$$V_T = \frac{kT}{q}, \tag{7.2}$$

where q = electron charge: 1.6022×10^{-19} coulombs (C)
        T = absolute temperature in Kelvin (K = 273 + °C)
        k = Boltzmann's constant: 1.3806×10^{-23} J per K

At a junction temperature of 25°C (77°F, 298 K), Eq. (7.2) gives

$$V_T = \frac{kT}{q} = \frac{1.3806 \times 10^{-23} \times (273 + 25)}{1.6022 \times 10^{-19}} \approx 25.8 \text{ mV}$$

At a specified temperature, the leakage current I_S remains constant for a given diode. For small-signal diodes, the typical value of I_S is 10^{-14} A.

The diode characteristic of Fig. 7.1 can be divided into three regions:

Forward-bias region, where the forward voltage $V_D > 0$
Reverse-bias region, where the reverse voltage $V_R < 0$
Breakdown region, where the breakdown voltage $V_R < -V_{BR}$

7.2.1 Forward-Bias Region

In the forward-bias region, $V_D > 0$. The diode current I_D is very small if the diode voltage V_D is less than a specific value V_{TD} (typically 0.7 V). The diode conducts fully if V_D is higher than this value V_{TR}, which is referred to as the *threshold voltage*, the *cut-in voltage*, or the *turn-on voltage*. Thus, the threshold voltage is a voltage at which the diode conducts.

For $V_D > 0.1$ V, which is usually the case, $I_D \gg I_S$, and Eq. (7.1) can be approximated to

$$I_D = I_S[e^{V_D/\eta V_T} - 1] \approx I_S e^{V_D/\eta V_T} \tag{7.3}$$

Taking natural (base e) logarithm on both sides, we get

$$\ln I_D = \ln I_S + \frac{V_D}{\eta V_T}$$

which, after simplification, gives the diode voltage V_D as

$$V_D = \eta\, V_T \ln\left(\frac{I_D}{I_S}\right) \tag{7.4}$$

Converting the natural log of base e to the logarithm of base 10, Eq. (7.4) becomes

$$V_D = 2.3\,\eta\, V_T \log\left(\frac{I_D}{I_S}\right) \tag{7.5}$$

If I_{D1} is the diode current corresponding to diode voltage V_{D1}, Eq. (7.4) gives

$$V_{D1} = 2.3\,\eta\, V_T \log\left(\frac{I_{D1}}{I_S}\right) \tag{7.6}$$

Similarly, if V_{D2} is the diode voltage corresponding to the diode current I_{D2}, we get

$$V_{D2} = 2.3\,\eta\, V_T \log\left(\frac{I_{D2}}{I_S}\right) \tag{7.7}$$

Therefore, the difference in diode voltages can be expressed by

$$V_{D2} - V_{D1} = 2.3\,\eta\, V_T \log\left(\frac{I_{D2}}{I_S}\right) - 2.3\,\eta\, V_T \log\left(\frac{I_{D1}}{I_S}\right)$$

$$= 2.3\,\eta\, V_T \log\left(\frac{I_{D2}}{I_{D1}}\right) \tag{7.8}$$

Thus, for a decade (factor of 10) change in diode current, that is, $I_{D2} = 10\, I_{D1}$, the diode voltage would change by $2.3\,\eta\, V_T$.

7.2.2 Reverse-Bias Region

In the reverse-bias region, the reverse voltage $V_R < 0$. If V_R is negative, and $|V_R| \gg V_T$, which occurs for $V_R < -0.1$, the exponential term in Eq. (7.1) becomes negligibly small compared to unity, and the reverse diode current I_R becomes,

$$I_R = I_S[e^{-|V_R|/\eta V_T} - 1] \approx -I_S \tag{7.9}$$

which indicates that the reverse diode current I_R is almost constant and equal to $-I_S$. In practical diodes, however, the reverse current I_R increases with the reverse voltage V_R.

7.2.3 Breakdown Region

In the breakdown region, the reverse voltage V_R has a high magnitude, usually greater than 100 V. If the magnitude of the reverse voltage exceeds a specified voltage known as the *reverse breakdown voltage* V_{BR}, the magnitude of the reverse current I_R increases rapidly with a small change in reverse voltage. The reverse current corresponding to V_{BR} is denoted as the reverse current I_{BV}.

7.3   ANALYSIS OF DIODE CIRCUITS

Diodes are used in electronic circuits for signal processing. Consider the diode circuit of Fig. 7.2, in which a small-amplitude sinusoidal voltage v_s is superimposed on a dc source V_{DD}. The source V_{DD} sets the operating point (Q-point) by coordinates V_D and I_D. The operating point will vary with time as the ac signal v_s varies. If this variation is small, then the superposition theorem can be applied to find the instantaneous diode current i_D from

$$i_D = I_D + i_d$$

where I_D is the dc current due to V_{DD} only ($v_s = 0$)
  i_d is the small-signal current due to v_s only, and $V_{DD} = 0$.

Thus, the analysis of a diode circuit will consist of

  dc analysis and
  small-signal ac analysis

7.3.1   Dc Analysis

Assuming that $v_s = 0$, Fig. 7.2 is reduced to Fig. 7.3(a). From Kirchhoff's voltage law (*KVL*),

$$V_{DD} = V_D + R_L I_D$$

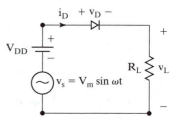

FIGURE 7.2

Diode circuit with a small signal superimposed on a dc source.

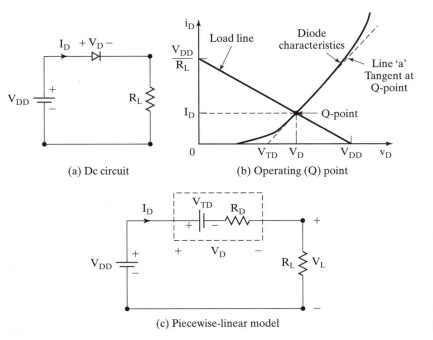

(a) Dc circuit

(b) Operating (Q) point

(c) Piecewise-linear model

FIGURE 7.3

Piecewise-linear model.

the diode current can be expressed as

$$I_D = -\frac{V_D}{R_L} + \frac{V_{DD}}{R_L} \tag{7.10}$$

Since the diode is forward biased, I_D is related to the diode voltage V_D by

$$I_D = I_S[e^{V_D/\eta V_T} - 1] \tag{7.11}$$

I_D depends on V_D, which in turn depends on I_D. Equations (7.10) and (7.11) can be solved for V_D and I_D at the operating point (Q-point) either graphically or iteratively.

The diode characteristic can be represented, approximately, by a fixed voltage drop V_{TD} and a straight line, as shown in Fig. 7.3(b). The straight line a takes into account the current dependency of the voltage drop, and it represents a fixed resistance R_D. The line a is tangent to the diode characteristic at the estimated Q-point. This representation is known as a *piecewise-linear model* consisting of two parts: a fixed part V_{TD} and a current-dependent part R_D. The resistance is given, approximately, by

$$R_D \approx \frac{\eta V_T}{I_D} = \frac{0.0258\,\eta}{I_D}\bigg|_{\text{at } Q\text{-point}} \tag{7.12}$$

If the diode is replaced by its piecewise-linear model, Fig. 7.3(a) can be represented by a linear circuit, as shown in Fig. 7.3(c). The diode current can be calculated from

$$I_D = \frac{V_{DD} - V_{TD}}{R_L + R_D} \tag{7.13}$$

and the diode voltage can be found from

$$V_D = V_{TD} + R_D I_D \tag{7.14}$$

7.3.2   Small-Signal Ac Analysis

Replacing the diode by its piecewise-linear model, we can find the instantaneous diode current i_D from Fig. 7.4(a). The diode will exhibit a junction capacitance C_D due to the time-variant signal v_s. Since the dc sources V_{DD} and V_{TD} in Fig. 7.4(a) will offer zero impedances to v_s, we can find the small-signal diode current i_d from the small-signal circuit of Fig. 7.4(b).

The value of diode resistance R_D will be low if the diode is forward biased and high if it is reverse biased. The junction capacitance will also vary depending upon its basing condition. The small-signal ac model of a diode is shown in Fig. 7.5. R_s is included to represent the contact and bulk resistance of the diode. C_D is the diffusion capacitance, and its value depends on the forward diode current I_D. C_j is the depletion capacitance, and its value depends on the reverse voltage according to the relation

$$C_j = \frac{C_{jo}}{(1 - V_D/V_j)^m} \quad \text{for } V_D < 0 \tag{7.15}$$

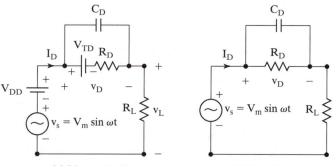

(a) Linear circuit

(b) Small-signal equivalent

FIGURE 7.4

Small-signal equivalent circuit.

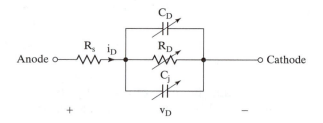

FIGURE 7.5

Small-signal ac diode model.

where m is a junction-gradient coefficient, ranging from 0.33 to 0.5;

V_j is the *junction potential* (for a silicon diode, $V_j \approx 0.5$ V to 0.9 V, and for a germanium diode, $V_j \approx 0.2$ V to 0.6 V)

V_D is negative in the reverse direction and positive in the forward direction

C_{jo} is the zero-bias junction capacitance.

We can conclude that the analysis of a diode circuit having a dc source and an ac source requires the following steps:

1. Find the dc operating point (V_D and I_D) of a diode.
2. Determine the parameters of the small-signal ac model for the diode.
3. Find the small-signal ac voltage and current of the diode.
4. Find the instantaneous diode voltage v_D and diode current i_D by superimposing the small-signal ac diode voltage v_d and diode current i_d on the voltage V_D and the dc current I_D, respectively.

7.4 DIODE MODEL

The PSpice model for a reverse-biased diode is shown in Fig. 7.6 [1,8]. The small-signal model and the static model, which are generated by PSpice, are shown in Figs. 7.7 and 7.8, respectively. In the static model, the diode current that depends on its voltage is represented by a current source. The small-signal parameters are generated by PSpice from the operating point.

PSpice generates a complex model for diodes. The model equations that are used by PSpice are described in the manual [8] and in Antognetti [1]. In many cases, especially the level at which this book is aimed, such complex models are not necessary. Many model parameters can be ignored by the users, and PSpice assigns default values for the parameters.

The model statement of a diode has the general form

```
.MODEL DNAME D (P1=A1 P2=A2 P3=A3 . . . PN=AN)
```

where DNAME is the model name. DNAME can begin with any character, but its word size is normally limited to eight characters. D is the type symbol for diodes.

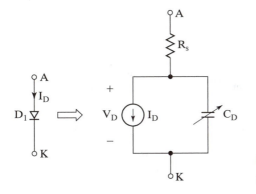

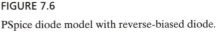

FIGURE 7.6

PSpice diode model with reverse-biased diode.

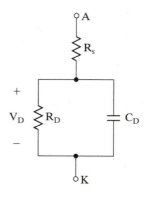

FIGURE 7.7

Small-signal diode model.

FIGURE 7.8

Static diode model with reverse-biased diode.

$P_1, P_2, \ldots$ and $A_1, A_2, \ldots$ are the model parameters and their values, respectively. The model parameters are listed in Table 7.1.

An *area factor* is used to determine the number of equivalent parallel diodes of a specified model. The model parameters that are affected by the area factor are marked by an asterisk (*) in the descriptions of the model parameters.

TABLE 7.1 Parameters of Diode Model

Name	Area	Model Parameter	Units	Default	Typical
IS	*	Saturation current	Amps	1E–14	1E–14
RS	*	Parasitic resistance	Ohms	0	10
N		Emission coefficient		1	
TT		Transit time	seconds	0	0.1NS
CJO	*	Zero-bias p-n capacitance	Farads	0	2PF
VJ		Junction potential	Volts	1	0.6
M		Junction-gradient coefficient		0.5	0.5
EG		Activation energy	Electron volts	1.11	11.1
XTI		IS temperature exponent		3	3
KF		Flicker noise coefficient		0	
AF		Flicker noise exponent		1	
FC		Forward-bias depletion capacitance coefficient		0.5	
BV		Reverse breakdown voltage	Volts	∞	50
IBV	*	Reverse breakdown current	Amps	1E–10	

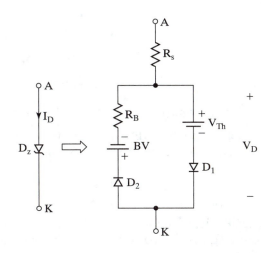

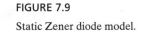

FIGURE 7.9

Static Zener diode model.

The diode is modeled as an ohmic resistance (value = RS/area) in series with an intrinsic diode. The resistance is attached between node NA and an internal anode node. The [(area) value] scales IS, RS, CJO, and IBV, and defaults to 1. IBV and BV are both specified as positive values.

The dc characteristic of a diode is determined by the reverse saturation current IS, the emission coefficient N, and the ohmic resistance RS. The charge storage effects are modeled by the transit time TT, a nonlinear depletion layer capacitance, which depends on the zero-bias junction capacitance CJO, the junction potential VJ, and grading coefficient M. The temperature of the reverse saturation current is defined by the gap-activation energy (or gap energy) EG and saturation-temperature exponent XTI.

In order to simulate a Zener diode, the model in Fig. 7.9 can be used. Diode D_1 and the threshold voltage source V_{Th} represent the normal forward voltage and reverse behavior of a Zener diode. Diode D_2, the voltage source BV, and resistance R_B define the breakdown region. Diode D_2 does not conduct until $V_D = -BV$. If the reverse voltage is increased, then diode D_2 becomes forward biased, and the reverse current flows through R_B.

Reverse breakdown is modeled by an exponential increase in the reverse diode current and is determined by the reverse breakdown voltage, BV, and the current at breakdown voltage, I_{BV}.

7.5 DIODE STATEMENT

The symbol for a diode is D. The name of a diode must start with D, and it takes the general form

```
D⟨name⟩  NA  NK  DNAME  [(area) value]
```

where NA and NK are the node and cathode nodes, respectively. The current flows from anode node NA through the diode to cathode node NK. DNAME is the model name. For describing the diode model and the parameters, see Fig. 7.12.

Some Statements for Diodes

```
DM    5   6    DNAME
.MODEL DNAME   D (IS=0.5UA   RS=6 BV=5.20 IBV=0.5UA)
D15   33   35    SWITCH   1.5
.MODEL SWITCH D (IS=100E-15   RS=16 CJO=2PF TT=12NS BV=100 IBV=100E-15)
DCLAMP   0   8    DMOD
.MODEL DMOD   D (IS=100E-15   RS=16 CJO=2PF TT=12NS BV=100 IBV=100E-15)
```

Note. Diode DM, having model name DNAME, is a Zener diode with a Zener break-down voltage of 5.2 V; the current at the Zener break is 0.5 μA.

7.6 DIODE PARAMETERS

Some versions of SPICE (e.g., PSpice) support device library files that give the model parameters. The library file EVAL.LIB contains the list of device library files and the device model statements that are available with the student's version. The software PARTS of PSpice can generate SPICE models from the data-sheet parameters of diodes. The SPICE model parameters are also supplied by some manufacturers. However, some model parameters of a diode can be determined from the data sheet. The data sheet of diode 1N914, which is shown in Fig. 7.10, gives the following:

Reverse recovery time t_r = 4 ns   (From diode data)

Reverse breakdown voltage BV = 100 V

Forward voltage V_D = 0.4 V at I_D = 25 μA   (From the curve of I_F versus V_F)

V_D = 0.6 V at I_D = 2 mA

Reverse current I_R = 25 nA at V_R = 20 V   (From diode data)

I_R = 5 μA at V_R = 75 V

Capacitance with zero reverse voltage (from the curve of capacitance versus reverse voltage) C_{jo} = 1.7 pF.

Assuming that V_T = 25.8 mV = 0.0258, V_{D1} = 0.4 V at I_{D1} = 25 μA = 0.025 mA, and V_{D2} = 0.6 V at I_{D2} = 2 mA, we can apply Eq. (7.8) to find the *emission coefficient* η as in

$$0.6 - 0.4 = 2.3\ \eta \times 0.0258 \log\left(\frac{2}{0.025}\right)$$

which gives η = 1.77.

For n = 1.77, and V_D = 0.6 V at I_{D2} = 2 mA, we can apply Eq. (7.1) to find the saturation current I_S:

$$2\text{ mA} = I_S[e^{0.6/(1.77 \times 0.0258)} - 1], \quad \text{which gives } I_S = 3.93\text{E} - 9\text{ A}$$

Assuming that I_R = 5 μA at V_R = 75 V is equal to the reverse breakdown current, then I_{BV} = 5 μA.

The transit time τ_T can be calculated approximately from

$$\tau_T = \frac{Q_{RR}}{I_{FM}}$$

National Semiconductor

Diode Data

Computer Diodes (Glass Package)

Device No.	Package No.	V_{RRM} V Min	I_R nA Max	@	V_R V	V_F V Min	V_F V Max	@	I_F mA	C pF Max	t_{rr} ns Max	Test Cond.	Proc. No.
1N625	DO-35	30	1000		20		1.5		4		1000	(Note 1)	D4
1N914	DO-35	100	25 5000		20 75		1.0		10		4	(Note 2)	D4
1N914A	DO-35	100	25 5000		20 75		1.0		20		4	(Note 2)	D4
1N914B	DO-35	100	25 5000		20 75		0.72 1.0		5 100		4	(Note 2)	D4
1N916	DO-35	100	25 5000		20 75		1.0		10		4	(Note 2)	D4
1N916A	DO-35	100	25 5000		20 75		1.0		20		4	(Note 2)	D4
1N916B	DO-35	100	25 5000		20 75		0.73 1.0		5 30		4	(Note 2)	D4
1N3064	DO-35	75	100		50		0.575 0.650 0.710 1.0		0.250 1.0 2.0 10.0	2	4	(Note 3)	D4
1N3600	DO-35	75	100		50	0.54 0.66 0.76 0.82 0.87	0.62 0.74 0.86 0.92 1.0		1.0 10.0 50.0 100.0 200.0	2.5	4	(Note 4)	D4
1N4009	DO-35	35	100		25		1.0		30	4	2	(Note 2)	D4
1N4146	DO-35	See Data for 1N914A/914B											
1N4147	DO-35	See Data for 1N914A/914B											
1N4148	DO-35	See Data for 1N914											
1N4149	DO-35	See Data for 1N916											
1N4150	DO-35	See Data for 1N3600											
1N4151	DO-35	75	50		50		1.0		50	4	2	(Note 2)	D4
1N4152	DO-35	40	50		30	0.49 0.53 0.59 0.62 0.70 0.74	0.55 0.59 0.67 0.70 0.81 0.88		0.1 0.25 1.0 2.0 10.0 20.0	4	2	(Note 2)	D4
1N4153	DO-35	75	50		50	See 1N4152				4	2	(Note 2)	D4
1N4154	DO-35	35	100		25		1.0		30	4	2	(Note 2)	D4

FIGURE 7.10

Data sheet for diode 1N914 (Courtesy of National Semiconductor, Inc.).

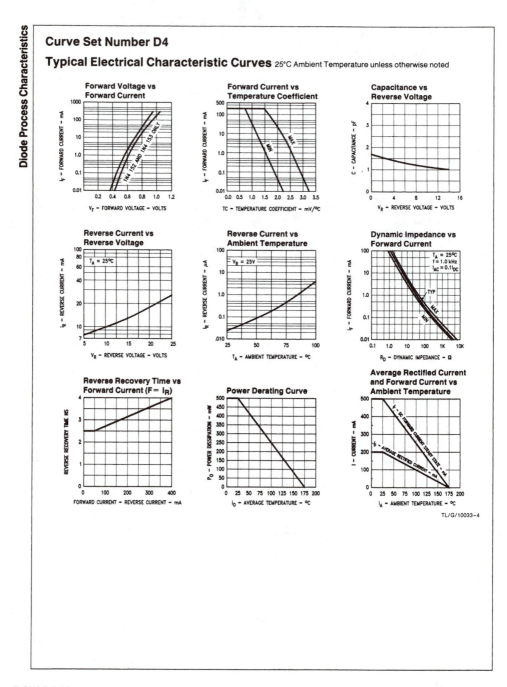

Diode Process Characteristics

Curve Set Number D4

Typical Electrical Characteristic Curves 25°C Ambient Temperature unless otherwise noted

TL/G/10033–4

FIGURE 7.10

(*Continued*)

where Q_{RR} = reverse recovery charge
I_{FM} = reverse recovery current.

Since Q_{RR} is not specified on the data sheet, it can be related approximately to t_r and I_{FM} by

$$Q_{RR} = \frac{1}{2}t_r I_{FM} \qquad (7.16)$$

Thus, the transit time $TT = \tau_T = t_r/2 = 4/2 = 2$ ns.

Assuming that RS = 1 Ω, the model statement for diode 1N914 is

```
.MODEL DMOD D(IS=3.93E-9 RS=1 BV=100V IBV=5E-6 CJO=1.7PF TT=2NS)
```

Note. The model parameters that were calculated previously from the data sheet are approximate and will differ from those supplied by the diode manufacturer and SPICE/PSpice library.

7.7 EXAMPLES OF DC ANALYSIS

In dc circuits, the sources are dc. The diode is specified by its model parameters. If the model parameters are not specified, PSpice assumes the default values as indicated in Table 7.1. After calculating the steady-state dc voltage and current of a diode, PSpice determines the small-signal model parameters, which can be printed by the .OP command. The following examples illustrate the PSpice simulation of dc diode circuits.

Example 7.1: Describing diode model parameters and plotting diode charateristics

A diode circuit is shown in Fig. 7.11. Plot the $v-i$ characteristic of the diode for forward voltage from 0 to 2 V and for temperatures of 50°, 100°, and 150°. The diode is of type D1N914, and the model parameters are IS = 3.93E − 9, RS = 1, BV = 100V, IBV = 5E − 6, CJO = 1.7PF, and TT = 2NS.

Solution The model name Dbreak1, as shown in Fig. 7.12(b), is changed from the edit model as shown in Fig. 7.12(a), which is selected from the edit menu. The model parameters are changed from the edit model text menu as shown in Fig. 7.12(c). The temperature values are varied by nested sweep within the main dc sweep, which is selected from the dc sweep menu of the analysis setup. The listing of the circuit file follows.

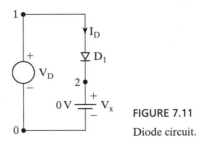

FIGURE 7.11

Diode circuit.

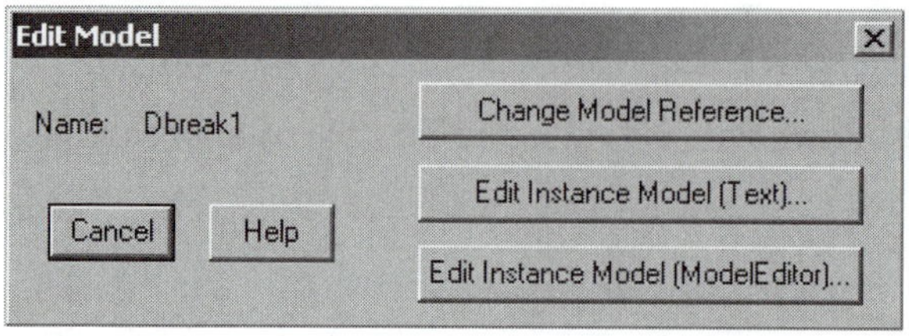

Edit Model   ×

Name: Dbreak1   Change Model Reference...

                        Edit Instance Model (Text)...

[Cancel] [Help]   Edit Instance Model (ModelEditor)...

(a) Edit model

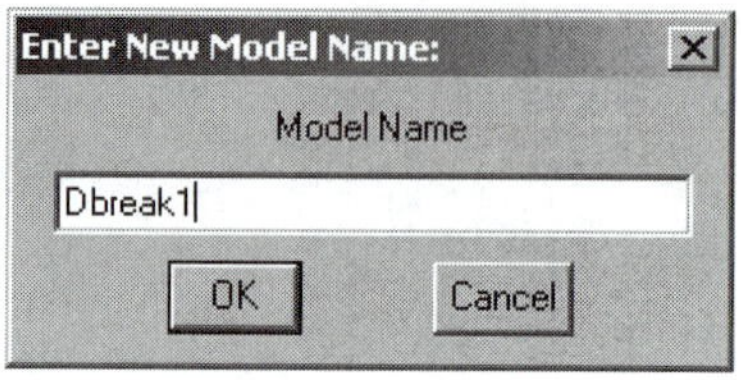

Enter New Model Name: ×

              Model Name

Dbreak1|

     [OK]   [Cancel]

(b) Change model name

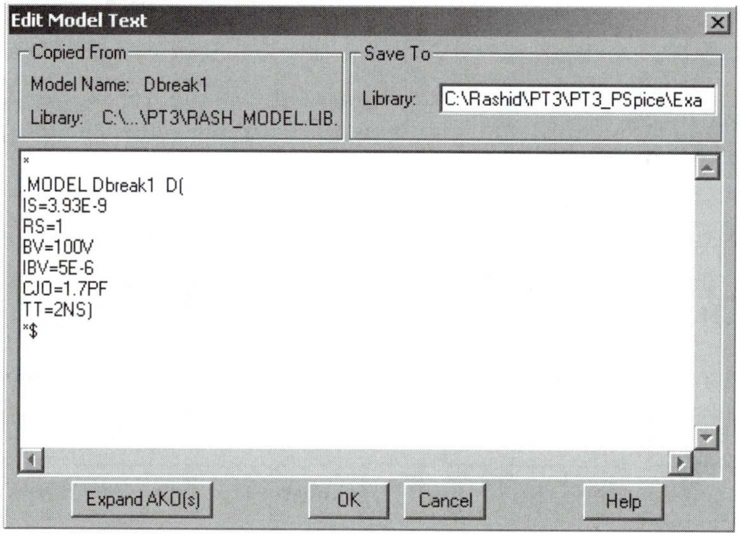

Edit Model Text   ×

┌─ Copied From ──────────────┐ ┌─ Save To ──────────────────┐
Model Name: Dbreak1   Library: C:\Rashid\PT3\PT3_PSpice\Exa
Library: C:\...\PT3\RASH_MODEL.LIB.

```
*
.MODEL Dbreak1 D(
IS=3.93E-9
RS=1
BV=100V
IBV=5E-6
CJO=1.7PF
TT=2NS)
*$
```

[Expand AKO(s)]   [OK]   [Cancel]   [Help]

(c) Exit model text

FIGURE 7.12

Editing diode model name and parameters.

Example 7.1   Diode characteristic

```
▲  VD   1   0   DC   0V ; Dc input voltage is overridden during dc sweep
▲▲ D1   1   2   D1N914      ; Diode with model D1N914
     *  anode cathode model
    VX   2   0   DC   0V        ; Measures the diode current ID
     *   Diode model defines the model parameters
    .MODEL  D1N914  D (IS=3.93E−9 RS=1 BV=100V IBV=5E−6 CJO=1.7PF TT=2NS)
    .TEMP  50  100  150    ; Operating temperatures: 50, 100, and 150 degrees
    .OPTIONS NOPAGE  NOECHO        ; Options
▲▲▲    .DC  VD  0  1V  0.01V ; Dc sweep from 0 to 2 V with 0.01 V increment
        .PLOT   DC   I(VX)    ; Plots the diode current in the output file
        .PROBE                      ; Graphical waveform analyzer
    .END                         ; End of circuit file
```

The $v-i$ characteristic that is obtained by varying the input voltage is shown in Fig. 7.13.

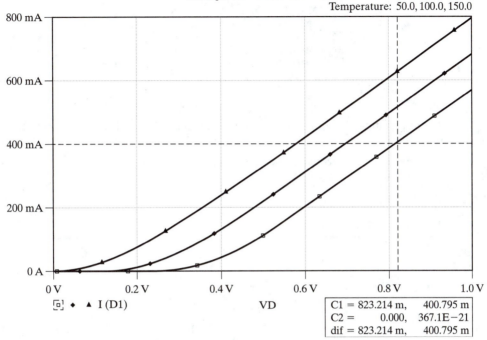

FIGURE 7.13

Diode forward characteristics for Example 7.1.

Notes

1. The diode drop decreases with the temperature. That is, the diode current increases with the temperature.
2. Using the Dbreak diode permits user-defined model parameters.

Example 7.2: Describing the diode characteristic of tabular data

The diode circuit of Fig. 7.14(a) has the characteristics $V_{DD} = 15$ V and $R_L = 500\ \Omega$. Use PSpice to calculate the operating point (V_D, I_D). The diode characteristic is represented by the following table:

I_D (mA)	0	10	20	30	40	50	60	70
V_D (V)	0.5	0.87	0.98	1.058	1.115	1.173	1.212	1.25

Solution The diode is replaced by a voltage-controlled voltage source, shown in Fig. 7.14(b), to represent the table of the current-voltage characteristic. The PSpice schematic is shown in Fig. 7.15(a). The diode is modeled by an ETABLE from the *abm.slb* library. The current is related to the voltage in a table form as shown in Fig. 7.15(b). The listing of the circuit file follows.

Example 7.2   Diode circuit

```
▲     VD  1   0    DC   15V      ; Dc voltage of 15 V
▲▲    VX  3   2    DC   0V       ; Measures the diode current I_D
      RL  2   0    500
      ED  1   3    TABLE {I (VX) } =
      +   (0. 0,  0.5)    (10E-3,  0.87)  (20E-3, 0.98)   (30E-3, 1.058)
      +   (40E-3, 1.115)  (50E-3,  1.173) (60E-3, 1.212)  (70E-3, 1.250)
▲▲▲   .OP                         ; Prints the details of the operating point
      .END                        ; End of circuit file
```

The information about the operating point, which is obtained from the output file EX7.2.OUT, is as follows:

```
****     OPERATING POINT INFORMATION TEMPERATURE - 27.000 DEG C
    NAME                ED
 V-SOURCE      1.042E+00
 I-SOURCE      2.792E-02
```

Note: The diode current is listed in terms of (i, v) data points.

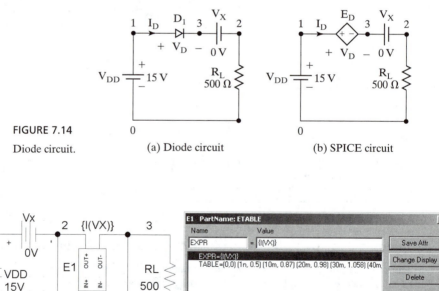

FIGURE 7.14

Diode circuit.

(a) Diode circuit

(b) SPICE circuit

(a) ETABLE

(b) ETABLE parameters

FIGURE 7.15

PSpice schematic for Example 7.2.

Example 7.3: Transfer characteristic of a symmetrical Zener limiter

A Zener voltage regulator is shown in Fig. 7.16. Plot the dc transfer characteristic if the input voltage is varied from -15 V to 15 V with an increment of 0.5 V. The Zener voltages of the diodes are the same, and $V_Z = 5.2$ V; the current at the Zener breakdown is $I_Z = 0.5 \mu A$. The model parameters are IS $= 0.5$UA, RS $= 1$, BV $= 5.20$, and IBV $= 0.5$UA. The operating temperature is 50°C. V_{in} has a normal voltage of 10 V (dc). The details of the operating point are to be printed.

Solution A Zener diode is implemented by setting the model parameters BV $= V_Z = 5.2$ V, and IBV $= I_Z = 0.5 \mu A$. The parameters for model Dbreak2 are changed from the edit model text menu as shown in Fig. 7.17. The listing of the circuit file follows.

Example 7.3 Zener regulator

```
▲  VIN 1  0  DC  10V        ; Dc input voltage is overridden during dc sweep.
▲▲  R1  1  2    500
     D1  2  3   DNAME        ; Diode with model name DNAME
     D2  0  3   DNAME        ; Diode with model name DNAME
     RL  2  0    1K
     *   Model DNAME defines the parameters of Zener diodes.
    .MODEL DNAME  D (IS=0.5UA RS=1 BV=5.20 IBV=0.5UA)
    .TEMP    50                ; Operating temperature 50 degrees
    .OPTIONS NOPAGE NOECHO    ; Options
▲▲▲    .DC VIN -15  15V  0.5V ; DC sweep from -15 to 15 V with 0.5 V increment
       .PRINT DC V(2)     ; Plots the load voltage in the output file
       .PROBE             ; Graphical waveform analyzer
       .OP
.END                           ; End of circuit file
```

The information about the operating point, which is obtained from the output file EX7.3.OUT, is as follows:

****	OPERATING POINT INFORMATION		TEMPERATURE = 50.000 DEG C
	NAME	D1	D2
	MODEL	DNAME	DNAME
I_D	ID	3.19E−03	−3.19E−03
V_D	VD	1.56E−01	−5.45E+00
R_D	REQ	8.69E+00	8.76E+00
C_D	CAP	0.00+00	0.00E+00

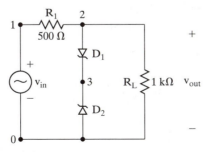

FIGURE 7.16

Zener regulator.

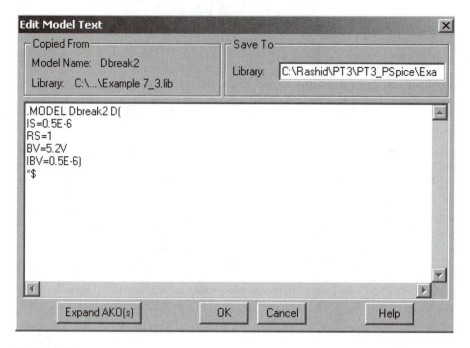

FIGURE 7.17

Model parameters of Zener diodes in Example 7.3.

The dc transfer characteristic is shown in Fig. 7.18.

Note: Due to identical Zener diodes, the transfer characteristic is symmetrical. The output voltage will be limited to ±5.2 V.

Example 7.4: Transfer characteristic of a diode waveform-shaping circuit

A diode waveform-shaping circuit is shown in Fig. 7.19, where the output is taken from Node 2. Plot the transfer characteristic between V_{in} and V(2) for values of V_{in} in the range of −15V to 30 V in steps of 0.5 V. The model parameters of the diodes are IS = 100E − 15, RS = 16, BV = 100, and IBV = 100E − 15. V_{in} has a normal voltage of 10 V (dc). The details of the operating point are to be printed.

Solution The listing of the circuit file follows.

Example 7.4 Diode waveform-shaping circuit

```
▲  *    The input voltage that is overridden by the dc sweep voltage is assumed
   *    to be 10 V.
      VIN   1   0   DC   −10V
      VCC   6   0   DC    15V
      VEE   10  0   DC   −15V
▲▲    R1    6   5   2K
      R2    5   4   2K
      R3    4   3   2K
```

```
R4  3  2  2K
R5  2  7  2K
R6  7  8  2K
R7  8  9  2K
R8  9 10  2K
*    All diodes have the same model name of DMOD.
D1  1  5  DMOD
D2  1  4  DMOD
D3  1  3  DMOD
D4  1  7  DMOD
D5  1  8  DMOD
D6  1  9  DMOD
*    Diode model DMOD defines diode parameters.
.MODEL DMOD D (IS=100E-15 RS=16 BV=100 IBV=100E-15)
*       Dc sweep from -15 V to 30 V with 0.5 V increment
.DC  VIN  -15   30   0.5
*    Plot the results of dc sweep: (V2) versus VIN.
.PLOT  DC  V(2)
*    Graphic post-processor
.OP                          ; Prints the details of the operating point
.PROBE
```
.END

Example 7.3   Zener regulator

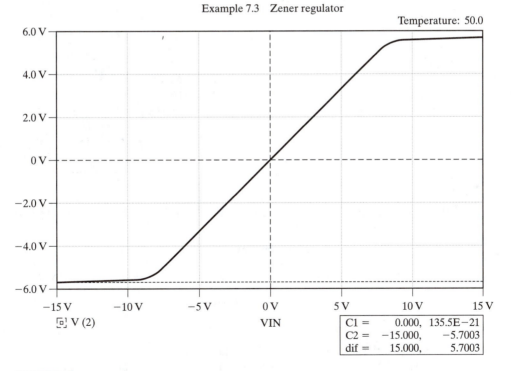

FIGURE 7.18

Dc transfer characteristic for Example 7.3.

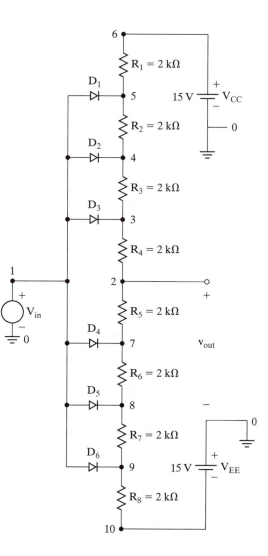

FIGURE 7.19

Diode waveform-shaping circuit.

The information about the operating point, which is obtained from the output file EX7.4.OUT, is as follows:

****	OPERATING	POINT	INFORMATION		TEMPERATURE = 27.000 DEG C	
	NAME	D1	D2	D3	D4	D5
	MODEL	D1N914	D1N914	D1N914	D1N914	D1N914
I_D	ID	−4.26E−12	−3.43E−12	−2.59E−12	−9.25E−13	6.08E−14
V_D	VD	−4.16E+00	−3.33E+00	−2.49E+00	−8.25E−01	1.05E−02
R_D	REQ	1.00E+12	1.00E+12	1.00E+12	1.00E+12	1.47E+11
C_D	CAP	0.00E+00	0.00E+00	0.00E+00	0.00E+00	0.00E+00

The results of the simulations are shown in Fig. 7.20.

Note: Each diode in Fig. 7.19 changes the slope of the transfer characteristic. There are a total of six breaks corresponding to the six diodes. The slopes can be viewed by plotting $d(V(2))$ or $d(V(R4:1))$.

Example 7.4   Diode waveform-shaping circuit

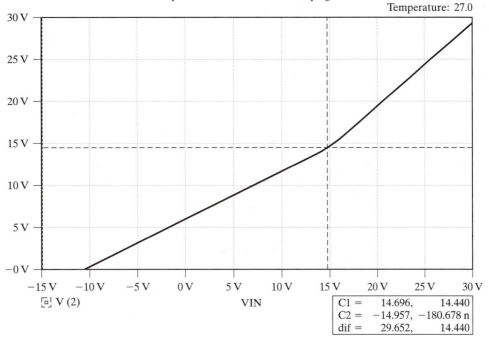

FIGURE 7.20

Dc transfer characteristic for Example 7.4.

7.8   EXAMPLES OF TRANSIENT AND AC ANALYSIS

The transient response is determined for diode circuits in which the input signal is time variant, usually sinusoidal. PSpice first determines the transient bias point, which is different from the regular (dc) point. The initial values of the circuit nodes are taken into account in calculating the transient bias point, along with the small-signal parameters of the nonlinear elements. The capacitors and inductors, which may have initial values, therefore, remain as parts of the circuit. The various nodes can be assigned to initial voltages by the .IC statement.

PSpice determines the small-signal model parameters from the transient bias point. The following examples illustrate the transient analysis of diode circuits.

Example 7.5:   Output voltage of a single-phase-full-wave rectifier

A full-wave rectifier is shown in Fig. 7.21, where the output is taken between terminals 4 and 3. Plot the transient response of the output voltage V(4, 3) for the time duration of 0 to 20 ms in steps of 0.1 ms. The peak voltage of the transformer primary is 120 V, 60 Hz. The ratio of primary to secondary windings is 10:1. The model parameters are the default values. Calculate and print the coefficients of the Fourier series.

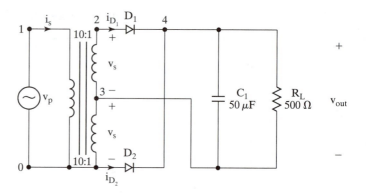

FIGURE 7.21

Rectifier with single-phase center-tapped transformer.

Solution   The transformer secondaries may be considered as a voltage-controlled voltage source, as shown in Fig. 7.22. The primary is represented by a voltage source with a very high resistance (e.g., 10 GΩ).

The current-controlled current-source $F1$ has a degree of 2 and it adjusts the current flow from the ac main supply so that the primary V_pI_p equals to the secondary V_sI_s and the relationship $V_p/V_s = I_s/I_p$ is maintained. $F1$ is dependent on the diode currents and is described by

$$F_1 = P_1I_{D1} + P_2I_{D2},$$

where $P_1 = P_2 = 20$.

The listing of the circuit file follows.

Example 7.5   Rectifier with single-phase center-tapped transformer

```
▲   *    Primary is modeled as a voltage source of 120 V peak at 60 Hz
    *    with zero offset voltage.
   VP    1   0   SIN (0 120 60HZ)
▲▲  *    Primary winding is assumed to have a very high resistance:
    *    R1 = 10 GΩ.
   R1    1   0   10GOHM
   F₁    1   0   POLY(2)  Vx  Vy  0  20  20
   Vx    5   2   DC  0V
   Vy    6   3   DC  0V
    *    Secondary winding is assumed as a voltage-controlled voltage source
    *    with a voltage gain of 0.1.
   E1    5   3   1   0   0.1.
   E2    6   0   1   0   0.1
   C1    4   3   50UF
   RL    4   3   500
    *    Diode D1 with model name DIODE
   D1    2   4   DIODE
   D2    0   4   DIODE
    *    Diode model with default values
   .MODEL DIODE D
▲▲▲ *    Transient analysis from 0 to 20 ms with 0.1-ms increment
   .TRAN    10us    33.33ms
    *    Plot the results of transient analysis for voltage across nodes
    *    4 and 3.
   .PLOT   TRAN   V(4,3)
   .FOUR   60HZ   V(4,3)
    *    Graphic postprocessor
   .OP                           ; Prints details of the operating point
   .PROBE
.END
```

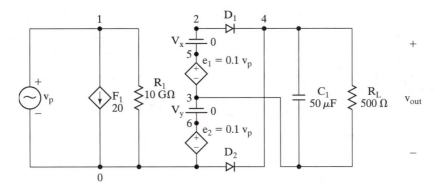

FIGURE 7.22

The equivalent circuit for Figure 7.22.

The transient response for Example 7.5 is shown in Fig. 7.23. The coefficients of the Fourier analysis follow:

```
****        FOURIER ANALYSIS                    TEMPERATURE = 27.000 DEG C
FOURIER COMPONENTS OF TRANSIENT RESPONSE V (4,3)
DC COMPONENT =  1.00866SE+01
```

HARMONIC NO	FREQUENCY (HZ)	FOURIER COMPONENT	NORMALIZED COMPONENT	PHASE (DEG)	NORMALIZED PHASE (DEG)
1	6.000E+01	2.112E−03	1.000E+00	−2.607E+00	0.000E+00
2	1.200E+02	9.843E−01	4.661E+02	1.182E+01	1.443E+01
3	1.800E+02	7.890E−04	3.736E−01	2.452E+01	2.713E+01
4	2.400E+02	3.833E−01	1.815E+02	2.131E+01	2.392E+01
5	3.000E+02	1.167E−03	5.528E−01	8.314E+01	8.575E+01
6	3.600E+02	1.669E−01	7.902E+01	3.761E+01	4.022E+01
7	4.200E+02	1.756E−03	8.314E−01	1.076E+02	1.102E+02
8	4.800E+02	7.140E−02	3.381E+01	6.687E+01	6.948E+01
9	5.400E+02	1.895E−03	8.971E−01	1.279E+02	1.305E+02

```
    TOTAL HARMONIC DISTORTION = 5.075342E+04 PERCENT
        JOB CONCLUDED
        TOTAL JOB TIME        22.62
```

Note: The final simulation time (e.g., 33.33 ms) should be a multiple of the period of the output waveform. Otherwise, the phase angles of the Fourier components will not be measured with respect to the input reference voltage.

Example 7.6:   Diode clamping circuit

A clamping circuit is shown in Fig. 7.24, where the output is taken from Node 2. Plot the transient response of the output voltage V(2) for the time duration of 0 to 3 ms in steps of 20 μs. The initial capacitor voltage is −15 V. The model parameters of the diode are the default values.

Solution   The listing of the circuit file follows.

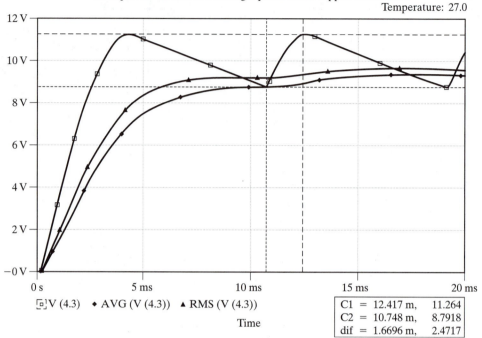

Example 7.5   Rectifier with single-phase center-tapped transformer

Temperature: 27.0

□ V (4.3)   ◆ AVG (V (4.3))   ▲ RMS (V (4.3))

Time

C1	=	12.417 m,	11.264
C2	=	10.748 m,	8.7918
dif	=	1.6696 m,	2.4717

FIGURE 7.23

Output voltage for Example 7.5.

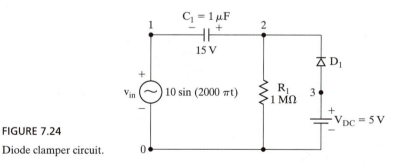

FIGURE 7.24

Diode clamper circuit.

Example 7.6   Diode clamper circuit

```
▲ *    Input voltage of a 10-V peak at 1 kHz and zero offset voltage is
   *    connected between nodes 1 and 0.
▲▲    VIN 1  0  SIN (0 10  1KHZ)
      *    Capacitor with an initial voltage of −15 V
      C1  1   2   1UF IC=−15V
      R1  2   0   1MEG
      VDC 3   0   DC 5
      *    Diode with model name DIODE is connected between nodes 3 and 2.
      D1  3   2   DIODE
```

```
      *    Diode model with default values of parameters
      .MODEL DIODE  D
▲▲▲   *    Transient analysis for 0 to 3 ms with 20-µs increment with UIC
      .TRAN  20US  3MS  UIC
      *    Plot transient voltages at nodes 1 and 2.
      .PLOT TRAN  V(2)  V(1)
      .OP                          ; Prints details of the operating point
        .PROBE
  .END
```

The information about the operating point, which is obtained from the output file EX7.6.OUT, is as follows:

```
****        OPERATING POINT INFORMATION          TEMPERATURE = 27.000 DEG C
            NAME          D1
            MODEL         DIODE
  I_D       ID            4.48E-06
  V_D       VD            5.15E-01
  R_D       REQ           5.77E+03
  C_D       CAP           0.00E+00
```

The input and output voltages of the diode clamper circuit are shown in Fig. 7.25. With the PROBE command, there is no need for the .PLOT statement.

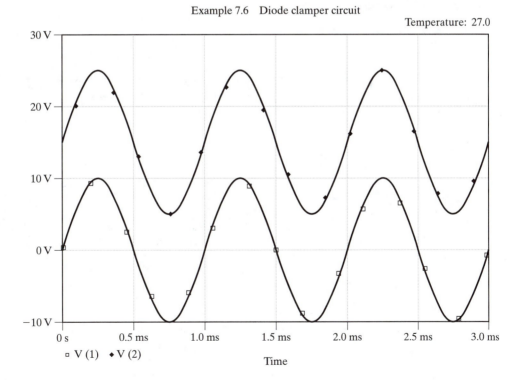

Example 7.6   Diode clamper circuit

Temperature: 27.0

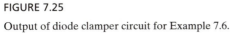

FIGURE 7.25

Output of diode clamper circuit for Example 7.6.

Note: The instantaneous output voltage is described by

$$v_o(t) = V_m \sin \omega t + V_c(t = 0) = 10 \sin \omega t + V_m + V_{DC} \quad \text{for} \quad \omega t \geq 3\pi/2$$
$$= 10 \sin \omega t + 10 + 5 = 10 \sin \omega t + 15$$

Under the steady-state conditions, the capacitor will change to a peak voltage of $V_m + E_1 (= 10 + 5 = 15 \text{ V})$ when the diode D_1 is turned on during the negative half-cycle of the input signal.

Example 7.7: Diode circuit with a small-signal superimposed on a dc biasing voltage

A diode circuit is shown in Fig. 7.26(a). The ac input voltage is $v_{in} = 10 \times 10^{-3} \sin(2\pi \times 10^3 t)$. Print the bias point and the small-signal parameters of the diode. Plot the output voltage from 0 to 200 μs with 2-μs increments. If the frequency of the ac voltage is varied from 1 Hz to 1 kHz, plot the magnitude and phase angle of the output voltage. The model parameters are IS = 100E − 15, RS = 16, CJO = 2PF, TT = 12NS, BV = 100, and IBV = 100E − 15.

Solution The equivalent circuit for calculating the bias point and small-signal parameters is shown in Fig. 7.26(b). Capacitor C_1 blocks any dc signal to the output load. For the ac small-signal, C_1 will practically be shorted, and the output voltage v_{out} will due to the input signal v_{in}. The listing of the circuit file follows.

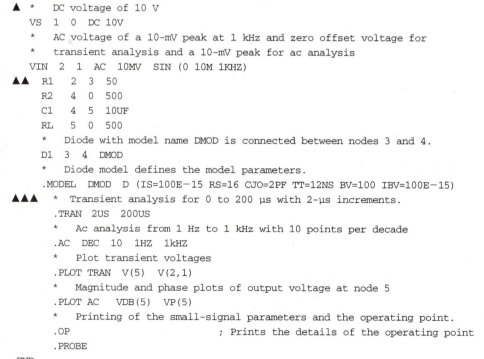

Example 7.7   Diode circuit

```
▲ *    DC voltage of 10 V
  VS   1  0   DC 10V
  *    AC voltage of a 10-mV peak at 1 kHz and zero offset voltage for
  *    transient analysis and a 10-mV peak for ac analysis
  VIN  2  1   AC  10MV  SIN (0 10M 1KHZ)
▲▲ R1   2  3  50
   R2   4  0  500
   C1   4  5  10UF
   RL   5  0  500
  *    Diode with model name DMOD is connected between nodes 3 and 4.
  D1  3  4  DMOD
  *    Diode model defines the model parameters.
  .MODEL  DMOD  D (IS=100E−15 RS=16 CJO=2PF TT=12NS BV=100 IBV=100E−15)
▲▲▲ *   Transient analysis for 0 to 200 μs with 2-μs increments.
  .TRAN  2US  200US
  *    Ac analysis from 1 Hz to 1 kHz with 10 points per decade
  .AC   DEC  10  1HZ  1kHZ
  *    Plot transient voltages
  .PLOT TRAN  V(5)  V(2,1)
  *    Magnitude and phase plots of output voltage at node 5
  .PLOT AC   VDB(5)  VP(5)
  *    Printing of the small-signal parameters and the operating point.
  .OP                        ; Prints the details of the operating point
  .PROBE
  .END
```

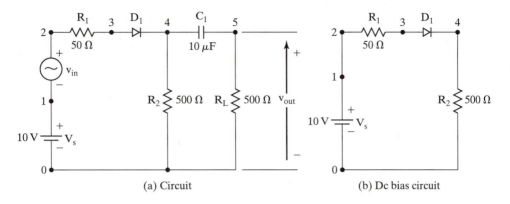

FIGURE 7.26

Diode circuit.

Note that with the .PROBE command, there is no need for the .PLOT statement. The transient response is shown in Fig. 7.27, and the frequency response is shown in Fig. 7.28. The details of the dc bias point and the small-signal parameters are as follows:

```
****    SMALL-SIGNAL BIAS SOLUTION          TEMPERATURE = 27.000 DEG C
NODE     VOLTAGE       NODE     VOLTAGE       NODE     VOLTAGE       NODE     VOLTAGE
(  1)    10.0000       (  2)    10.0000       (  3)    9.1756        (  4)    8.2438
(  5)     0.0000
         VOLTAGE SOURCE CURRENTS
         NAME            CURRENT
         VS            -1.649E-02
         VIN           -1.649E-02
         TOTAL POWER DISSIPATION    1.65E-01    WATTS

****    OPERATING POINT INFORMATION          TEMPERATURE = 27.000 DEG C
         NAME          D1
         MODEL         DMOD
         ID            1.65E-02
         VD            9.32E-01
         REQ           1.57E+00
         CAP           7.65E-09
```

Notes

1. The output voltage can be found from

$$v_o(t) = \frac{R_L \| R_2}{R_1 + r_d + R_L \| R_2} v_{in}$$

where r_d is the small-signal ac resistance of the diode such that $r_d = \dfrac{25.8 \text{ mV}}{I_D}$ and I_D is the dc biasing current of the diode.

2. Figure 7.27 shows the reduction in the output voltage and $v_o(t) = 0.78226\, v_{in}$.

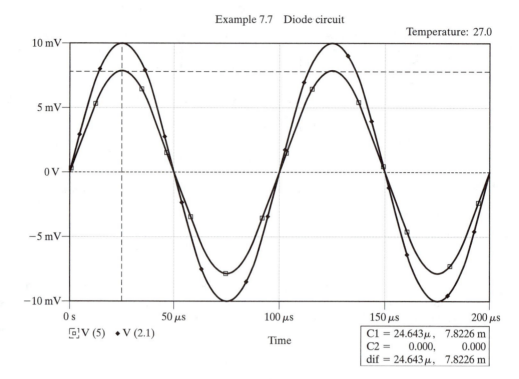

Example 7.7 Diode circuit

FIGURE 7.27

Transient response for Example 7.7.

3. Figure 7.28 shows the frequency response of the circuit. For an input frequency of $f > 300$ Hz, C_1 offers a very low impedance and has negligible effect on the output voltage.

Example 7.8: Worst-case analysis of a diode circuit

Use PSpice to perform the worst-case analysis for the transient response of the diode circuit in Fig. 7.26. The model parameters are $R = 1$ for resistors, $C = 1$ for the capacitor, and I_S for the diode. The circuit parameters having uniform deviations are

$$R_1 = 50 \pm 10\%$$
$$R_2 = 500 \ \Omega \pm 15\%$$
$$R_L = 500 \ \Omega \pm 20\%$$
$$C_1 = 10 \ \mu\text{F} \pm 15\%$$
$$I_S = 100\text{E} - 15 \pm \text{DEV} = 10\%$$

The maximum value of the output voltage is to be printed.

Solution The PSpice schematic is shown in Fig. 7.29(a). The tolerances are specified in the break models. The worst-case analysis is selected from the analysis setup. The worst-case maximum plot of the output voltage and its nominal value are shown in Fig. 7.29(b). The listing of the circuit file follows.

Example 7.8   Worst-case analysis of diode circuit

```
▲  VS  1  0   DC  10V                    ; DC voltage of 10 V
   VIN 2  1   AC  10MV  SIN (0 10M 10KHZ) ; sinusoidal voltage of 10 mV
▲▲ R1  2  3   RMOD1   50
   R2  4  0   RMOD2   500
   RL  5  0   RMODL   500
   C1  4  5   CMOD  10UF
   .MODEL RMOD1  RES (R=1 DEV=10%)
   .MODEL RMOD2  RES (R=1 DEV=20%)
   .MODEL RMODL  RES (R=1 DEV=25%)
   .MODEL CMOD   CAP (C=1 DEV=15%)
   D1  3 4  D1N914                       ; Diode with model name D1N914
   *  Diode model defines the model parameters
   .MODEL D1N914 D(IS=100E−15 DEV=10% RS=1 CJO=2PF TT=12NS BV=100
   +IBV=100E−15)
▲▲▲  .TRAN  2US  200US ; Transient analysis for 0 to 200 μs with 2 μs
                       ; increment
     .WCASE   TRAN  V(5)  MAX
     .OP                  ; Prints details of the operating point
     .PROBE               ; Graphics postprocessor
.END                     ; End of circuit file
```

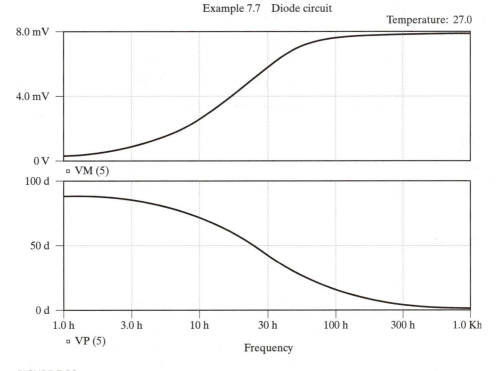

Example 7.7   Diode circuit

Temperature: 27.0

FIGURE 7.28

Frequency response for Example 7.7.

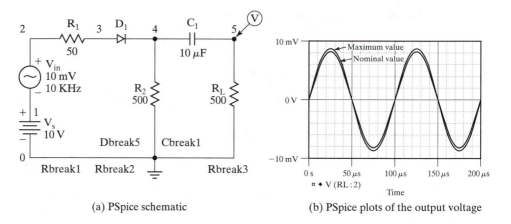

(a) PSpice schematic

(b) PSpice plots of the output voltage

FIGURE 7.29

PSpice schematic for Example 7.8.

The results of the worst-case analysis, which are obtained from the output file EX7.8.OUT, are as follows:

```
****      SORTED DEVIATIONS OF V(5)              TEMPERATURE = 27.000 DEG C
                                    WORST-CASE SUMMARY
      RUN               MAXIMUM VALUE
      ALL DEVICES       8.6367E-03 at T = 125.1210E-06
                        (104.82% of Nominal)
      NOMINAL           8.2395E-03 at T = 125.1210E-06
```

Note: The tolerance for the diode parameter IS is specified immediately after IS (e.g., $IS = 100E - 15\ DEV = 10\%$).

SUMMARY

The statements for diodes are

```
D⟨name⟩   NA   NK   DNAME   [(area) value]
.MODEL    DNAME    D (P1=A1   P2=A2   P3=A3  . . .  PN=AN)
```

REFERENCES

[1]   P. Antognetti, *Power Integrated Circuits*. New York: McGraw-Hill, 1986.

[2]   Clifton G. Fonstad, *Microelectronic Devices and Circuits*. New York: McGraw-Hill, 1994.

[3]   M. S. Ghausi, *Electronic Devices and Circuits: Discrete and Integrated*, pp. 3–7. New York: Holt, Reinehart and Winston, 1985.

[4]   Paul R. Gray, Paul J. Hurst, Stephen H. Lewis, and Robert G. Meyer, *Electronic Devices and Circuits: Discrete and Integrated*, 4th ed. New York: John Wiley & Sons, 2001.

[5]   Allan R. Hambley, *Electronics—A Top-Down Approach to Computer-Aided Circuit Design*. New York: Macmillan Publishing, 1994.

[6]   A. Laha and D. Smart, "A Zener diode model with application to SPICE2," *IEEE Journal of Solid-State Circuits*, Vol. SC-16, No. 1, pp. 21–22, 1981.

[7] Donald A. Neamen, *Semiconductor Physics & Devices*, 3d ed. New York: McGraw-Hill, 2002.

[8] *PSpice Manual.* Irvine, California: MicroSim Corporation, 1988.

[9] M. H. Rashid, *SPICE for Circuits and Electronics Using PSpice.* Englewood Cliffs, New Jersey: Prentice Hall, 1995.

[10] M. H. Rashid, *Electronics Analysis and Design Using Electronics Workbench.* Boston: PWS Publishing, 1997.

[11] M. H. Rashid, *Microelectronic Circuits: Analysis and Design.* Boston: PWS Publishing, 1999.

PROBLEMS

7.1 For the diode circuit in Fig. P7.1, print the bias point and the small-signal parameters of the diode. Use the default values of model parameters.

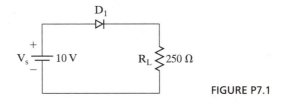

FIGURE P7.1

7.2 If the input voltage to the circuit in Fig. 7.16 is $v_{in} = 15 \sin(2000\pi t)$, plot the transient response of the output voltage for a time duration of 0 to 2 ms with a time increment of 10 μs. Print the details of the transient analysis bias point. The Zener voltages of the diodes are the same, $V_z = 5.2$ V, and the current at the Zener breakdown is $I_z = 0.5$ μA. The model parameters are IS = 0.5 UA, RS = 6, CJO = 2PF, TT = 12NS, BV = 5.20, and IBV = 0.5UA. The operating temperature is 50°C.

7.3 If the input voltage of the circuit in Fig. 7.19 is $V_{in} = 10$ V DC, print the details of the dc operating point. Print the voltage gain (V_{out}/V_{in}), the input resistance, and the output resistance.

7.4 A full-wave bridge rectifier is shown in Fig. P7.4. Plot the transient response of the output voltage for the time duration of 0 to 20 ms in steps of 0.1 ms. The model parameters are the default values. Print the details of the transient analysis bias point and the coefficients of the Fourier series.

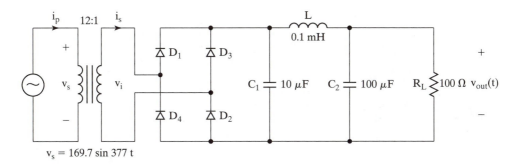

FIGURE P7.4

7.5 For the diode circuit in Fig. P7.5, plot the dc transfer characteristic between v_{in} and v_{out} for values of v_{in} in the range of -18 V to 18 V in steps of 0.5 V. The model parameters of the diodes are the default values.

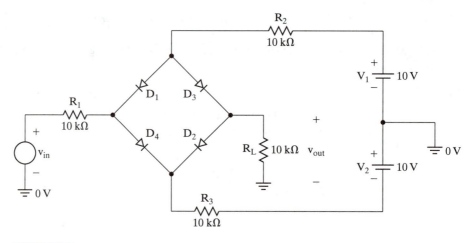

FIGURE P7.5

7.6 For the diode circuit in Fig. P7.6, plot the input current against the input voltage for values of V_{in} in the range of -10 V to 10 V in steps of 0.25 V. The model parameters are IS = 0.5UA, RS = 6, BV = 5.20, and IBV = 0.5UA. The operating temperature is 50°C.

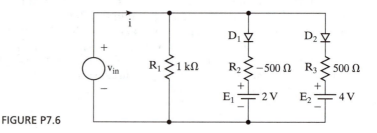

FIGURE P7.6

7.7 Repeat Example 7.6 if the direction of the diode D_1 is reversed.

7.8 Repeat Example 7.5 if the diodes are represented by voltage-controlled switches. The model parameters of the switches are RON = 0.25, ROFF = 1E + 6, VON = 0.25, and VOFF = 0.

7.9 A demodulator circuit is shown in Fig. P7.9. Plot the transient output voltage for the time duration of 0 to 100 μs with an increment of 0.5 μs. The model parameters are IS = 0.5UA, RS = 6, CJO = 2PF, TT = 12NS, BV = 5.20, and IBV = 0.5UA. The input voltage is given by

$$v_{in} = 10[1 + 0.5\sin(2\pi \times 10 \times 10^3 t)]\sin(2\pi \times 20 \times 10^6 t).$$

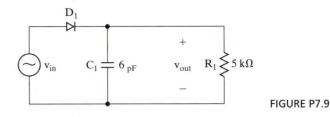

FIGURE P7.9

7.10 Use PSpice to perform a Monte Carlo analysis for six runs and for the transient response of Problem 7.4. The model parameters are $R = 1$ for the resistors, L for the inductor, and $C = 1$ for the capacitors. The circuit parameters having uniform deviations are

$$R_L = 100 \ \Omega \pm 5\%$$
$$C_1 = 10 \ \mu F \pm 15\%$$
$$C_2 = 100 \ \mu F \pm 10\%$$
$$L = 0.1 \ mH \pm 20\%$$

(a) The greatest difference from the nominal run is to be printed.
(b) The maximum value of the output voltage is to be printed.
(c) The minimum value of the output voltage is to be printed.
(d) The first occurrence of the output voltage crossing below 5 V is to be printed.

7.11 Use PSpice to perform the worst-case analysis for Problem 7.10.
7.12 Use PSpice to perform a Monte Carlo analysis for five runs and for the dc sweep of Problem 7.5. The model parameter for resistors is $R = 1$. The circuit parameters having uniform deviations are

$$R_1 = 10 \ k\Omega \pm 20\%$$
$$R_2 = 10 \ k\Omega \pm 5\%$$
$$R_3 = 10 \ k\Omega \pm 10\%$$
$$R_L = 10 \ \Omega \pm 15\%$$

(a) The greatest difference from the nominal run is to be printed.
(b) The maximum value of the output voltage is to be printed.
(c) The minimum value of the output voltage is to be printed.
(d) The first occurrence of the output voltage crossing below -2 V is to be printed.

7.13 Use PSpice to perform the worst-case analysis for Problem 7.12.
7.14 Use PSpice to perform a Monte Carlo analysis for five runs and for the dc sweep of Problem 7.6. The model parameters for resistors are $R = 1$ and I_S for diodes. The circuit parameters having uniform deviations are

$$R_1 = 1 \ k\Omega \pm 20\%$$
$$R_2 = 500 \ \Omega \pm 15\%$$
$$R_3 = 500 \ \Omega \pm 20\%$$
$$I_S = 0.5E - 6 + 10\%$$

(a) The greatest difference from the nominal run is to be printed.
(b) The maximum value of the input current is to be printed.
(c) The minimum value of the input current is to be printed.

7.15 Use Pspice to perform the worst-case analysis for Problem 7.14.

7.16 Use PSpice to find the operating current of the diode circuit in Fig. 7.15 if the diode is described by the following v-i relation:

$$i_D = 5 \times 10^{-4} v_D^2$$

7.17 Use PSpice to find the worst-case minimum and maximum output voltages $V_{out(max)}$ and $V_{out(min)}$ for the circuit as shown in Fig. 7.19. Assume uniform tolerances of $\pm 10\%$ for all resistances and an operating temperature of 25°C. Also assume that $V_{in} = 5V$.

C H A P T E R   8

Bipolar Junction Transistors

After completing this chapter, students should be able to

- Model SPICE BJTs and identify the SPICE model parameters
- Derive the model parameters from the data sheets of BJTs
- Plot the BJT characteristics in SPICE
- Find the dc biasing conditions of BJT amplifiers
- Perform the dc analysis, the transient analysis, and the ac analysis of BJT amplifiers and circuits.

8.1   INTRODUCTION

A bipolar junction transistor (BJT) may be specified by a device statement in conjunction with a model statement. Similar to diode models, the BJT model incorporates an extensive range of characteristics: for example, dc and small-signal behavior, temperature dependency, and noise generation. The model parameters take into account temperature effects, various capacitances, and the physical properties of semiconductors.

8.2   BJT MODEL

PSpice generates a complex model for BJTs. The model equations that are used by PSpice are described in Gummel and Poon [6] and Getreu [4]. If a complex model is not necessary, the model parameters can be ignored by the users, and PSpice assigns default values to the parameters.

The PSpice model, which is based on the integral charge-control model of Gummel and Poon [4,6], is shown in Fig. 8.1. The small-signal and static models that are generated by PSpice are shown in Figs. 8.2 and 8.3, respectively.

The model statement for NPN transistors has the general form

        .MODEL QNAME NPN (P1=A1 P2=A2 P3=A3 ... PN=AN)

and the general form for PNP transistors is

        .MODEL QNAME PNP (P1=A1 P2=A2 P3=A3 ... PN=AN)

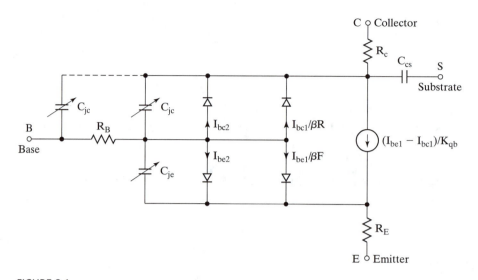

FIGURE 8.1

PSpice BJT model.

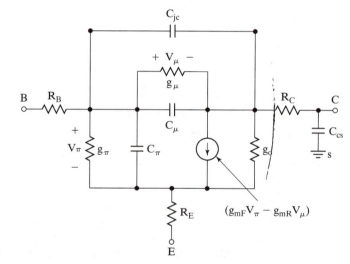

FIGURE 8.2

Small-signal BJT model.

where QNAME is the name of the BJT model. NPN and PNP are the type symbols for NPN and PNP transistors, respectively. QNAME, which is the model name, can begin with any character, and its word size is normally limited to eight characters. P1, P2, ... and A1, A2, ... are the parameters and their values, respectively. Table 8.1 shows the model parameters of BJTs. If certain parameters are not specified, PSpice assumes the simple Ebers–Moll model [3], which is shown in Fig. 8.4(a). The large signal model is shown in Fig. 8.4(b).

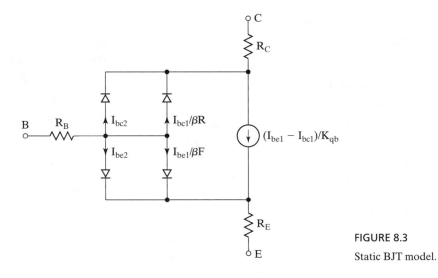

FIGURE 8.3

Static BJT model.

As with diodes, an *area factor* is used to determine the number of equivalent parallel BJTs of a specified model. The model parameters that are affected by the area factor are marked by an asterisk (*) in Table 8.1. RC, RE, and RB represent the contact and bulk resistances per unit area of the collector, emitter, and base, respectively. The bipolar transistor is modeled as an intrinsic device. The area value, which is the relative device area, is specified in the .MODEL statement (Section 8.3) and changes the actual resistance values. The area has a default value of 1.

Some parameters have alternate names, such as VAF and VA. One can use either name, VAF or VA. These are indicated by parentheses in Table 8.1.

The parameters ISE (C2) and ISC (C4) can be either greater than or less than 1. If they are less than 1, they represent the absolute currents. If they are greater than 1, they act as the multipliers of IS instead of absolute currents. That is, the value of ISE becomes ISE*IS for ISE > 1 and that of ISC becomes ISC*IS for ISC > 1.

The dc model is defined (1) by parameters BF, C2, IK, and NE, which determine the forward-current gain; (2) by BR, C4, IKR, and VC, which determine the reverse-current gain characteristics; (3) by VA and VB, which determine the output conductance for forward and reverse regions; and (4) by the reverse saturation current IS.

Base-charge storage is modeled (1) by forward and reverse transit times TF and TF, and nonlinear depletion-layer capacitances, which are determined by CJE, PE, and ME for a b–e junction, and (2) by CJC, PC, and MC for a b–c junction. CCS is a constant collector-substrate capacitance.

The temperature dependence of the saturation current is determined by the energy gap EG and the saturation-current temperature exponent PT.

TABLE 8.1   Model Parameters of BJTS

Name	Area	Model Parameters	Units	Default	Typical
IS	*	$p–n$ saturation current	Amps	1E–16	1E–16
BF		Ideal maximum forward beta		100	100
NF		Forward current emission coefficient		1	1
VAF(VA)		Forward Early voltage	Volts	∞	100
IKF(IK)		Corner for forward beta high-current roll-off	Amps	∞	10m
ISE(C2)		Base-emitter leakage saturation current	Amps	0	1000
NE		Base-emitter leakage emission coefficient		1.5	2
BR		Ideal maximum reverse beta		1	0.1
NR		Reverse current emission coefficient		1	
VAR(VB)		Reverse Early voltage	Volts	∞	100
IKR	*	Corner for reverse beta high-current roll-off	Amps	∞	100m
ISC(C4)		Base-collector leakage saturation current	Amps	0	1
NC		Base-collector leakage emission coefficient		2	2
RB	*	Zero-bias (maximum) base resistance	Ohms	0	100
RBM		Minimum base resistance	Ohms	RB	100
IRB		Current at which RB falls halfway to RBM	Amps	∞	
RE	*	Emitter ohmic resistance	Ohms	0	1
RC	*	Collector ohmic resistance	Ohms	0	10
CJE	*	Base-emitter zero-bias $p–n$ capacitance	Farads	0	2P
VJE(PE)		Base-emitter built-in potential	Volts	0.75	0.75
MJE(ME)		Base-emitter $p–n$ grading factor		0.33	0.33
CJC	*	Base-collector zero-bias $p–n$ capacitance	Farads	0	1P
VJC(PC)		Base-collector built-in potential	Volts	0.75	0.75
MJC(MC)		Base-collector $p–n$ grading factor		0.33	0.33
XCJC		Fraction of C_{bc} connected internal to R_B		1	
CJS(CCS)		Collector-substrate zero-bias $p–n$ capacitance	Farads	0	2PF
VJS(PS)		Collector-substrate built-in potential	Volts	0.75	
MJS(MS)		Collector-substrate $p–n$ grading factor		0	
FC		Forward-bias depletion capacitor coefficient		0.5	
TF		Ideal forward transit time	Seconds	0	0.1NS
XTF		Transit-time bias dependence coefficient		0	
VTF		Transit-time dependency on V_{bc}	Volts	∞	
ITF		Transit-time dependency on I_c	Amps	0	
PTF		Excess phase at $1/(2\pi * TF)$Hz	Degrees	0	30°
TR		Ideal reverse transit time	Seconds	0	10NS
EG		Bandgap voltage (barrier height)	Electron-volts	1.11	1.11
XTB		Forward and reverse beta temperature coefficient		0	
XTI(PT)		IS temperature-effect exponent		3	
KF		Flicker noise coefficient		0	6.6E–16
AF		Flicker noise exponent		1	1

8.3   BJT STATEMENTS

The symbol for a bipolar junction transistor (BJT) is Q. The name of a bipolar transistor must start with Q, and it takes the general form

```
Q⟨name⟩ NC NB NE NS QNAME [(area) value]
```

where NC, NB, NE, and NS are the collector, base, emitter, and substrate nodes, respectively. QNAME could be any name of up to eight characters. The substrate node is

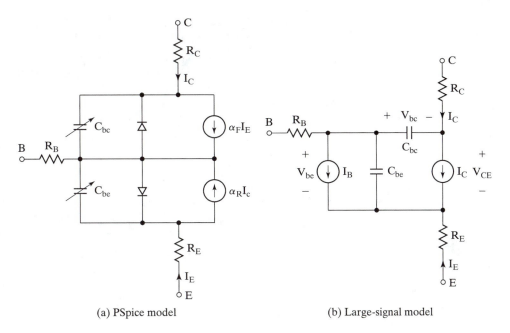

(a) PSpice model   (b) Large-signal model

FIGURE 8.4

Ebers–Moll BJT model.

optional; if not specified, it defaults to ground. Positive current is the current that flows into a terminal. That is, the current flows from the collector node through the device to the emitter node for an NPN BJT.

Some Statements for BJTs

```
QIN   5  7  8    2N2222
Q5    2  4  5    2N2907    1.5
QX    1  4  9    NMOD
.MODEL 2N2222 NPN(IS=3.108E-15 XTI=3 EG=1.11 VAF=131.5 BF=217.5
+   NE=1.541 ISE=190.7E-15 IKF=1.296 XTB=1.5 BR=6.18 NC=2 ISC=0 IKR=0
+   RC=1 CJC=14.57E-12 VJC=.75 MJC=.3333 FC=.5 CJE=26.08E-12 VJE=.75
+   MJE=.3333 TR=51.35E-9 TF=451E-12 ITF=.1 VTF=10 XTF=2)
.MODEL 2N2907 PNP (IS=9.913E-15 XTI=3 EG=1.11 VAF=90.7 BF=197.8
+   NE=2.264 ISE=6.191E-12 IKF=.7322 XTB=1.5 BR=3.369 NC=2 ISC=0 IKR=0
+   RC=1 CJC=14.57E-12 VJC=.75 MJC=.3333 FC=.5 CJE=20.16E-12 VJE=.75
+   MJE=.3333 TR=29.17E-9 TF=405.7E-12 ITF=.4 VTF=10 XTF=2)
.MODEL NMOD NPN
```

Note: A + (plus) sign at the first column indicates the continuation of the statement preceding it.

The PSpice schematic is shown in Fig. 8.5(a). The PSpice specifies the model parameters of a known BJT (e.g., Q2N2222) as shown in Fig. 8.5(b). If the specific BJT is not available in the PSpice library, the user can specify the model parameters of a Qbreak BJT.

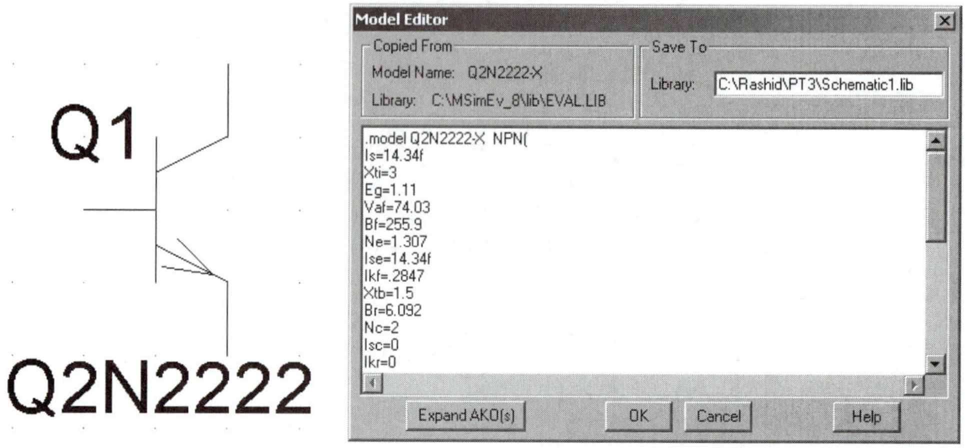

(a) Schematic   (b) Model parameters

FIGURE 8.5

Schematic and model parameters of BJTs.

8.4 BJT PARAMETERS

The data sheet for NPN transistor of type 2N2222A is shown in Fig. 8.6. SPICE parameters are not quoted directly in the data sheet. Some versions of SPICE (e.g., PSpice) support device library files that give the model parameters. The library file EVAL.LIB contains the list of devices and their model statements in the student's version of PSpice. The software PARTS of PSpice can generate SPICE models from the datasheet parameters of diodes. SPICE model parameters are also supplied by some manufacturers. However, some parameters that significantly influence the performance of a transistor can be determined from the data sheet [12,15].

The diode characteristic described by Eq. (7–1) can be applied to an NPN transistor by selecting appropriate subscripts, as in the following equation:

$$I_C = I_S[e^{V_{BE}/\eta V_T} - 1] \tag{8.1}$$

Using Eq. (7–8), the difference in the base-emitter voltages can be expressed by

$$V_{BE2} - V_{BE1} = 2.3 \, \eta V_T \log\left(\frac{I_{C2}}{I_{C1}}\right) \tag{8.2}$$

From the data sheet for V_{BD} versus I_C, we get $V_{BE1} = 0.6$ V at $I_{C1} = 0.2$ mA, and $V_{BE2} = 0.7$ V at $I_{C2} = 20$ mA. Assuming that $V_T = 25.8$ mV $= 0.0258$, we can apply Eq. (8.2) to find the *emission coefficient* η from the formula

$$0.7 - 0.6 = 2.3 \, \eta \times 0.0258 \log\left(\frac{20}{0.2}\right)$$

which gives $\eta = 0.843$. Since we did not include any contact and bulk resistance RE of the emitter, we got the value $\eta < 1$; its practical value is $\eta \geq 1$. Let us assume that

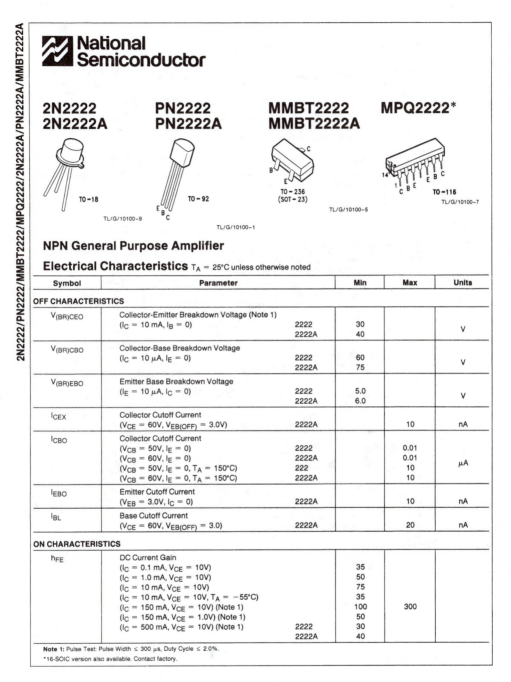

FIGURE 8.6

Data sheet for transistor of type 2N2222A (Courtesy of National Semiconductor, Inc.).

NPN General Purpose Amplifier (Continued)

Electrical Characteristics T_A = 25°C unless otherwise noted (Continued)

Symbol	Parameter		Min	Max	Units
ON CHARACTERISTICS (Continued)					
$V_{CE(sat)}$	Collector-Emitter Saturation Voltage (Note 1)				
	(I_C = 150 mA, I_B = 15 mA)	2222		0.4	
		2222A		0.3	
	(I_C = 500 mA, I_B = 50 mA)	2222		1.6	V
		2222A		1.0	
$V_{BE(sat)}$	Base-Emitter Saturation Voltage (Note 1)				
	(I_C = 150 mA, I_B = 15 mA)	2222	0.6	1.3	
		2222A	0.6	1.2	V
	(I_C = 500 mA, I_B = 50 mA)	2222		2.6	
		2222A		2.0	
SMALL-SIGNAL CHARACTERISTICS					
f_T	Current Gain—Bandwidth Product (Note 3)				
	(I_C = 20 mA, V_{CE} = 20V, f = 100 MHz)	2222	250		MHz
		2222A	300		
C_{obo}	Output Capacitance (Note 3)				
	(V_{CB} = 10V, I_E = 0, f = 100 kHz)			8.0	pF
C_{ibo}	Input Capacitance (Note 3)				
	(V_{EB} = 0.5V, I_C = 0, f = 100 kHz)	2222		30	pF
		2222A		25	
rb'C_C	Collector Base Time Constant				
	(I_E = 20 mA, V_{CB} = 20V, f = 31.8 MHz)	2222A		150	ps
NF	Noise Figure				
	(I_C = 100 μA, V_{CE} = 10V, R_S = 1.0 kΩ, f = 1.0 kHz)	2222A		4.0	dB
Re(h_{ie})	Real Part of Common-Emitter High Frequency Input Impedance				
	(I_C = 20 mA, V_{CE} = 20V, f = 300 MHz)			60	Ω
SWITCHING CHARACTERISTICS					
t_D	Delay Time	(V_{CC} = 30V, $V_{BE(OFF)}$ = 0.5V,	except		10	ns
t_R	Rise Time	I_C = 150 mA, I_{B1} = 15 mA)	MPQ2222		25	ns
t_S	Storage Time	(V_{CC} = 30V, I_C = 150 mA,	except		225	ns
t_F	Fall Time	I_{B1} = I_{B2} = 15 mA	MPQ2222		60	ns

Note 1: Pulse Test: Pulse Width < 300 μs, Duty Cycle ≤ 2.0%.
Note 2: For characteristics curves, see Process 19.
Note 3: f_T is defined as the frequency at which $|h_{fe}|$ extrapotates to unity.
Note 4: 2N also available in JAN/TX/V series.

FIGURE 8.6

(*Continued*)

Process 19

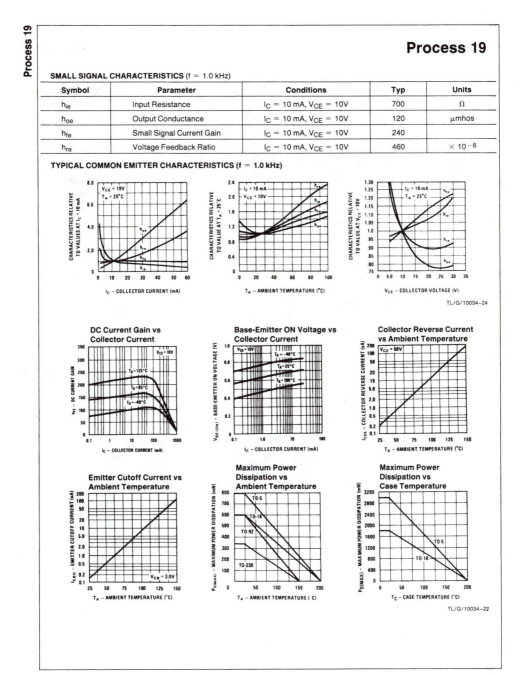

SMALL SIGNAL CHARACTERISTICS (f = 1.0 kHz)

Symbol	Parameter	Conditions	Typ	Units
h_{ie}	Input Resistance	$I_C = 10$ mA, $V_{CE} = 10$V	700	Ω
h_{oe}	Output Conductance	$I_C = 10$ mA, $V_{CE} = 10$V	120	μmhos
h_{fe}	Small Signal Current Gain	$I_C = 10$ mA, $V_{CE} = 10$V	240	
h_{re}	Voltage Feedback Ratio	$I_C = 10$ mA, $V_{CE} = 10$V	460	$\times 10^{-6}$

TYPICAL COMMON EMITTER CHARACTERISTICS (f = 1.0 kHz)

TL/G/10034–24

TL/G/10034–22

FIGURE 8.6

(*Continued*)

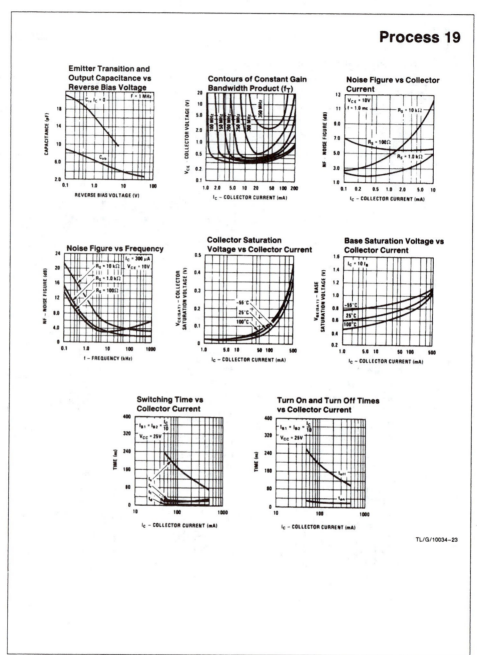

TL/G/10034–23

FIGURE 8.6

(*Continued*)

$\eta = 1$. For $\eta = 1$ and $V_{BE2} = 0.7$ V at $I_{C2} = 20$ mA, we can apply Eq. (8.1) to find the saturation current from

$$20 \text{ mA} = I_S[e^{0.7/(1 \times 0.0258)} - 1]$$

which gives $I_S = 3.295E{-}14$ A. The dc current gain at 150 mA is $h_{FE} = 100$ to 300. Taking the geometric mean gives $BF = \sqrt{(100 \times 300)} \approx 173$.

The input capacitance at the base-emitter junction is $C_{ibo} = 25$ pF at $V_{EB} = 0.5$ V, $I_C = 0$ (reverse-biased). Since $C_{je} = C_{ibo}$, then C_{jeo} can be found from

$$C_{je} = \frac{C_{jeo}}{(1 + V_{EB}/V_{je})^{M_{je}}} \tag{8.3}$$

where $M_{je} = MJE \approx 1/3$, and $V_{je} = VJE \approx 0.75$ V. Equation (8.3) gives $C_{jeo} = CJE = 29.6$ pF at $V_{BE} = 0$ V.

The output capacitance is $C_{obo} = 8$ pF at $V_{CB} = 10$ V, $I_E = 0$ (reverse-biased). Since $C_{\mu} = C_{obo}$, $C_{\mu o}$ can be found from

$$C_{\mu} = \frac{C_{\mu o}}{(1 + V_{CB}/V_{jc})^{M_{je}}} \tag{8.4}$$

where $M_{jc} = MJC \approx 1/3$ and $V_{jc} \approx VJC = 0.75$ V. From Eq. (8.4), $C_{\mu o} = CJC = 19.4$ pF at $V_{CB} = 0$ V.

The transition frequency $f_{T(\min)} = 300$ MHz at $V_{CE} = 20$ V, $I_C = 20$ mA. The transition period is $\tau_T = 1/2\pi f_T = 1/(2\pi \times 300 \text{ MHz}) = 530.5$ ps. Assuming that $V_{BE} = 0.7$ V, we have $V_{CB} \approx V_{CE} - V_{BE} = 20 - 0.7 = 19.3$ V, and Eq. (8.4) gives $C_{\mu} = 6.49$ pF.

Since the transition frequency $f_{T(\min)} = 300$ MHz is specified at $I_C = 20$ mA, we need to find the transconductance g_m (at $I_C = 20$ mA), which is given by

$$\begin{aligned} g_m &= \frac{I_C}{V_T} \\ &= 20 \text{ mA}/25.8 \text{ mV} = 775.2 \text{ mA/V}. \end{aligned} \tag{8.5}$$

The transition period τ_T is related to forward transit time τ_F by

$$\tau_T = \tau_F + \frac{C_{je}}{g_m} + \frac{C_{\mu}}{g_m} \tag{8.6}$$

or

$$530.5 \text{ ps} = \tau_F + \frac{25 \text{ pF}}{0.7752} + \frac{6.49 \text{ pF}}{0.7752}$$

which gives $\tau_F = TF = 489.88$ ps.

The output conductance h_{oe} of a transistor is related to the collector current I_C and the *Early voltage* V_A by

$$\frac{1}{h_{oe}} = \frac{V_A}{I_C} \tag{8.7}$$

From the data sheet, $h_{oe} = 120\ \mu$mhos at $I_C = 10$ mA and $V_{CE} = 10$ V. From Eq. (8.7), the Early voltage becomes

$$V_A = \text{VA} = I_C/h_{oe} = 10\ \text{mA}/120\ \mu\text{mhos} = 83.3\ \text{V}.$$

The reverse transit time can be approximated to $\tau_R = \text{TR} = 10\tau_F = 4.9$ ns.
    The model statement for transistor 2N2222A is

```
.MODEL   Q2N2222A NPN (IS=3.295E-14 BF=173 VA=83.3V CJE=29.6PF CJC=19.4PF
+        TF=489.88PS TR=4.9NS)
```

This model is used to plot the characteristics of the BJT as illustrated in Example 8.1. It may be necessary to modify the parameter values to conform to the actual characteristics.

Note.   If a model parameter is not specified in the model statement, SPICE assumes its default value. However, it should be noted that some default values represent ideal conditions (e.g., TR = 0 and VAF = ∞). More accurate results can be obtained by using the typical values (e.g., TR = 0.1NS and VAF = 100V) rather than the default ones.

8.5   EXAMPLES OF BJT CIRCUITS

The PSpice simulation of BJT circuits requires specifying the BJT model parameters. If the model parameters are not specified, PSpice assumes the default values that are given in Table 8.1. The following examples illustrate the PSpice simulation of BJT circuits.

Example 8.1:   Plotting the BJT characteristics

For the NPN BJT transistor of Fig. 8.7, plot the output characteristics (I_C versus V_{CE}) if V_{CE} is varied from 0 to 10 V in steps of 0.02 V and I_B is varied from 0 to 1 mA in steps of 200 μA. Use the model parameters that were determined in Section 8.4. Print the details of the small-signal parameters at the operating point for $I_B = 1$ mA and $V_{CE} = 12$ V.

Solution   The PSpice schematic is shown in Fig. 8.8(a). The model parameters are changed from the edit model text menu as shown in Fig. 8.8(c). The model name, QbreakN1 as shown in Fig. 8.8(d), is changed from the edit model as shown in Fig. 7.8(c), which is selected from the edit menu. The temperature values are varied by nested sweep within the main dc sweep, which is selected from the dc sweep menu of the analysis setup. The parameters of model QbreakN1 are specified by a model statement in file RASH_MODEL.LIB.

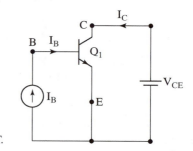

FIGURE 8.7

A circuit with an NPN BJT.

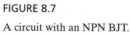

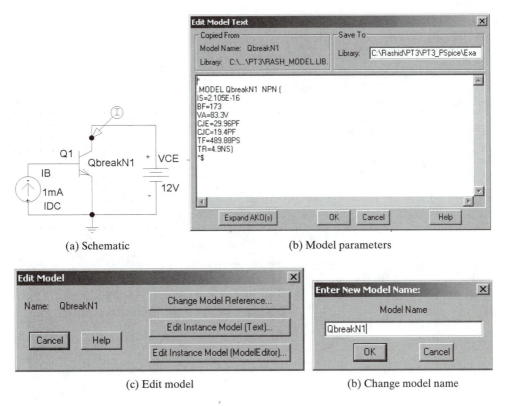

(a) Schematic   (b) Model parameters

(c) Edit model   (b) Change model name

FIGURE 8.8

PSpice schematic and BJT model parameters for Example 8.1.

The dc sweep in PSpice schematic that is set from the analysis setup is shown in Fig. 8.9(a) and the nested sweep is shown in Fig. 8.9(b) with nested sweep enabled. The listing of the circuit file follows.

Example 8.1   NPN BJT characteristics

```
▲  IB    0   1   DC   1MA              ; Base current
   VCE   2   0   DC   12V              ; Collector-emitter voltage
▲▲ Q1        2   1   0      Q2N2222A    ; BJT statement
   .MODEL  Q2N2222A  NPN (IS=2.105E-16 BF=173 VA=83.3V CJE=29.6PF CJC=19.4PF
   +       TF=489.88PS TR=4.9NS)       ; Model parameters
▲▲▲  .DC  VCE  0  10V   0.02V  IB 0  1MA  200UA  ; Dc sweep for VCE and IB
     .PROBE                            ; Graphics postprocessor

.END
```

The output characteristics, which are plots of I_C versus V_{CE}, are shown in Fig. 8.10. The students are encouraged to compare the transistor characteristics with those obtained by using the model parameters of transistor 2N2222A, which is listed in the PSpice library file EVAL.LIB. This can be done by replacing the .MODEL statement in the preceding circuit file by the .LIB EVAL.LIB statement.

The small-signal parameters of the transistor at the operating point, which are obtained from the output file EX8.1.OUT, are as follows:

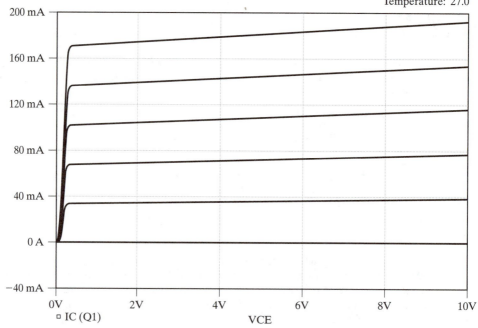

(a) Main dc sweep   (b) Nested sweep

FIGURE 8.9

Dc sweep setup for Example 8.1.

Example 8.1   NPN–BJT characteristics

Temperature: 27.0

FIGURE 8.10

Output characteristics of the BJT in Example 8.1.

****   OPERATING   POINT   INFORMATION   TEMPERATURE = 27.000 DEG C

NAME	Q1
MODEL	Q2N2222A
IB	1.00E−03
IC	1.96E−01
VBE	8.88E−01
VBC	−1.11E+01
VCE	1.20E+01
BETADC	1.96E+02
GM	7.58E+00
RPI	2.59E+01
RX	0.00E+00
RO	4.82E+02
CBE	3.77E−09
CBC	7.80E−12
CBX	0.00E+00
CJS	0.00E+00
BETAAC	1.96E+02
FT	3.19E+08

Note: The collector current i_C shown in Fig. 8.10 increases with the base current i_B. For $V_{CE} > 0.5\text{V}$, the i_C-V_{CE} characteristic remains almost constant at a low value. That is, the output resistance r_o of the transistor is high.

Example 8.2:   Sensitivity analysis of BJT amplifier

A bipolar transistor circuit is shown in Fig. 8.11(a), where the output is taken from Node 4. Calculate and print the sensitivity of the collector current with respect to all parameters.

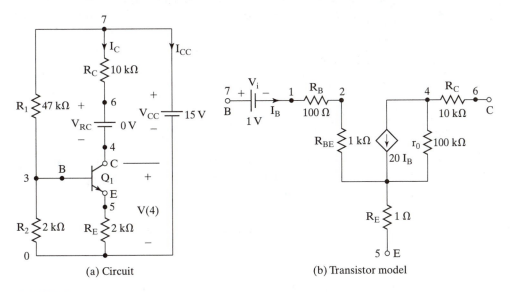

(a) Circuit

(b) Transistor model

FIGURE 8.11

Bipolar transistor circuit.

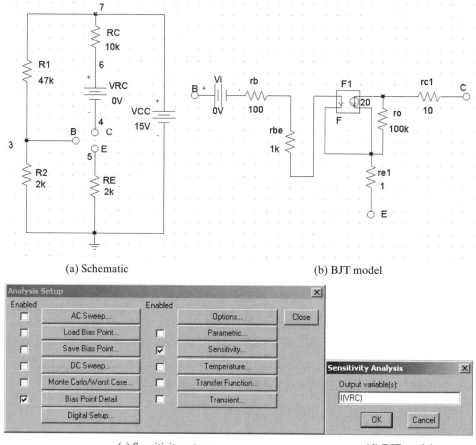

(a) Schematic   (b) BJT model

(c) Sensitivity setup   (d) BJT model

FIGURE 8.12

Sensitivity for Example 8.2.

Print the details of the bias point. The equivalent circuit for transistor Q_1 is shown in Fig. 8.11(b).

Solution The PSpice schematic is shown in Fig. 8.12(a) and the BJT model in Fig. 8.12(b). The sensitivity analysis that is set from the analysis setup is shown in Fig. 8.12(c). The output variable is defined as shown in Fig. 8.12(c).

The listing of the circuit file follows.

Example 8.2   Biasing sensitivity of bipolar transistor amplifier

```
▲  *   Supply voltage is 15 V DC.
   VCC  7  0  DC 15V
   *   A dummy voltage source of 0 V to measure the collector current
   VRC  6  4  DC  0V
```

▲▲ R1   7 3   47K
     R2   3 0   2K
     RC   7 6   10K
     RE   5 0   2K
     *   Subcircuit call for transistor model QMOD; the substrate is
     *   connected to ground by default.
     XQ1 4 3 5   QMOD
     * Subcircuit definition for QMOD
     .SUBCKT   QMOD 6 7 5
     RB 1 2   100
     RE 3 5   1
     RC 4 6   10
     RBE 2 3   1K
     RO 4 3   100K
     *   A dummy voltage source of 0 V to measure the controlling current
     VI 7 1   DC 0V
     F1 4 3 VI 20
     *   End of subcircuit definition
     .ENDS QMOD
▲▲▲   .OPTIONS NOPAGE NOECHO
       *   Sensitivity of collector current (which is the current through voltage
       *   source VRC)
        .SENS   I (VRC)
   .END

The .SENS command does not require a .PRINT command for printing the output. The output for the sensitivity analysis and the bias point follow:

****   **SMALL-SIGNAL BIAS SOLUTION**   **TEMPERATURE =**   **27.000 DEG C**

NODE	VOLTAGE	NODE	VOLTAGE	NODE	VOLTAGE	NODE	VOLTAGE
(3)	.5960	(4)	12.1520	(5)	.5864	(6)	12.1520
(7)	15.0000	(XQ1.1)	.5960	(XQ1.2)	.5952	(XQ1.3)	.5867
(XQ1.4)	12.1500						

       VOLTAGE SOURCE CURRENTS

NAME	CURRENT
VCC	$-5.912E-04$
VRC	$2.848E-04$
XQ1.VI	$8.456E-06$

TOTAL POWER DISSIPATION 8.87E−03 WATTS

****   **DC SENSITIVITY ANALYSIS**   **TEMPERATURE =**   **27.000 DEG C**

DC SENSITIVITIES OF OUTPUT I (VRC)

ELEMENT NAME	ELEMENT VALUE	ELEMENT SENSITIVITY (AMPS/UNIT)	NORMALIZED SENSITIVITY (AMPS/PERCENT)
R1	4.700E+04	$-5.481E-09$	$-2.576E-06$
R2	2.000E+03	$1.252E-07$	$2.505E-06$
RC	1.000E+04	$-3.134E-10$	$-3.134E-08$
RE	2.000E+03	$-1.288E-07$	$-2.576E-06$
XQ1.RB	1.000E+02	$-3.705E-09$	$-3.705E-09$
XQ1.RE	1.000E+00	$-1.288E-07$	$-1.288E-09$
XQ1.RC	1.000E+01	$-3.134E-10$	$-3.134E-11$

```
XQ1.RBE      1.000E+03     -3.705E-09     -3.705E-08
XQ1.RO       1.000E+05     -1.273E-10     -1.273E-07
VCC          1.500E+01      1.898E-05      2.848E-06
VRC          0.000E+00     -1.101E-06      0.000E+00
XQ1.VI       0.000E+00     -4.381E-04      0.000E+00
JOB CONCLUDED
TOTAL JOB TIME        2.52
```

Note: The sensitivity analysis gives the effects of variations of a particular element on the desired output. For example, any increase in V_{CC} and R_1 will increase and decrease the collector current, respectively.

Example 8.3:   Transfer function analysis of a BJT amplifier

A bipolar Darlington pair amplifier is shown in Fig. 8.13. Calculate and print the voltage gain, the input resistance, and the output resistance. The input voltage is 5 V. The model parameters of the bipolar transistors are BF = 100, BR = 1, RB = 5, RC = 1, RE = 0, VJE = 0.8, and VA = 100.

Solution   The PSpice schematic is shown in Fig. 8.14(a). The transfer-function analysis is set from the analysis setup menu. The output variable is defined as shown in Fig. 8.14(b).
    The listing of the circuit file follows.

Example 8.3   Darlington pair

```
▲   .OPTIONS  NOPAGE  NOECHO
    VCC  2  0  DC  12V
    VIN  1  0  DC  5V
▲▲    *  BJTs with model QM
        Q1  2  1  3  QM
        Q2  2  3  4  QM
        RB  2  1  47K
        RE  4  0  4.7K
        *  Model QM for NPN BJTs
        .MODEL  QM  NPN (BF=100 BR=1 RB=5 RC=1 RE=0 VJE=0.8 VA=100)
▲▲▲  *  Transfer-function analysis to calculate dc gain, input
       *  resistance, and output resistance
        .TF  V(4)  VIN
  .END
```

The results of the transfer-function analysis follow:

```
****    SMALL-SIGNAL BIAS SOLUTION               TEMPERATURE = 27.000 DEG C
    NODE    VOLTAGE        NODE    VOLTAGE      NODE    VOLTAGE     NODE    VOLTAGE

(   1)    5.0000     (   2)    12.0000    (   3)  4.3560   (   4)   3.5909

        VOLTAGE SOURCE CURRENTS
        NAME        CURRENT

        VCC        -9.129E-04
        VIN         1.489E-04
        TOTAL POWER DISSIPATION    1.02E-02 WATTS
```

```
****        SMALL-SIGNAL CHARACTERISTICS
    V(4)/VIN =   9.851E-01
    INPUT RESISTANCE AT VIN =    4.696E+04
    OUTPUT RESISTANCE AT V(4) =  6.679E+01
        JOB CONCLUDED
        TOTAL JOB TIME        2.97
```

Note: The transfer function analysis gives the input resistance, output resistance, and gain of the circuit. In other words, it gives the values of the Thevenin's equivalent circuit.

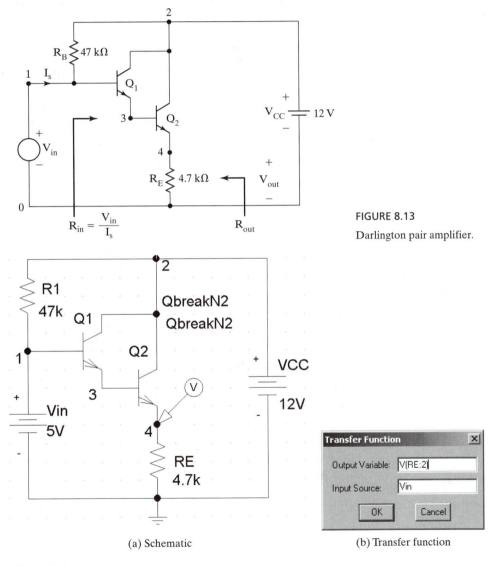

FIGURE 8.13

Darlington pair amplifier.

(a) Schematic

(b) Transfer function

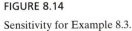

FIGURE 8.14

Sensitivity for Example 8.3.

Example 8.4: Transient and frequency response of a BJT amplifier

A bipolar transistor amplifier circuit is shown in Fig. 8.15. The output is taken from Node 6. Calculate and plot the magnitude and phase of the voltage gain for frequencies from 1 Hz to 10 kHz with a decade increment and with 10 points per decade. The input voltage for ac analysis is 10 mV. Calculate and plot the transient response of voltages at Nodes 4 and 6 for an input voltage of $v_{in} = 0.01 \sin(2\pi \times 1000t)$ and for a duration of 0 to 2 ms in steps of 50 μs. The details of ac and transient analysis operating points should be printed. The model parameters of the PNP BJT are IS = 2E –16, BF = 100, BR = 1, RB = 5, RC = 1, RE = 0, TF = 0.2NS, TR = 5NS, CJE –0.4PF, VJE = 0.8, ME = 0.4, CJC = 0.5PF, VJC = 0.8, CCS = 1PF, and VA = 100.

Solution   The PSpice schematic is shown in Fig. 8.16(a). The ac analysis is set from the analysis setup menu as shown in Fig. 8.16(b) and the transient analysis in Fig. 8.16(c).

The listing of the circuit file follows.

Example 8.4   Bipolar transistor amplifier

```
▲ *  Input voltage is 10-mV peak for ac analysis and for transient response:
    *  It is a 10-mV peak at 1 kHz with zero-offset value.
    VIN  1   0   AC  10MV  SIN(0  10MV  1KHZ)
    VCC  0   7   DC  15V
▲▲ RS  1   2   500
    R1   7   3   47K
    R2   3   0   5K
    RC   7   4   10K
    RE   5   0   2K
    RL   6   0   20K
    C1   2   3   1UF
    C2   4   6   1UF
    CE   5   0   10UF
    *  Transistor Q1 with model QM
    Q1   4   3   5   0   QM
    * Model QM for PNP transistors
    .MODEL  QM  PNP (IS=2E-16 BF=100 BR=1 RB=5 RC=1 RE=0 TF=0.2NS TR=5NS
    +          CJE=0.4PF VJE=0.8 ME=0.4 CJC=0.5PF VJC=0.8 CCS=1PF VA=100)
▲▲▲ *  Plot the results of transient analysis for voltages at nodes 4, 6, and 1.
    .PLOT  TRAN  V(4)  V(6)  V(1)
    *    Plot the results of the ac analysis for the magnitude and phase angle
    *    of output voltage at node 6.
    .PLOT  AC    VM(6)   VP(6)
    .OPTIONS  NOPAGE NOECHO
    *    Transient analysis for 0 to 2 ms with 50-μs increment
    *    Print details of transient analysis operating point.
    .TRAN/OP  50US  2MS
    *  AC analysis from 1 Hz to 10 KHz with a decade increment and 10 points
    *  per decade
    .AC  DEC  10  1HZ  10KHZ
    *  Print the details of the ac analysis operating point.
    .OP
    .PROBE
.END
```

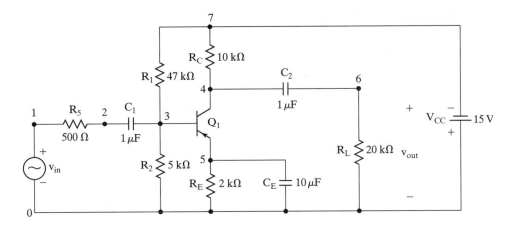

FIGURE 8.15

Bipolar transistor amplifier circuit.

Note: PLOT statements generate graphical plots in the output file. If the .PROBE command is included, there is no need for the .PLOT command.

The determination of the operating point is the first step in analyzing a circuit with nonlinear devices (e.g., bipolar transistors). The equivalent circuit for determining the ac analysis (or dc analysis) bias point of the amplifier in Fig. 8.15 is shown in Fig. 8.17, where the capacitors are open circuited. The details of the bias point follow:

```
****    SMALL-SIGNAL BIAS SOLUTION          TEMPERATURE =   27.000 DEG C

NODE      VOLTAGE     NODE     VOLTAGE     NODE     VOLTAGE     NODE      VOLTAGE


(   1)    0.0000    (   2)    0.0000    (   3)    -1.4280   (   4)     -11.5240
(   5)    -.7016    (   6)    0.0000    (   7)    -15.0000


    VOLTAGE SOURCE CURRENTS
    NAME           CURRENT

    VIN        0.000E+00
    VCC       -6.364E-04
    TOTAL POWER DISSIPATION     9.55E-03 WATTS
```

Once the dc bias point is determined, PSpice generates a small-signal model of the BJT. This model is similar to that in Fig. 8.18. PSpice replaces the transistor with this circuit model. It should be noted that this model is valid only at the operating

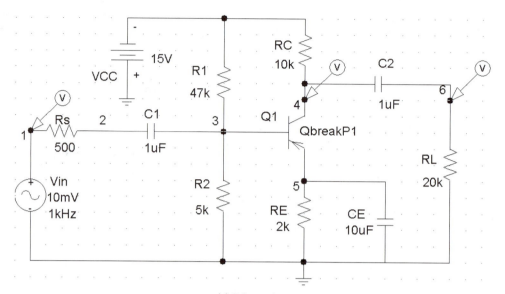

(a) Schematic

(b) Setup for ac sweep

(c) Setup for transient

FIGURE 8.16

PSpice schematic for Example 8.4.

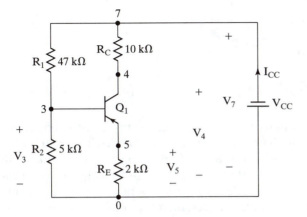

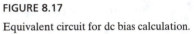

FIGURE 8.17

Equivalent circuit for dc bias calculation.

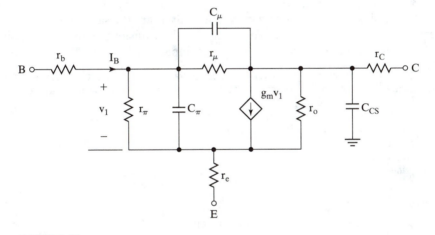

FIGURE 8.18

Small-signal equivalent circuit of bipolar transistors.

point. The details of the operating point and model values follow:

```
****         OPERATING POINT INFORMATION        TEMPERATURE = 27.000 DEG C
**** BIPOLAR JUNCTION TRANSISTORS
   NAME          Q1
   MODEL         QM
   IB            -3.16E-06
   IC            -3.48E-04
   VBE           -7.26E-01
   VBC            1.01E+01
   VCE           -1.08E+01
   BETADC         1.10E+02
   GM             1.34E-02
   RPI            8.19E+03
   RX             5.00E+00
```

```
RO              3.17E+05
CBE             3.39E-12
CBC             2.11E-13
CBX             0.00E+00
CJS             1.00E-12
BETAAC          1.10E+02
FT              5.94E+08
```

Prior to the transient analysis, PSpice determines the small-signal parameters of the nonlinear devices and the potentials of the various nodes. The method for the calculation of the transient analysis bias point differs from that of the dc analysis bias point because, in transient analysis, all the nodes have to be assigned initial values, and the nonlinear sources may have transient values at the beginning of transient analysis. The capacitors, which may have initial values, therefore remain as parts of the circuit. The equivalent circuit for determining the transient analysis bias point for the circuit in Fig. 8.15 is shown in Fig. 8.19. Since the capacitors in Fig. 8.15 do not have any initial values, the bias points for dc and transient analysis are the same. There, the small-signal parameters are also the same. The details of the transient analysis bias point and the small-signal parameters are given next to compare with those of dc analysis:

```
****      INITIAL   TRANSIENT   SOLUTION            TEMPERATURE = 27.000 DEG C
  NODE    VOLTAGE     NODE     VOLTAGE    NODE     VOLTAGE    NODE     VOLTAGE

(    1)   0.0000   (    2)    0.0000   (   3)    -1.4280  (    4)   -11.5240
(    5)   -.7016   (    6)    0.0000   (   7)   -15.0000
```

```
VOLTAGE SOURCE CURRENTS
  NAME          CURRENT

VIN        0.000E+00
VCC       -6.364E-04
TOTAL POWER DISSIPATION 9.55E-03 WATTS
```

```
****      OPERATING POINT INFORMATION            TEMPERATURE = 27.000 DEG C
**** BIPOLAR JUNCTION TRANSISTORS
```

```
NAME            Q1
MODEL           QM
IB              -3.16E-06
IC              -3.48E-04
VBE             -7.26E-01
VBC              1.01E+01
VCE             -1.08E+01
BETADC           1.10E+02
GM               1.34E-02
RPI              8.19E+03
RX               5.00E+00
RO               3.17E+05
CBE              3.39E-12
CBC              2.11E-13
CBX              0.00E+00
CJS              1.00E-12
BETAAC           1.10E+02
FT               5.94E+08
```

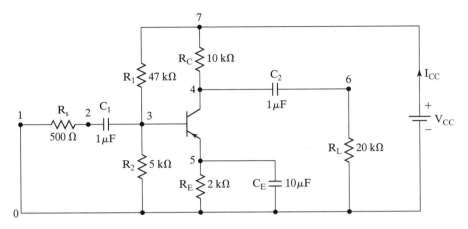

FIGURE 8.19

Equivalent circuit for the transient analysis bias point.

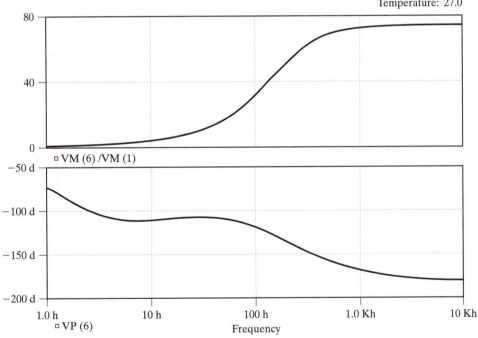

FIGURE 8.20

Frequency response for Example 8.4.

The frequency and transient responses are shown in Figs. 8.20 and 8.21, respectively.

Note: The output signal of an amplifier is an amplified version (inverters) of the input signal. The collector voltage composed of the output signal and a dc biasing voltage. Capacitors C_1 and C_2 isolate the load R_L and input signal v_s from the dc biasing circuit, consisting of R_1, R_2, R_C, and R_E.

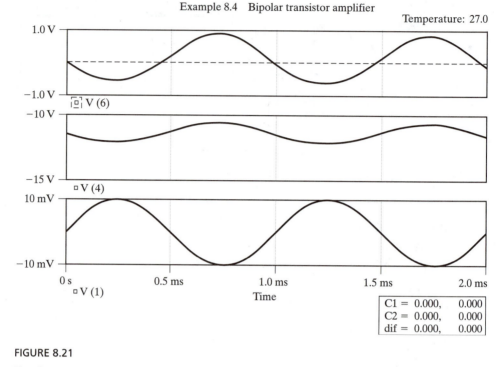

FIGURE 8.21

Transient response for Example 8.4.

Example 8.5:   Transient and frequency response of a BJT amplifier

If the transistor in Fig. 8.15 is replaced by the equivalent circuit of Fig. 8.22, repeat Example 8.4. There is no need to print the details of the operating point.

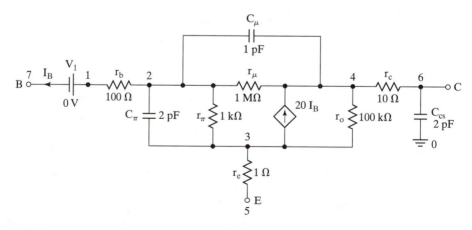

FIGURE 8.22

Subcircuit for PNP bipolar transistor.

Solution The PSpice schematic is shown in Fig. 8.23(a) and the subcircuit model of the transistor is shown in Fig. 8.23(b). The listing of the circuit file follows.

Example 8.5 **Bipolar transistor amplifier**

```
▲ *  Input voltage is 10 mV peak for ac analysis and for transient response:
  *  It is 10 mV peak at 1 kHz with zero-offset value.
  VIN  1  0  AC  1mV SIN(0 0.01 1KHZ)
  VCC  0  7  DC  15V
▲▲ RS  1  2  500
  R1   7  3  47K
  R2   3  0  5K
  RC   7  4  10K
  RE   5  0  2K
  RL   6  0  20K
  C1   2  3  1UF
  C2   4  6  1UF
  CE   5  0  10UF
  *  Calling subcircuit for transistor model TRANS
  XQ1 4  3  5    TRANS
  * Subcircuit definition for TRANS
  .SUBCKT TRANS 6 7 5
  RB   1  2  100
  RE   3  5  1
  RC   4  6  10
  RPI  2  3  1K
  CPI  2  3  2PF
  RU   2  4  1MEG
  CU   2  4  1PF
  RO   4  3  100K
  CCS 6  0  2PF
  * A dummy voltage source of 0 V through which the controlling current flows
  VI  1  7  DC  OV
  * The collector current is controlled by the current through source VI.
  F1  3  4  VI  20
  *  End of subcircuit definition
  .ENDS TRANS
▲▲▲ .OPTIONS  NOPAGE  NOECHO
     *   Transient analysis for 0 to 2 ms with 50-µs increment
     .TRAN  50US  2MS
     * Ac analysis from 1 Hz to 10 KHz with a decade increment and
     * 10 points per decade
     .AC  DEC  10  1HZ  10KHZ
     *  Plot the results of transient analysis for voltages at nodes 4,6, and 1.
     .PLOT  TRANS  V(4)  V(6)  V(1)
     *  Plot the results of ac analysis for the magnitude and phase angle
     *  of voltage at node 6.
     .PLOT  AC  VM(6)  VP(6)
     .PROBE
  .END
```

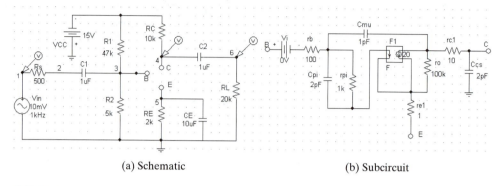

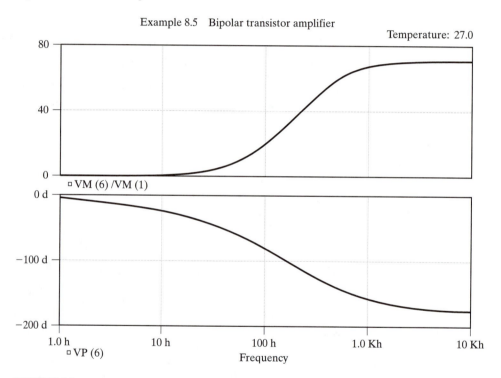

(a) Schematic   (b) Subcircuit

FIGURE 8.23

PSpice schematic for Example 8.5.

FIGURE 8.24

Frequency response for Example 8.5.

The frequency and transient responses are shown in Figs. 8.24 and 8.25, respectively. The .PLOT statements generate graphical plots in the output file. If the .PROBE command is included, there is no need for the .PLOT command.

Note: The isolating capacitors C_1, C_2, and C_3 set the low cut-off frequency, whereas the transistor capacitances C_π and C_μ set the high cut-off frequency. A typical amplifier exhibits a band-pass characteristic.

Example 8.5 Bipolar transistor amplifier

Temperature: 27.0

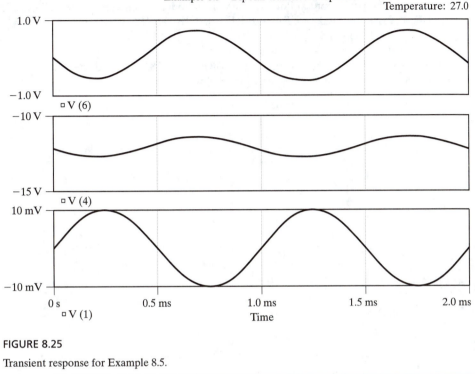

FIGURE 8.25

Transient response for Example 8.5.

Example 8.6: Frequency response of a BJT amplifier

A two-stage bipolar transistor amplifier is shown in Fig. 8.26. The output is taken from Node 9. Plot (a) the magnitude and phase angle of the voltage gain and (b) the magnitude of input

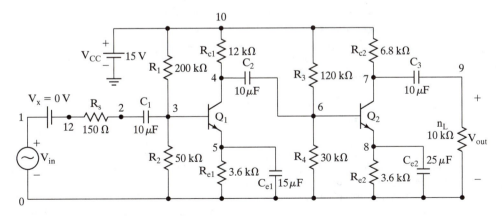

FIGURE 8.26

Two-stage BJT amplifier.

impedance for frequencies from 10 Hz to 10 MHZ with a decade increment and 10 points per decade. The peak input voltage is 1 mV. The model parameters of the BJTs are IS = 2E –16, BF = 50, BR = 1, RB = 5, RC = 1, RE = 0, CJE = 0.4PF, VJE = 0.8, ME = 0.4, CJC = 0.5PF, VJC = 0.8, CCS = 1PF, and VA = 100.

Solution   The PSpice schematic is shown in Fig. 8.27(a) and the setup for ac analysis in Fig. 8.27(b). The listing of the circuit file follows.

Example 8.6   Two-Stage BJT amplifier

```
▲ VCC   10   0   DC   15V
  *       Input voltage is 1 mV peak for frequency response.
  VIN    1   0   AC   1MV
  * A dummy voltage source of 0 V to measure the input current
  VX   1    12   DC   0V
▲▲  RS   12   2    150
    C1   2    3    10UF
    R1   10   3    200K
    R2   3    0    50K
    *   Transistors Q1 and Q2 have model name QM.
    Q1   4   3   5   0   QM
    Q2   7   6   8   0   QM
    RC1  10   4   12K
    RE1  5    0   3.6K
    CE1  5    0   15UF
    C2   4    6   10UF
    R3   10   6   120K
    R4   6    0   30K
    RC2  10   7   6.8K
    RE2  8    0   3.6K
    CE2  8    0   25UF
    C3   7    9   10UF
    RL   9   0   10K
     * Model statement for NPN transistors whose model name is QM
     .MODEL QM NPN (IS=2E—16 BF=50 BR=1 RB=5 RC=1 RE=0 CJE=0.4PF
     +              VJE=0.8 ME=0.4 CJC=0.5PF VJC=0.8 CCS=1PF VA=100)
▲▲▲  * Ac analysis from 10 Hz to 10 MHz with a decade increment and 10
     * points per decade
     .AC   DEC   10   10HZ   10MEGHZ
     .PLOT   AC   VM(9)   VP(9)
     .PROBE
 .END
```

The results of the frequency response are shown in Fig. 8.28. If the .PROBE command is included, there is no need for the .PLOT command.

Note: The isolating capacitors C_1, C_2, and C_3 set the low cut-off frequency, whereas the transistor capacitances C_π and C_μ set the high cut-off frequency. A typical amplifier exhibits a band-pass characteristic. The cut-off frequency is defined as the frequency at which gain is 70.7% of the maximum value.

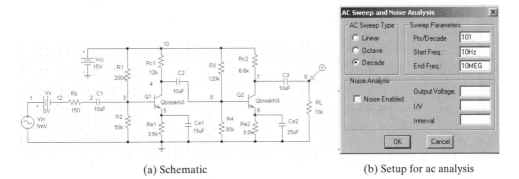

(a) Schematic   (b) Setup for ac analysis

FIGURE 8.27

PSpice schematic for Example 8.6.

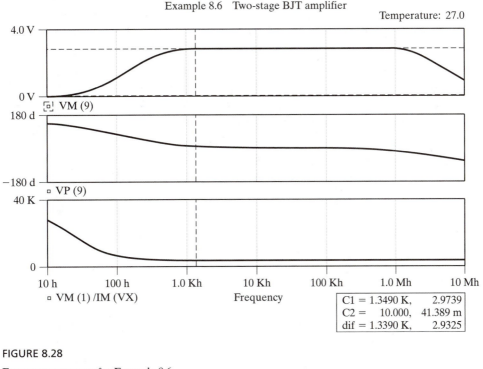

FIGURE 8.28

Frequency response for Example 8.6.

Example 8.7: Frequency response of a two-stage BJT amplifier with shunt-series feedback

A two-stage amplifier with shunt-series feedback is shown in Fig. 8.29. Plot (a) the magnitude and phase angle of voltage gain and (b) the magnitude of the input impedance if the frequency is varied from 10 Hz to 10 MHz in decade steps with 10 points per decade. The peak input voltage is 10 mV. The model parameters of the BJTs are IS = 2E − 16, BF = 50, BR = 1, RB = 5,

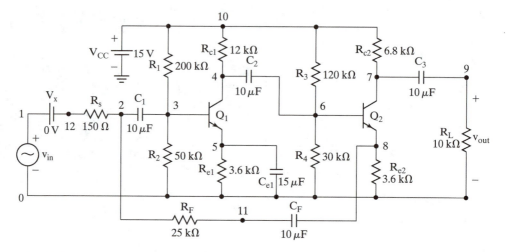

FIGURE 8.29

Two-stage BJT amplifier with shunt-series feedback.

RC = 1, RE = 0, CJE = 0.4PF, VJE = 0.8, ME = 0.4, CJC = 0.5PF, VJC = 0.8, CCS = 1PF, and VA = 100.

Solution   The PSpice schematic is shown in Fig. 8.30(a) and the setup for ac analysis in Fig. 8.30(b).

Example 8.7   Two-stage BJT amplifier with shunt-series feedback

```
▲ VCC   10  0   DC   15V
  *     Input  voltage of 10 mV peak for frequency response
  VIN   1   0   AC   10MV
  *     A dummy voltage source of 0 V
  VX    1   12  DC   0V
▲▲ RS   12  2     150
   C1   2   3     10UF
   R1   10  3     200K
   R2   3   0     50K
   *    Substrate of BJTs with model QM is connected to node 0.
   Q1   4   3   5   0   QM
   Q2   7   6   8   0   QM
   RC1  10  4     12K
   RE1  5   0     3.6K
   CE1  5   0     15UF
   C2   4   6     10UF
   R3   10  6     120K
   R4   6   0     30K
   RC2  10  7     6.8K
   RE2  8   0     3.6K
   CF   11  8     10UF
   RF   2   11    25K
   C3   7   9     10UF
   RL   9   0     10K
   *  Model statement for NPN transistors with model name QM
   .MODEL QM NPN (IS=2E-16 BF=50 BR=1 RB=5 RC=1 RE=0 CJE=0.4PF
   +             VJE=0.8 ME=0.4 CJC=0.5PF VJC=0.8 CCS=1PF VA=100)
```

```
▲▲▲  *  Ac analysis for 10 Hz to 10 MHz with a decade increment and 10
     *  points per decade
     .AC  DEC  10  10  10MEGHZ
     .PLOT  AC  VM(9)  VP(9)
.END
```

The results of the frequency response are shown in Fig. 8.31. If the .PROBE command is included, there is no need for the .PLOT command.

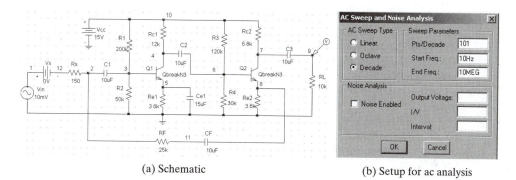

(a) Schematic (b) Setup for ac analysis

FIGURE 8.30

PSpice schematic for Example 8.7.

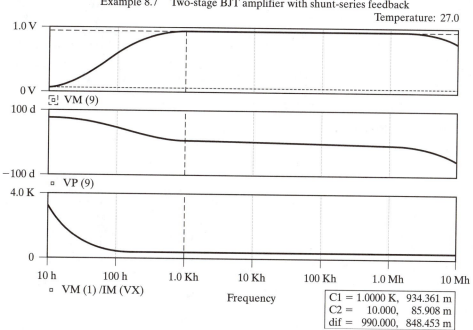

Example 8.7 Two-stage BJT amplifier with shunt-series feedback

Temperature: 27.0

C1 = 1.0000 K,	934.361 m
C2 = 10.000,	85.908 m
dif = 990.000,	848.453 m

FIGURE 8.31

Frequency response for Example 8.7.

Note: Feedback increases the bandwidth, but reduces the gain by the same amount. The series-shunt feedback particularly increases the input resistance and reduces the output resistance by the same proportion.

Example 8.8: Transient response of an astable multivibrator

An astable multivibrator is shown in Fig. 8.32. The output is taken from Nodes 1 and 2. Plot the transient responses of voltages at Nodes 1 and 2 from 0 to 15 μs in steps of 0.1 μs. The initial voltages of nodes 1 and 3 are 0. The CPU time should be limited to 1.22E2 s. The model parameters of the BJTs are IS = 2E − 16, BF = 50, BR = 1, RB = 5, RC = 1, RE = 0, TF = 0.2NS, and TR = 5NS.

Solution Due to the regenerative nature of the circuit, the solution may not converge, and the simulation will continue for a very long time. The CPU time is limited so that the circuit does not run for a long time. The run time should be less than the CPU time itself if the circuit converges. The PSpice schematic is shown in Fig. 8.33. The listing of the circuit file follows.

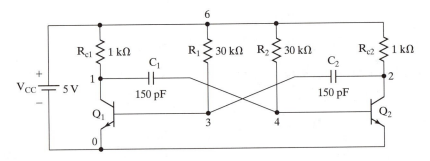

FIGURE 8.32

Astable multivibrator.

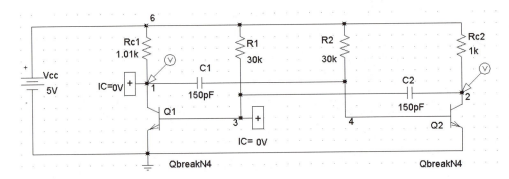

FIGURE 8.33

PSpice schematic for Example 8.8.

Example 8.8   Astable multivibrator

```
▲ VCC   6   0   DC   5V
▲▲ RC1   6   1    1K
   RC2   6   2    1K
   R1    6   3    30K
   R2    6   4    30K
   C1    1   4    150PF
   C2    2   3    150PF
   *   Q1 and Q2 with model QM and substrate connected to ground by
   +default
   Q1    1   3   0   QM
   Q2    2   4   0   QM
   *   Model statement for NPN transistors
   .MODEL QM NPN (IS=2E−16 BF=50 BR=1 RB=5 RC=1 RE=0 TF=0.2NS TR=5NS)
   *   CPU time is limited.
   .OPTIONS  NOPAGE  NOECHO  CPTIME=1.2E2
   *    Node voltages are set to defined values to break the tie-in
   +condition.
   . NODESET V(1)=0 V(3)=0
▲▲▲ *  Transient analysis from 0 to 10 µs with 0.1-µs increment
    .TRAN/OP  0.1US  15US
    *    Plot the results of transient analysis: voltages at nodes 1 and 2.
    .PLOT TRAN V(1) V(2)
    .OPTIONS ABSTOL=1.0N RELTOL=10M VNTOL=1M ITL5=40000
    .PROBE
  .END
```

Example 8.8   Astable multivibrator

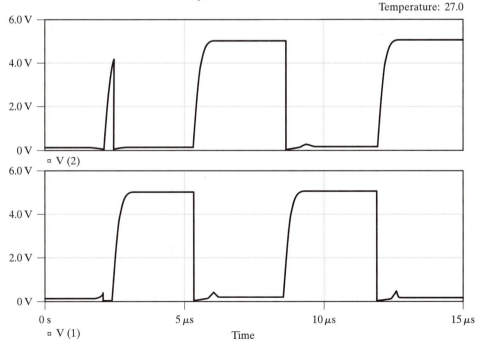

FIGURE 8.34

Transient responses for Example 8.8.

The transient responses are shown in Fig. 8.34. If the .PROBE command is included, there is no need for the .PLOT command.

Note: An astable multivibrator is a free-running oscillator.

Example 8.9:   Transient response and dc transfer characteristic of a simple TTL inverter

A TTL inverter circuit is shown in Fig. 8.35(a). The output is taken from node 4. Plot the dc transfer characteristic V(4) versus V_{in} if the input voltage is varied from 0 to 2 V with a step of 0.01 V. If the input is a pulse voltage with a period of 60 μs, as shown in Fig. 8.35(b), plot the transient response of voltage at Node 4 from 0 to 80 ns in steps of 1 ns. The model parameters of the BJTs are   BF = 50, RB = 70, RC = 40, CCS = 2PF, TF = 0.1NS, TR = 10NS, VJC = 0.85, and VAF = 50.

Solution   The PSpice schematic is shown in Fig. 8.36. The listing of the circuit file follows.

Example 8.9   TTL inverter

```
▲ *   Pulsed input voltage
    VIN  1  0  PULSE (0  5  1NS  1NS  1NS  38NS  60NS)
    VCC  6  0  DC 5V
▲▲ *  BJTs with model QN and substrate connected to ground by default
    Q1  3  2  1  QN
    Q2  4  3  5  QN
    Q3  4  5  0  QN
    *  Model for NPN BJTs with model QN
    .MODEL QN NPN (BF=50 RB=70 RC=40 CCS=2PF TF=0.1NS TR=10NS VJC=0.85
    +VAF=50)
    R1  6  2  4K
    R2  5  0  1K
    R3  6  4  1K
▲▲▲ * DC sweep for 0 to 2 with 0.01-V increment
    .DC  VIN  0  2  0.01
    *  Transient analysis for 0 to 80 ns with 1-ns increment
    .TRAN  0.5NS  80NS
    *  Plot the results of dc sweep: voltage at node 4 versus VIN.
    .PLOT  DC  V(4)
    *  Plot the results of transient analysis voltage at nodes 4 and 1.
    .PLOT  TRAN  V(4)  V(1)
    .PROBE
.END
```

The results of the dc sweep and transient analyses are shown in Figs. 8.37 and 8.38, respectively. If the .PROBE command is included, there is no need for the .PLOT command.

Notes:

1. The transition, as shown in Fig. 8.37, from the high-state at 4 V (or higher) does not occur until at an input voltage of 1.442 V. The logic high-output is at $V_{oH} \geq 4$ V and the logic low-output is $V_{oL} \leq 1$ V (approximately).
2. Due to the BJT storage time and internal capacitances, the output voltage rises exponentially rather than a sharp rise and a fall similar to the input voltage.

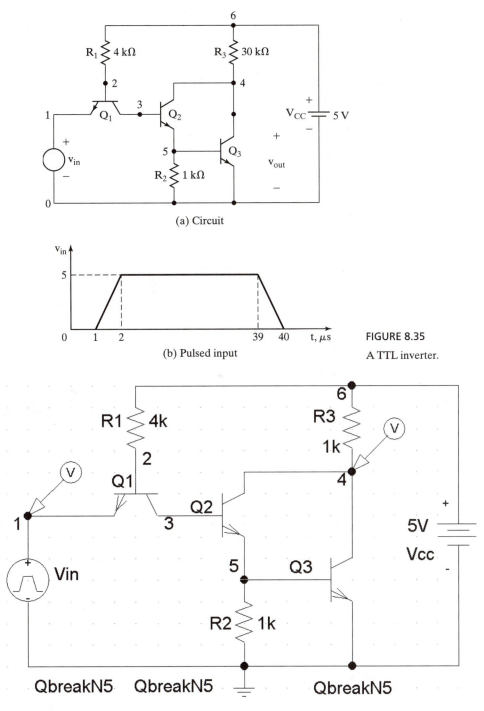

(a) Circuit

(b) Pulsed input

FIGURE 8.35

A TTL inverter.

FIGURE 8.36

PSpice schematic for Example 8.9.

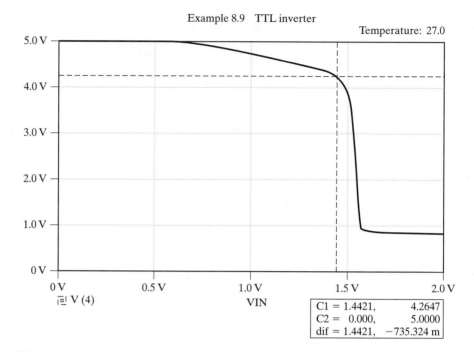

Example 8.9 TTL inverter

Temperature: 27.0

C1 = 1.4421,	4.2647
C2 = 0.000,	5.0000
dif = 1.4421,	−735.324 m

FIGURE 8.37

Dc transfer characteristic for Example 8.9.

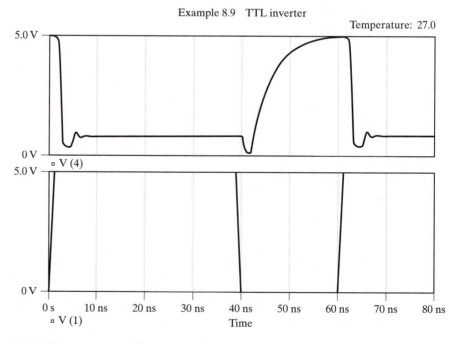

Example 8.9 TTL inverter

Temperature: 27.0

FIGURE 8.38

Transient response for Example 8.9.

Example 8.10: Transient response and dc transfer characteristic of a TTL inverter

A TTL inverter circuit is shown in Fig. 8.39(a). Plot the dc transfer characteristic between nodes 1 and 9 for values of V_{in} in the range of 0 to 2 V in steps of 0.01 V. If the input is a pulsed waveform of period 80 μs, as shown in Fig. 8.39(b), plot the transient response from 0 to 100 ns with steps of 1 ns. The model parameters of the BJTs are BF = 50, RB = 70, RC = 40, TF = 0.1NS, TR = 10NS, VJC = 0.85, and VAF = 50. The model parameters of the diodes are RS = 40 and TT = 0.1NS.

Solution The PSpice schematic is shown in Fig. 8.40. The listing of the circuit file follows.

Example 8.10 TTL inverter

```
▲  *  Pulse input voltage
   VIN  1  0  PULSE (0 3.5V 1NS 1NS 1NS 38NS 80NS)
   VCC  13  0  5V
▲▲ RS  1  2  50
   RB1 13  3  4K
   RC2 13  5  1.4K
   RE2  6  0  1K
   RC3 13  7  100
   RB5 13  10 4K
   *  BJTs with model QNP and substrate connected to ground by default
   Q1  4  3  2  QNP
   Q2  5  4  6  QNP
   Q3  7  5  8  QNP
   Q4  9  6  0  QNP
   Q5  11 10 9  QNP
   *  Diodes with model DIODE
   D1  8  9    DIODE
   D2  11 12   DIODE
   D3  12 0    DIODE
   *  Model of NPN transistors with model QNP
   .MODEL QNP NPN (BF=50 RB=70 RC=40 TF=0.1NS TR=10NS VJC=0.85 VAF=50)
   *  Diodes with model DIODE
   .MODEL DIODE D (RS=40 TT=0.1NS)
▲▲▲ *  Dc sweep from 0 to 2 V with 0.01-V increment
   .DC   VIN  0  2  0.01
   *  Transient analysis from 0 to 80 ns with 1-ns increment
   .TRAN  1NS  100NS
   *  Plot the results of dc sweep: voltage at node 9 against VIN.
   .PLOT  DC  V(9)
   *  Plot the results of transient analysis: voltage at node 9.
   .PLOT  TRAN  V(9)
   .PROBE
   .END
```

The results of the dc sweep and transient analyses are shown in Figs. 8.41 and 8.42, respectively. If the .PROBE command is included, there is no need for the .PLOT command.

Notes:

1. The transition, as shown in Fig. 8.41, from the high-state at 2.54 V (or higher) does not occur until at an input voltage of 1.449 V. The logic high-output is at $V_{oH} \geq 2.5$ V and the logic low-output is $V_{oL} \leq 0.2$ V (approximately).
2. The output voltage as shown in Fig. 8.42 follows the input more closely than the simple inverter as shown in Fig. 8.38.

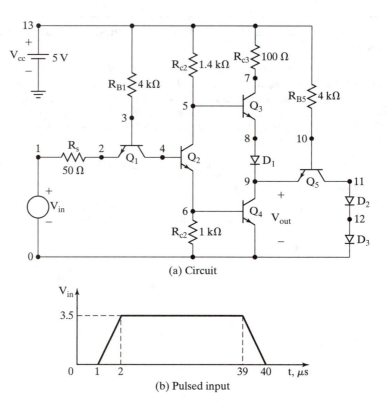

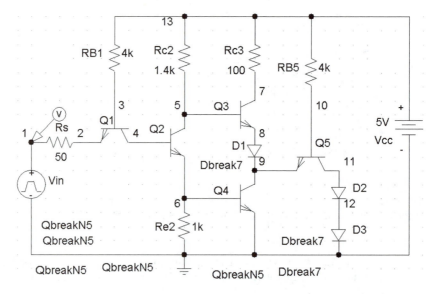

FIGURE 8.39

TTL inverter.

FIGURE 8.40

PSpice schematic for Example 8.10.

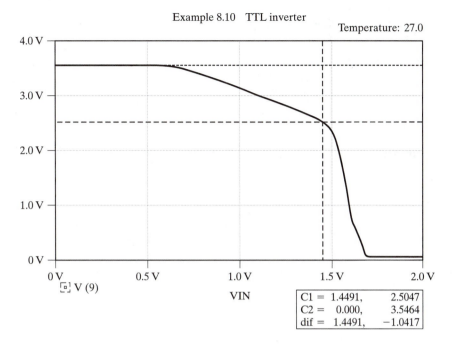

FIGURE 8.41

Dc transfer characteristic for Example 8.10.

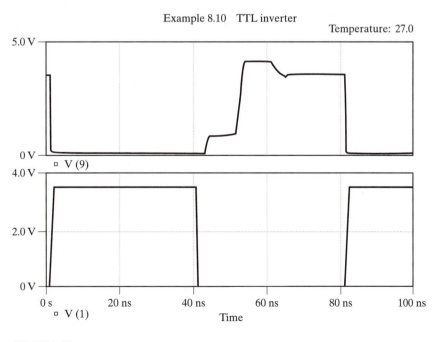

FIGURE 8.42

Transient response for Example 8.10.

Example 8.11: Transient response and dc transfer characteristic of an OR/NOR gate

The circuit diagram of an OR/NOR gate is shown in Fig. 8.43(a). The inputs to nodes 1 and 4 are pulses of period 60 μs, as shown in Fig. 8.43(b). Plot the transient responses of voltages at nodes 12, 13, and 1 from 0 to 100 ns in steps of 1 ns. The model parameters of the BJTs are BF = 50, RB = 70, RC = 40, TF = 0.1NS, TR = 10NS, VJC = 0.85, and VAF = 50. The parameters of the diodes are RS = 40 and TT = 0.1NS.

Solution The PSpice schematic is shown in Fig. 8.44. The listing of the circuit file follows.

Example 8.11 OR/NOR logic gate

```
▲ *  Pulsed input voltages
   VA  1  0   PULSE (0   -5   1NS   1NS   1NS   38NS   60NS)
   VB  4  0   PULSE (0   -5   1NS   1NS   1NS   38NS   60NS)
   VEE  0  14  DC  5.2V
▲▲ *   BJTs with model QN and substrate connected to ground by default
   Q1  2  4   3   QN
   Q2  7  8   3   QN
   Q3  2  1   3   QN
   Q4  0  9   8   QN
   Q5  0  2   13  QN
   Q6  0  7   12  QN
   .MODEL QN NPN (BF=50 RB=70 RC=40 TF=0.1NS TR=10NS VJC=0.85 VAF=50)
   *  Diodes with model DIODE
   D1   9    10  DIODE
   D2   10   11  DIODE
   .MODEL DIODE  D (RS=40 TT=0.1NS)
   R1   0   2   220
   R2   0   7   245
   R3   3   14   779
   R4   4   14   15K
   R5   1   14   15K
   R6   8   14   6.1K
   R7   0   9   907
   R8   11   14   4.98K
   RLO  12  14  10K
   RLN  13  14  10K
▲▲▲ *  Transient analysis from 0 to 80 ns with 1-ns increment
   .TRAN  0.5NS  120NS
   * Plot the results of transient analysis: voltages at nodes 12 and +13.
   .PLOT TRAN V(12) V(13) V(1)
   .PROBE
 .END
```

The results of the transient analysis are shown in Fig. 8.45. If the .PROBE command is included, there is no need for the .PLOT command.

Note: Any transit time and device capacitance will cause switching transients on the switching voltages as shown in Fig. 8.45.

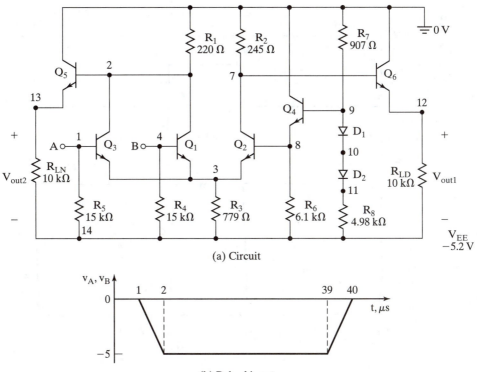

(a) Circuit

(b) Pulsed input

FIGURE 8.43

OR/NOR logic gate.

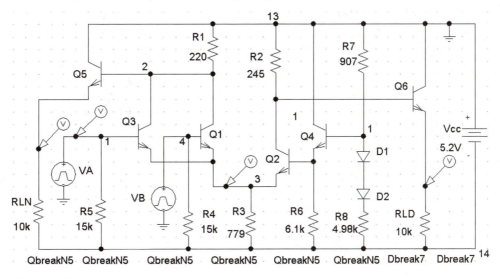

FIGURE 8.44

PSpice schematic for Example 8.11.

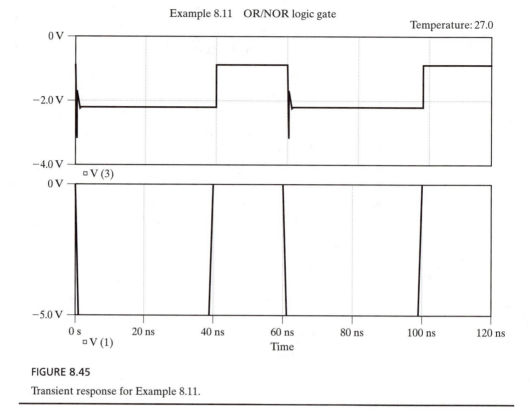

FIGURE 8.45

Transient response for Example 8.11.

Example 8.12: Finding the output current and resistance of a BJT active current source

The circuit diagram of a BJT current source is shown in Fig. 8.46. Use (a) the transfer function analysis to find the output current I_o and the output resistance R_o for a nominal current gain of $\beta_F = 255.9$, (b) the dc biasing currents, (c) the parametric sweep to find the output current I_O for current gains $\beta_F = 100, 400$, and (d) the sensitivity of the output current I_O with respect to the current gain β_F. Use the PSpice model parameters of the transistor 2N2222.

Solution The analysis types are selected from the analysis setup menu as shown in Fig. 8.47(a).

(a) The setup for the transfer function analysis as shown in Fig. 8.47(b) gives the output resistance and the output current whose values are found from the output file as follows:

I(V_Vy)/V_VCC = 1.736E-07

$$I_O = 1.736 \times 10^{-7} \times V_{CC} = 1.736 \times 10^{-7} \times 30 = 5.21 \ \mu A$$

INPUT RESISTANCE AT V_VCC = 5.730E+06

OUTPUT RESISTANCE AT I(V_Vy) = 7.585E+08 $R_O = 758.5 \ \text{M}\Omega$

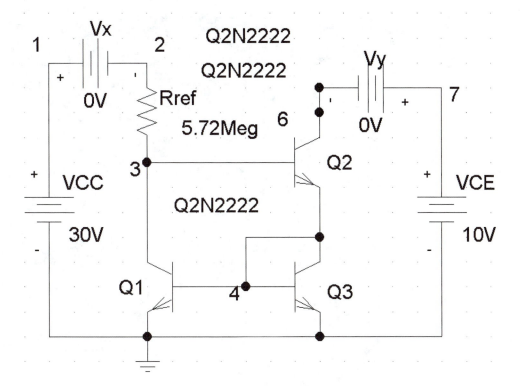

FIGURE 8.46

PSpice schematic for Example 8.12 [Ref. 14].

(b) From the output file, we get the currents through the voltage sources as follows:

Through V_Vx   5.067E-06   (5.067 μA)

Through V_Vy   5.044E-06   (5.044 μA)

(c) The setup for the parametric function analysis as shown in Fig. 8.47(d) gives the output resistance and the output current I_O for different values of the current gain β_F. From the output file, we get as follows:

For $\beta_F = 100$

I(V_Vy)/V_VCC = 1.738E-07

$$I_O = 1.738 \times 10^{-7} \times V_{CC} = 1.738 \times 10^{-7} \times 30 = 5.214 \ \mu A$$

INPUT RESISTANCE AT V_VCC = 5.730E + 06

OUTPUT RESISTANCE AT I(V_Vy) = 5.097E + 08 $R_O = 509.7 \ M\Omega$

For $\beta_F = 400$

I(V_Vy)/V_VCC = 1.736E-07

$$I_O = 1.736 \times 10^{-7} \times V_{CC} = 1.736 \times 10^{-7} \times 30 = 5.21 \ \mu A$$

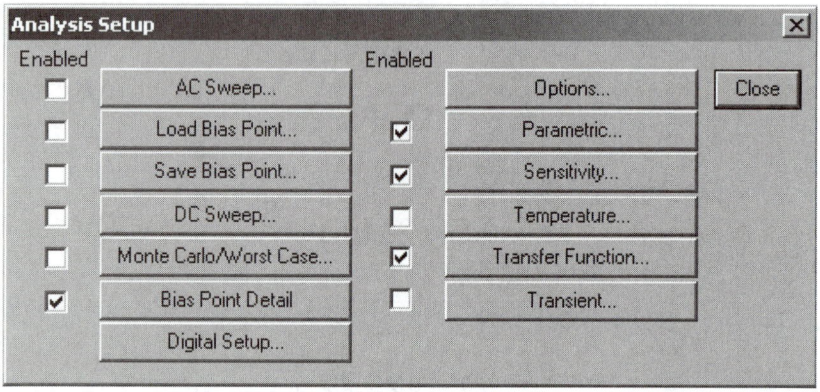

(a) Analysis setup

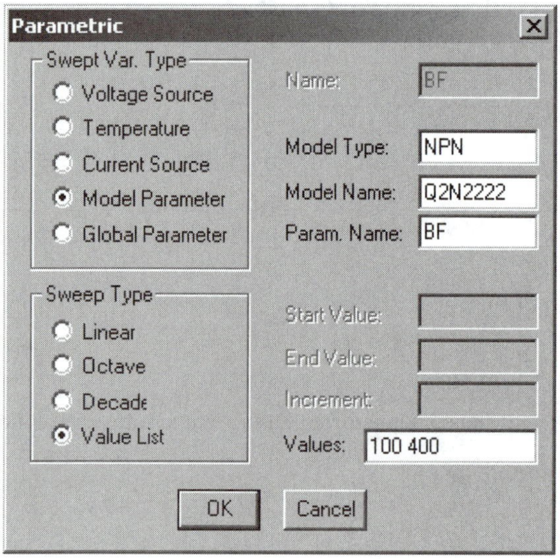

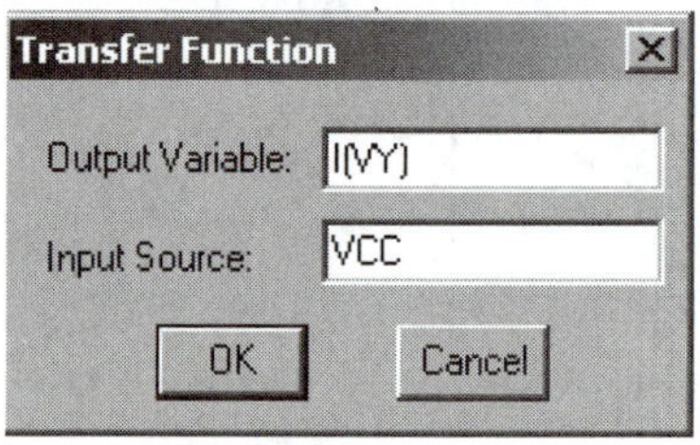

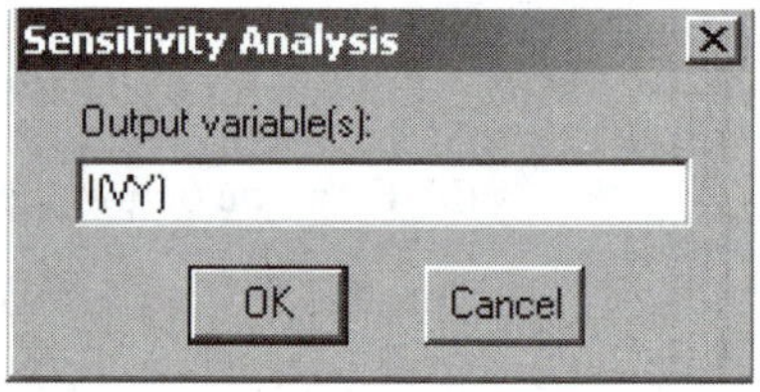

(d) Parametric-variation of β_F

(b) Transfer function

(c) Sensitivity analysis

FIGURE 8.47

Analysis setup.

INPUT RESISTANCE AT V_VCC = 5.730E + 06

OUTPUT RESISTANCE AT I(V_Vy) = 8.554E + 08 R_O = 855.4 MΩ

(d) The setup for the sensitivity analysis is shown in Fig. 8.47(c). From the PSpice output file, we get the sensitivity of the output current with respect to the current gain β_F as follows:

For Q1   −7.440E-11   amp/unit

For Q2   1.354E-10   amp/unit

For Q3   −7.411E-11   amp/unit

Notes: (a) $\beta_F = 255.9$. Using the KVL on the input side, we get

$$V_{CC} = I_{ref}R_{ref} + V_{BE2} + V_{BE1}$$

which gives the input reference current through R_{ref} as

$$I_{ref} \approx I_o = \frac{V_{CC} - V_{BE2} - V_{BE1}}{Rref} = \frac{30 - 0.7 - 0.7}{5.72 \times 10^6} = 5\ \mu A \text{ (Compared to 5.067 } \mu A \text{ by PSpice)}$$

(b) The effect of the current gain β_F on the output resistance and the output current are not significant. However, this example illustrates how to apply the paramertic variation of a model parameter on an output qualtity.

SUMMARY

The statements for BJTS are

```
Q⟨name⟩ NC  NB  NE  NS  QNAME    [(area) value]
.MODEL   QNAME   NPN (P1=V1 P2=V2 P3=V3........PN=VN)
.MODEL   QNAME   PNP (P1=V1 P2=V2 P3=V3........PN=VN)
```

REFERENCES

[1]   Phillip E. Allen and D. R. Holberg, *CMOS Analog Circuit*. New York: Holt, Rinehart and Winston, 1987.

[2]   Donald T. Comer and David J. Comer, "A new amplifier circuit with both practical and tutorial value." *IEEE Transactions on Education*, Vol. 43 (1), pp. 25–29, February 2000.

[3]   J. J. Ebers and J. J. Moll, "Large-signal behavior of junction transistors." *Proc. IRE*, Vol. 42, December 1954, pp. 1161–1172.

[4]   Ian Getreu, *Modeling the bipolar transistor—Part #062-2841-00*. Beaverton, Oregon: Tektronix, Inc., 1979.

[5]   A. S. Grove, *Physics and Technology of Semiconductor Devices*. New York: Wiley, 1967.

[6]   H. K. Gummel and H. C. Poon, "An integral charge control model for bipolar transistors." *Bell System Technical Journal*, Vol. 49, January 1970, pp. 827–852.

[7]   Mark N. Horenstein, *Microelectronic Circuits and Devices*, 2d ed. Upper Saddle River, New Jersey: Prentice Hall, 1996.

[8]   Roger T. Howe and C. G. Sodini, *Microelectronics—An Integrated Approach*. Upper Saddle River, New Jersey: Prentice Hall, 1997.

[9]   Richard Jaeger, *Microelectronic Circuit Design*. New York: McGraw-Hill, 1997.

[10]   David A. Johns and K. Martin, *Analog Integrated Circuit Design*. New York: John Wiley & Sons, 1997.

[11]   L. W. Nagel, *SPICE2—A computer program to simulate semiconductor circuits*, Memorandum no. ERL-M520, May 1975, Electronics Research Laboratory, University of California, Berkeley.

[12]   S. Natarajan, "An effective approach to obtain model parameters for BJTs and FETs from data books." *IEEE Transactions on Education*, Vol. 35, No. 2., 1992, pp. 164–169.

[13]   Donald A. Neamen, *Electronic Circuit Analysis and Design*. New York: McGraw-Hill, 2001.

[14]   M. H. Rashid, *Microelectronic Circuits: Analysis and Design*. Boston: PWS Publishing, 1999, Chapter 13.

[15] M. H. Rashid, *SPICE for Power Electronics and Electric Power*. Englewood Cliffs, New Jersey: Prentice Hall, 1993, Chapter 11.

[16] R. B. Schilling, "A bipolar transistor model for device and circuit design." *RCA Review*, Vol. 32, September 1971, pp. 339–371.

[17] Adel S. Sedra and K. C. Smith, *Microelectronic Circuits*, 4th ed. New York: Oxford University Press, 1997.

[18] Richard Spencer and Mohammed Ghausi, *Introduction to Electronic Circuit Design*. Upper Saddle River, New Jersey: Prentice Hall, 2003.

PROBLEMS

8.1 For Example 8.2, calculate the coefficients of a Fourier series for the output voltage.

8.2 For Example 8.6, calculate the equivalent input and output noise.

8.3 For Example 8.7, plot the output impedance and the current gain.

8.4 For Fig. 8.39, calculate the input and output noise for frequencies from 1 Hz to 10 kHz.

8.5 For Fig. 8.39, calculate and plot the frequency response of the output voltage from 10 Hz to 10 MHz in decade steps with 10 points per decade. Assume that the peak input voltage is 5 V. The model parameters of the BJTs are BF = 50, RB = 70, RC = 40, TF = 0.1NS, TR = 10NS, VJC = 0.85, and VAF = 50. The model parameters of the diodes are RS = 40 and TT = 0.1NS.

8.6 For the circuit in Fig. P8.6, calculate and plot (a) the magnitude and phase angle of voltage gain, (b) the magnitude of input impedance, and (c) the magnitude of output impedance. The frequency is varied from 1 Hz to 10 MHz in decade steps with 10 points per decade. The peak input voltage is 10 mV. The model parameters of the BJT are IS = 2E –16, BF = 50, BR = 1, RB = 5, RC = 1, RE = 0, CJE = 0.4PF, VJE =0.8, ME = 0.4, CJC = 0.5PF, VJC = 0.8, CCS = 1PF, and VA = 100.

8.7 Repeat Problem 8.6 for the circuit in Fig. P8.7.

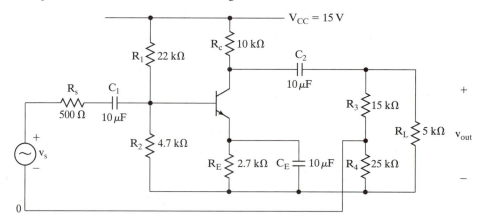

FIGURE P8.6

8.8 Repeat Problem 8.6 for the circuit in Fig. P8.8.

8.9 Repeat Problem 8.6 for the circuit in Fig. P8.9.

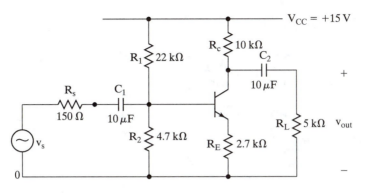

FIGURE P8.7

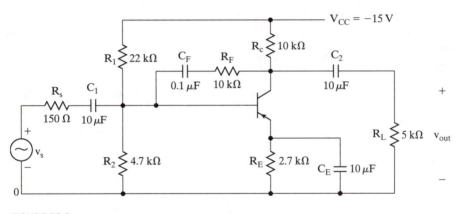

FIGURE P8.8

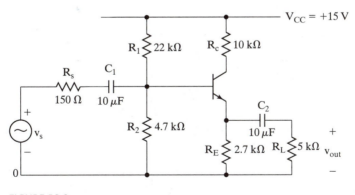

FIGURE P8.9

8.10 Repeat Problem 8.6 for the circuit in Fig. P8.10. Calculate the input and output noise.

8.11 For the circuit in Fig. P8.11, calculate and print the dc transfer function (the voltage gain, the input resistance, and the output resistance) between the output current and the input

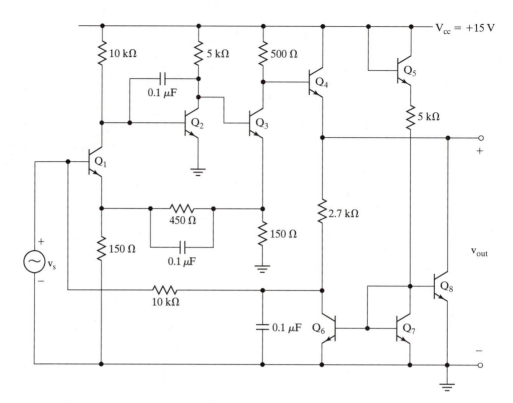

FIGURE P8.10

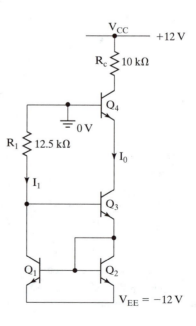

FIGURE P8.11

voltage V_{EE}. The model parameters of the BJTs are BF = 100, BR = 1, RB = 5, RC = 1, RE = 0, VJE = 0.8, and VA = 100.

8.12 For the circuit in Fig. P8.12, calculate and print the voltage gain, the input resistance, and the output resistance. The input voltage is 5 V dc. The model parameters of the BJTs are BF = 100, BR = 1, RB = 5, RC = 1, RE = 0, VJE = 0.8, and VA = 100.

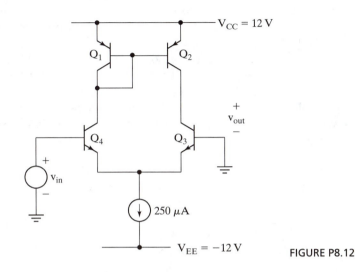

FIGURE P8.12

8.13 Use PSpice to perform a Monte Carlo analysis for six runs and for the dc sweep of Problem 8.11. The model parameter is $R = 1$ for resistors. The circuit and transistor parameters having uniform deviations are

$$R_1 = 12.5\,\text{k}\Omega \pm 5\%$$
$$R_C = 10\,\text{k}\Omega \pm 15\%$$
$$B_F = 100 \pm 50$$
$$V_A = 100 \pm 20$$

  (a) The greatest difference from the nominal run is to be printed.
  (b) The maximum value of the output voltage is to be printed.
  (c) The minimum value of the output voltage is to be printed.
  (d) The first occurrence of the output voltage crossing below 5 V is to be printed.

8.14 Use PSpice to perform the worst-case analysis for Problem 8.13.

8.15 Use PSpice to perform a Monte Carlo analysis for five runs and for the dc sweep of Problem 8.12. The transistor parameters having uniform deviations are

$$B_F = 100 \pm 50$$
$$V_A = 100 \pm 20$$

  (a) The greatest difference from the nominal run is to be printed.
  (b) The maximum value of the output voltage is to be printed.
  (c) The minimum value of the output voltage is to be printed.

8.16 Use PSpice to perform the worst-case analysis for Problem 8.15.

8.17 Use PSpice to find the worst-case minimum and maximum output voltages $V_{out(max)}$ and $V_{out(min)}$ for the circuit as shown in Fig. P8.12. Assume uniform tolerances of $\pm 20\%$ for the current gain β_F of all transistors and an operating temperature of 25°C. Also assume that $V_{in} = 2\,V$.

8.18 The values of the Comer [2] circuit elements in Fig. P8.18 are $R = 20\,k\Omega$, $R_A = 19\,k\Omega$, $R_B = 1\,k\Omega$, $R_X = R_Y = 20\,k\Omega$, $R_L = 10\,k\Omega$, $C = 5\,\mu F$ and $V_{CC} = -V_{EE} = 15\,V$. All transistors are identical and have a current gain of $\beta_F = 100$ (nominal). The input signal is $v_s = 0.1 \times \sin(2000\pi t)$. Use PSpice to plot the instantaneous output voltage $v_o(t)$ from 0 to 2 ms with an increment of 10 μs.

8.19 Use PSpice to plot the frequency response of the circuit in Problem 8.18 over the frequency range from 1 Hz to 100 kHz with a decade increment and 10 points per decade.

8.20 Use PSpice to plot the worst-case minimum and maximum values of the instantaneous output voltage $v_o(t)$ for Problem 8.18 if all circuit elements have uniform tolerances of $\pm 10\%$. The transistor β_F can also change by $\pm 20\%$. Assume that $V_{in} = 0.1\,V$ (*peak*) and the operating temperature is 25°C.

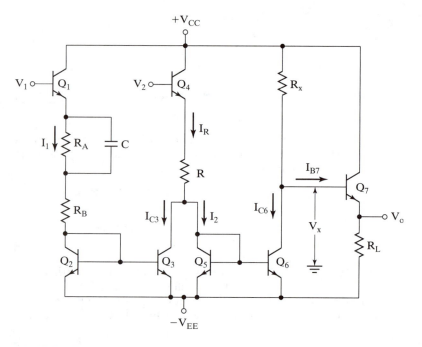

FIGURE P8.18 [Ref. 2, Comer]

CHAPTER 9

Field-Effect Transistors

After completing this chapter, students should be able to

- Mode SPICE FETs and identify the SPICE model parameters
- Derive the model parameters from the data sheets of FETs
- Plot the FET characteristics in SPICE
- Find the dc biasing conditions of FET amplifiers
- Perform the dc analysis, the transient analysis, and the ac analysis of FET amplifiers and circuits.

9.1   INTRODUCTION

A field-effect transistor (FET) may be specified by a device statement. PSpice generates complex models for FETs. These models are quite complex and incorporate an extensive range of device characteristics (e.g., dc and small-signal behavior, temperature dependency, and noise generation). If such complex models are not necessary, users can ignore many model parameters, and PSpice assigns default values to the parameters. The FETs are of three types:

> Junction field-effect transistors (JFETs)
> Metal-oxide silicon field-effect transistors (MOSFETs)
> Gallium arsenide MESFETs

9.2   JUNCTION FIELD-EFFECT TRANSISTORS

The PSpice JFET model is based on the FET model of Schichman and Hodges [10]. The model of an *n*-channel JFET is shown in Fig. 9.1. The small-signal model and the static (or dc) model, which are generated by PSpice, are shown in Figs. 9.2 and 9.3, respectively. The model parameters for a JFET device and the default values assigned by PSpice are given in Table 9.1. The model equations of JFETs that are used by PSpice are described in Schichman and Hodges [10], Vladimirescu and Liu [12], and the *PSpice Manual* [7].

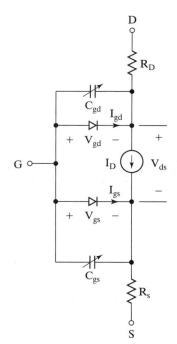

FIGURE 9.1

PSpice *n*-channel JFET.

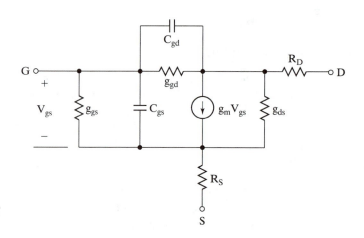

FIGURE 9.2

Small-signal *n*-channel JFET
model.

The model statement of an *n*-channel JFET has the general form

    .MODEL   JNAME   NJF   (P1=A1 P2=A2 P3=A3 . . . PN=AN)

and for a *p*-channel JFET, the statement has the form

    .MODEL   JNAME   PJF   (P1=A1 P2=A2 P3=A3 . . . PN=AN)

where JNAME is the model name; it can begin with any character and its word size is
normally limited to eight characters. NJF and PJF are the type symbols of *n*-channel

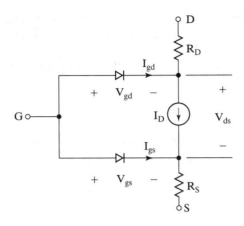

FIGURE 9.3

Static *n*-channel JFET model.

TABLE 9.1 Model Parameters of JFETS

Name	Area	Model Parameters	Units	Default	Typical
VTO		Threshold voltage	Volts	−2	−2
BETA	*	Transconductance coefficient	Amps/Volts^2	1E−4	1E−3
LAMBDA		Channel-length modulation	Volts^{-1}	0	1E−4
RD	*	Drain ohmic resistance	Ohms	0	100
RS	*	Source ohmic resistance	Ohms	0	100
IS	*	Gate *p–n* saturation current	Amps	1E−14	1E−14
PB		Gate *p–n* potential	Volts	1	0.6
CGD	*	Gate-drain zero-bias *p–n* capacitance	Farads	0	5PF
CGS	*	Gate-source zero-bias *p–n* capacitance	Farads	0	1PF
FC		Forward-bias depletion capacitance coefficient		0.5	
VTOTC		VTO temperature coefficient	Volts/°C	0	
BETATCE		BETA exponential temperature coefficient	percent/°C	0	
KF		Flicker noise coefficient		0	
AF		Flicker noise exponent		1	

and *p*-channel JFETs, respectively. P1, P2, ... and A1, A2, ... are the parameters and their values, respectively.

As with diodes and BJTs, an *area factor* is used to determine the number of equivalent parallel JFETs. The model parameters that are affected by the area factor are marked by an asterisk (*) in Table 9.1. The [(area) value] scales BETA, RD, RS, CGD, CGS, and IS; it defaults to 1.

RD and RS represent the contact and bulk resistances per unit area of the drain and source, respectively. The JFET is modeled as an intrinsic device. The area value, which is the relative device area, is specified in the .MODEL statement and changes the actual resistance values. The default value of the area is 1.

The dc characteristics that are represented by the nonlinear current source I_D are defined (1) by parameters VTO and BETA, which determine the variation of the drain current with the gate voltage; (2) by LAMBDA, which determines the output conductance; and (3) by IS, which determines the reverse saturation current of the two gate junctions. VTO is negative for depletion-mode JFETs, both for *n*-channel and *p*-channel types, and it is positive for enhancement-mode JFETs. VTO does not identify whether the JFET is *n*-channel or *p*-channel.

The symbol for a JFET is *J*. The name of a JFET must start with *J*, and it takes the general form

```
J⟨name⟩  ND  NG  JNAME  [(area) value]
```

where ND, NG, and NS are the drain, gate, and source nodes, respectively.

Some JFET Statements

```
JIM   5   6  8   JNAME
.MODEL JNAME  NJF
J15   3   9  12   SWITCH 1.5
.MODEL SWITCH NJF (IS=100E−14 RD=10 RS=10 BETA=1E−3 VTO=−5)
JQ    1   5  9   JMOD
.MODEL JMOD PJF (IS=100E−14 RD=10 RS=10 BETA=1E−3 CGD=5PF CGS=1PF
+VTO=5)
```

The PSpice schematic is shown in Fig. 9.4(a), and specifies the model parameters of a known JFET (e.g., J2N3819) as shown in Fig. 9.4(b). If the specific JFET is not available in the PSpice library, the user can specify the model parameters of a JbreakJFET.

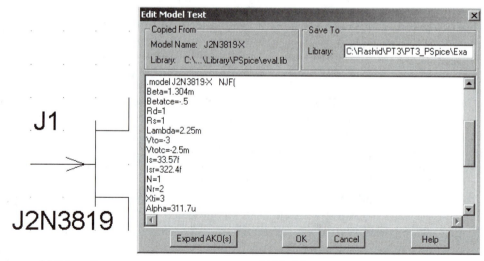

(a) Schematic   (b) Model parameters

FIGURE 9.4

Schematic and model parameters of JFETs.

9.3   JFET PARAMETERS

The library file EVAL.LIB of the student version of PSpice supports models for
n-channel JFETs: J2N3819 and J2N4393. As an example, we shall generate approximate values of some parameters [6, 9] from the data sheet of the *n*-channel JFET of
type 2N5459, shown in Fig. 9.5.

I_{DSS} = 4 to 16 mA at V_{GS} = 0 V, and V_{DS} = 15 V. Taking the geometric mean,

$$I_{DSS} = \sqrt{(4 \times 16)} = 8 \text{ mA}$$

The threshold voltage, $V_{Th} = V_{GS(off)}$ = −2 to −8 V. Taking the geometric mean
yields

$$V_{Th} = -\sqrt{(2 \times 8)} = -4 \text{ V}$$

That is, VTO = −4 V (for depletion-mode).

The transconductance coefficient is given by

$$\text{BETA} = \frac{I_{DSS}}{V_{Th}^2}$$

$$= \frac{8 \text{ mA}}{(-4)^2} = 0.5 \text{ mA/V}^2 \tag{9.1}$$

The gate reverse current I_{GSS} = −IS = −1 nA at V_{GS} = −15 V, and V_{DS} = 0.

The common-source reverse transfer capacitance, $C_{rss} = C_{gd}$ = 1.5 to 3 pF at
V_{DS} = 15 V, and V_{GS} = 0 V. Taking the geometric mean,

$$C_{rss} = C_{gd} = \sqrt{(1.5 \times 3)} \text{ pF} = 2.12 \text{ pF}$$

At $V_{DG} = V_{DS} - V_{GS}$ = 15 − 0 = 15 V, C_{gdo} can be found from

$$C_{gd} = \frac{C_{gdo}}{(1 + V_{DG}/V_{off})^{1/3}} \tag{9.2}$$

where V_{off} = 0.75 V. Equation (9.2) gives, C_{gdo} = 5.85 pF.

The common-source input capacitance, C_{iss} = 4.5 to 7 pF at V_{DS} = 15 V and
V_{GS} = 0 V. Taking the geometric mean,

$$C_{iss} = \sqrt{(4.5 \times 7)} \text{ pF} = 5.61 \text{ pF}$$

Since C_{iss} is measured at V_{GS} = 0 V, $C_{gs} = C_{gso}$. That is,

$$C_{iss} = C_{gso} + C_{gd}$$

which gives $C_{gso} = C_{iss} - C_{gd}$ = 5.61 − 2.12 = 3.49 pF.

The output admittance $|Y_{os}|$ = 10 to 50 μmhos at V_{DS} = 15 V, and V_{GS} = 0. Taking the geometric mean,

$$|Y_{os}| = \sqrt{(10 \times 50)} \text{ μmhos} = 22.36 \text{ μmhos}$$

**2N5460
thru
2N5465**

CASE 29-04, STYLE 7
TO-92 (TO-226AA)

**JFET
AMPLIFIER**

P-CHANNEL — DEPLETION

MAXIMUM RATINGS

Rating	Symbol	2N5460 2N5461 2N5462	2N5463 2N5464 2N5465	Unit
Drain-Gate Voltage	V_{DG}	40	60	Vdc
Reverse Gate-Source Voltage	V_{GSR}	40	60	Vdc
Forward Gate Current	$I_{G(f)}$	10		mAdc
Total Device Dissipation @ T_A = 25°C Derate above 25°C	P_D	310 2.82		mW mW/°C
Junction Temperature Range	T_J	−65 to +135		°C
Storage Channel Temperature Range	T_{stg}	−65 to +150		°C

ELECTRICAL CHARACTERISTICS (T_A = 25°C unless otherwise noted.)

Characteristic		Symbol	Min	Typ	Max	Unit
OFF CHARACTERISTICS						
Gate-Source Breakdown Voltage (I_G = 10 μAdc, V_{DS} = 0)	2N5460, 2N5461, 2N5462 2N5463, 2N5464, 2N5465	$V_{(BR)GSS}$	40 60	— —	— —	Vdc
Gate Reverse Current (V_{GS} = 20 Vdc, V_{DS} = 0) (V_{GS} = 30 Vdc, V_{DS} = 0) (V_{GS} = 20 Vdc, V_{DS} = 0, T_A = 100°C) (V_{GS} = 30 Vdc, V_{DS} = 0, T_A = 100°C)	2N5460, 2N5461, 2N5462 2N5463, 2N5464, 2N5465 2N5460, 2N5461, 2N5462 2N5463, 2N5464, 2N5465	I_{GSS}	— — — —	— — — —	5.0 5.0 1.0 1.0	nAdc μAdc
Gate Source Cutoff Voltage (V_{DS} = 15 Vdc, I_D = 1.0 μAdc)	2N5460, 2N5463 2N5461, 2N5464 2N5462, 2N5465	$V_{GS(off)}$	0.75 1.0 1.8	— — —	6.0 7.5 9.0	Vdc
Gate Source Voltage (V_{DS} = 15 Vdc, I_D = 0.1 mAdc) (V_{DS} = 15 Vdc, I_D = 0.2 mAdc) (V_{DS} = 15 Vdc, I_D = 0.4 mAdc)	2N5460, 2N5463 2N5461, 2N5464 2N5462, 2N5465	V_{GS}	0.5 0.8 1.5	— — —	4.0 4.5 6.0	Vdc
ON CHARACTERISTICS						
Zero-Gate-Voltage Drain Current (V_{DS} = 15 Vdc, V_{GS} = 0, f = 1.0 kHz)	2N5460, 2N5463 2N5461, 2N5464 2N5462, 2N5465	I_{DSS}	1.0 2.0 4.0	— — —	5.0 9.0 16	mAdc
SMALL-SIGNAL CHARACTERISTICS						
Forward Transfer Admittance (V_{DS} = 15 Vdc, V_{GS} = 0, f = 1.0 kHz)	2N5460, 2N5463 2N5461, 2N5464 2N5462, 2N5465	$|Y_{fs}|$	1000 1500 2000	— — —	4000 5000 6000	μmhos
Output Admittance (V_{DS} = 15 Vdc, V_{GS} = 0, f = 1.0 kHz)		$|Y_{os}|$	—	—	75	μmhos
Input Capacitance (V_{DS} = 15 Vdc, V_{GS} = 0, f = 1.0 MHz)		C_{iss}	—	5.0	7.0	pF
Reverse Transfer Capacitance (V_{DS} = 15 Vdc, V_{GS} = 0, f = 1.0 MHz)		C_{rss}	—	1.0	2.0	pF
FUNCTIONAL CHARACTERISTICS						
Noise Figure (V_{DS} = 15 Vdc, V_{GS} = 0, R_G = 1.0 Megohm, f = 100 Hz, BW = 1.0 Hz)		NF	—	1.0	2.5	dB
Equivalent Short-Circuit Input Noise Voltage (V_{DS} = 15 Vdc, V_{GS} = 0, f = 100 Hz, BW = 1.0 Hz)		e_n	—	60	115	nV/$\sqrt{Hz}$

FIGURE 9.5

Data sheet for the *n*-channel JFET of type 2N5459 (Courtesy of Motorola, Inc.).

Since $|Y_{os}|$ is given at $V_{GS} = 0$, the channel-modulation length λ (LAMBDA) can be found approximately from

$$\text{LAMBDA} = \frac{|Y_{os}|}{I_{DSS}} \approx \frac{22.36 \ \mu\text{mhos}}{8 \ \text{mA}} = 2.395\text{E}{-3}$$

The PSpice model statement for the JFET of type J2N5459 is

```
.MODEL J2N5459 NJF (IS=1N  VTO=−4  BETA=0.5M  CGDO=5.85PF
+                   CGSO=3.49PF  LAMBDA=2.395E−3)
```

9.4 EXAMPLES OF JFET AMPLIFIERS

The approximate values of JFET model parameters can be determined from the data sheet. If a model parameter is not specified, PSpice assumes its default value, as indicated in Table 9.1. The following examples illustrate the PSpice simulation of JFET circuits.

Example 9.1: Plotting the output characteristics of JFETs

For the n-channel JFET in Fig. 9.6, plot the output characteristics if V_{DD} is varied from 0 to 12 V in steps of 0.2 V and V_{GS} is varied from 0 to -4 V in steps of 1 V. The model parameters are IS = 100E−14, RD = 10, RS = 10, BETA = 1E−3, and VTO = −5.

Solution The PSpice schematic is shown in Fig. 9.7(a). The model parameters are changed from the edit model text menu as shown in Fig. 9.7(b). The model name, JbreakN1 as shown in Fig. 9.7(d), is changed from the edit model as shown in Fig. 9.7(c), which is selected from the edit menu. The gate-source voltage is varied by nested sweep within the main dc sweep, which is selected from the dc sweep menu of the analysis setup. The parameters of model JbreakN1 are specified by a model statement in file 'RASH_MODEL.LIB.'

The dc sweep in PSpice schematic that is set from the analysis setup is shown in Fig. 9.8(a), and the nested sweep is shown in Fig. 9.8(b) with nested sweep enabled. The listing of the circuit file follows.

Example 9.1 Output characteristics of an n-channel JFET

```
▲ *   Gate to source voltage of 0 V
    VGS  1  0  DC  0V
    *   A dummy voltage source of 0 V
    VX   3  2  DC  0V
    *   Dc supply voltage of 12 V
    VDD 3  0  DC  12V
▲▲ *  J1 with model JMOD
    J1   2  1  0  JMOD
    .MODEL  JMOD  NJF (IS=100E−14 RD=10 RS=10 BETA=1E−3 VTO=−5)
▲▲▲ *   VDD is swept from 0 to 12 V and VGS from 0 to −4 V.
    .DC  VDD  0  12  0.2  VGS  0  −4  1
    .PLOT  DC  I(VX)
    .PROBE
.END
```

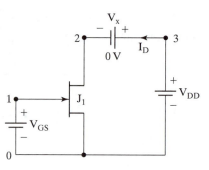

FIGURE 9.6

A circuit with an *n*-channel JFET.

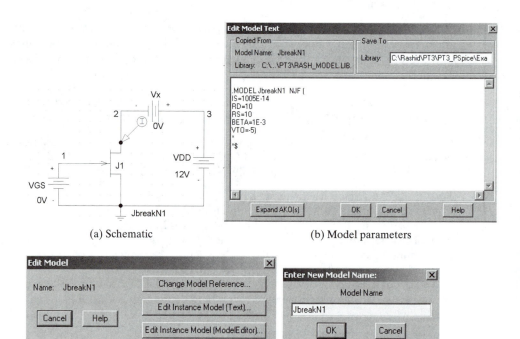

(a) Schematic

(b) Model parameters

(c) Edit model

(d) Change model name

FIGURE 9.7

PSpice schematic and JFET model parameters for Example 9.1.

The output characteristics, which are plots of I_D versus V_{DD}, are shown in Fig. 9.9.

Note: The drain current i_D decreases as the gate-source V_{GS} decreases from 0 to -4 V. For $V_{DS} > 5$ V, the slope of the i_D–V_{DS} characteristic remains almost constant at a low value. That is, the output resistance r_o of the JFET, is high.

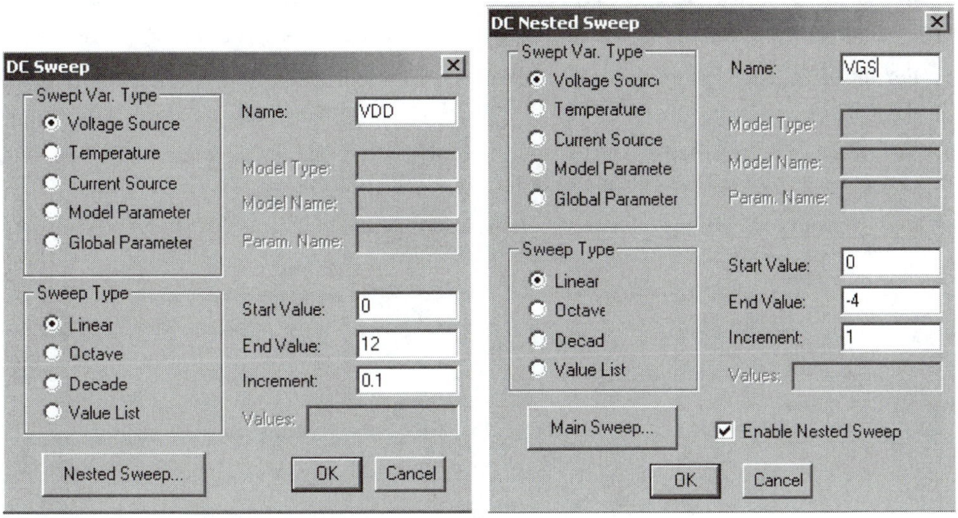

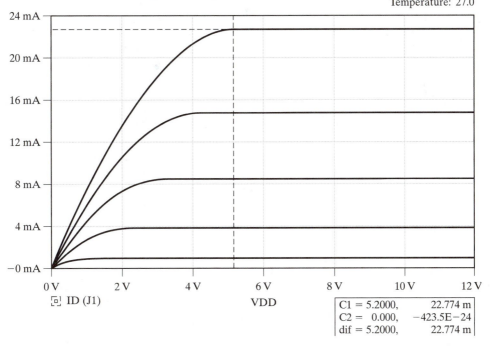

(a) Main dc sweep   (b) Nested sweep

FIGURE 9.8

Dc sweep setup for Example 9.1.

Example 9.1   Output characteristics of an n-channel JFET
Temperature: 27.0

FIGURE 9.9

Output characteristics of the JFET in Example 9.1.

Example 9.2: Plotting the transfer characteristics of JFETs

For the JFET in Example 9.1, plot the transfer characteristic if V_{GS} is varied from 0 to -5 V in steps of 0.1 V and $V_{DD} = 10$ V.

Solution   The transfer characteristic is obtained by potting the drain current i_D against the gate-source voltage V_{GS}. The listing of the circuit file follows.

Example 9.2   Transfer characteristics of an *n*-channel JFET

```
▲   VGS   1   0   DC   0V
    VX    3   2   DC   0V
    *     Dc supply voltage of 10 V
    VDD   3   0   DC   10V
▲▲  *  J1 with model JMOD
      J1    2  1  0   JMOD
      .MODEL JMOD   NJF  (IS=100E−14 RD=10 RS=10 BETA=1E−3 VTO=−5)
▲▲▲ *  VGS is swept from 0 to −5 V.
      .DC  VGS   0   −5V   0.1V
      .PLOT   DC   I(VX)
      .PROBE
  .END
```

The transfer characteristic, which is a plot of I_D versus V_{GS}, is shown in Fig. 9.10.

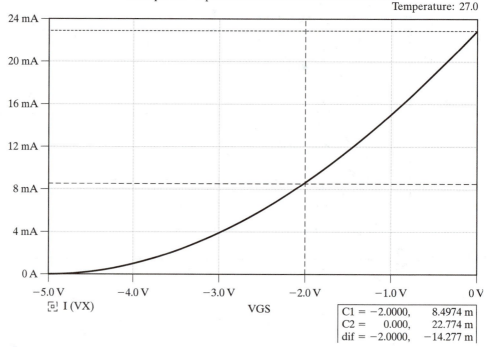

FIGURE 9.10

Input characteristic for Example 9.2.

Note: The slope of the i_D–V_{GS} characteristic gives the small-signal transconductance g_m of the JFET as given by

$$g_m = \frac{di_D}{dv_{GS}} = -\frac{I_{DSS}}{V_P}\left(1 - \frac{v_{GS}}{V_p}\right)_{\text{at Q-point } V_{GS}}$$

Example 9.3: Frequency response of a JFET amplifier

A JFET transistor amplifier circuit is shown in Fig. 9.11. The output is taken from Node 6. If the input voltage is $v_{in} = 0.5\sin(2000\pi t)$, use ac analysis to calculate and print the magnitudes and phase angles of the output voltage, the input current, and the load current. Plot the transient responses of the voltages at Nodes 1, 4, and 6 from 0 to 1 ms in steps of 10 μs. The model parameters of the JFET are IS = 100E–14, RD = 10, RS = 10, BETA = 1E–3, CGD = 5PF, CGS = 1PF, and VTO = −5. The details of the dc analysis and transient analysis operating points should be printed.

Solution   The PSpice schematic is shown in Fig. 9.12(a). The transient analysis is set from the analysis setup menu as shown in Fig. 9.12(b). The listing of the circuit file follows.

Example 9.3   An *n*-channel JFET amplifier

```
▲ .OPTIONS  NOPAGE  NOECHO
    *  Input voltage has 0.5 V peak at 1 kHz with zero offset value for
    *  transient response and 0.5-V peak for frequency response.
    VIN  1  0  AC  0.5V  SIN (0  0.5V  1KHZ)
    VDD  7  0  DC  20V
    *  Dummy voltage source of 0 V
    VI  8  2  DC  0V
    VX  6  9  DC  0V
▲▲ RRS  1  8  50
    RG  3  0  0.5MEG
    RD  7  4  3.5K
    RS  5  0  1.5K
    RL  9  0  20K
    C1  2  3  1UF
    C2  4  6  1UF
    CS  5  0  10UF
    *  n-channel JFET with model JMOD
    J1  4  3  5  JMOD
    .MODEL JMOD NJF (IS=100E-14 RD=10 RS=10 BETA=1E-3 CGD=5PF CGS=1PF VTO=-5)
▲▲▲ * Ac analysis at 1 kHz with a linear increment  and only 1 point
    .AC  LIN  1  1KHZ  1KHZ
    *  Transient analysis with details of transient analysis operating point
    .TRAN/OP  10US  1MS
    *   Print the details of the ac analysis operating point.
    .OP
    * Print the results of the ac analysis for the magnitudes of voltages at
    * Nodes 6 and 1 and for the magnitude of current through resistance RRS
    * and the current through VX.
    .PRINT AC  VM(6)  VP(6)  IM(RRS)  IP(RRS)
    .PRINT AC  IM(VI)  IP(VI)  IM(VX)  IP(VX)
    *  Plot transient response.
    .PLOT TRAN  V(6)  V(1)  V(4)
    .PROBE
    .END
```

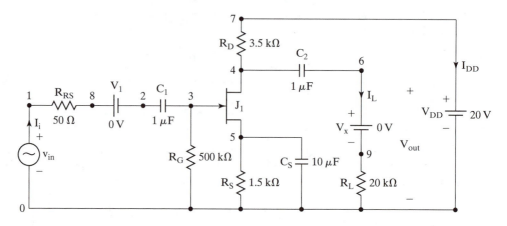

FIGURE 9.11

An *n*-channel JFET amplifier circuit.

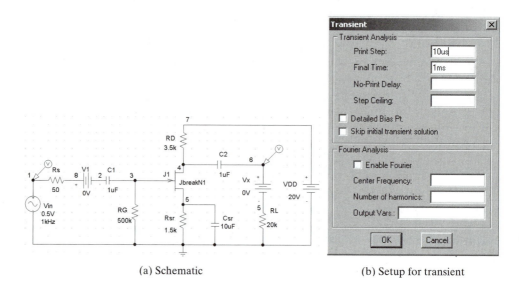

(a) Schematic   (b) Setup for transient

FIGURE 9.12

PSpice schematic for Example 9.3.

The equivalent circuit for determining the dc bias point is shown in Fig. 9.13. The details of the dc bias are as follows:

```
****    SMALL-SIGNAL BIAS SOLUTION              TEMPERATURE = 27.000 DEG C
NODE      VOLTAGE       NODE      VOLTAGE       NODE      VOLTAGE       NODE      VOLTAGE
(  1)     0.0000      (   2)     0.0000      (   3)     8.694E-06     (   4)     11.9300
(  5)     3.4585      (   6)     0.0000      (   7)    20.0000        (   8)     0.0000
(  9)     0.0000
```

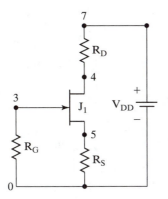

FIGURE 9.13

Equivalent circuit for dc bias calculation.

```
VOLTAGE SOURCE CURRENTS
NAME            CURRENT
VIN             0.000E+00
VDD            -2.306E-03
VI              0.000E+00
VX              0.000E+00
TOTAL POWER DISSIPATION    4.61E-02 WATTS
```

Once the dc bias points are determined, the small-signal parameters of the JFET in Fig. 9.11 are calculated. The details of the operating points are as follows:

```
****           OPERATING POINT INFORMATION           TEMPERATURE = 27.000 DEG C
****    JFETS
NAME           J1
MODEL          JMOD
ID             2.31E-03
VGS           -3.46E+00
VDS            8.47E+00
GM             3.04E-03
GDS            0.00E+00
CGS            4.72E-13
CGD            1.39E-12
```

The outputs at a frequency of 1 kHz are as follows:

```
FREQ         VM(6)        VP(6)        IM(RRS)      IP(RRS)
1.000E+03    4.382E+00   -1.769E+02    9.990E-07    2.555E+00
FREQ         IM(VI)       IP(VI)       IM(VX)       IP(VX)
1.000E+03    9.990E-07    2.555E+00    2.191E-04   -1.769E+02
```

The equivalent circuit for determining the transient analysis bias point is shown in Fig. 9.14. The transient analysis bias point and the operating point are the same as those of the dc analysis because the capacitors do not have any initial voltages. The transient responses are shown in Fig. 9.15.

Note: The output signal of an amplifier is an amplified version (inverters) of the input signal. The collector voltage composed of the output signal and a dc biasing voltage. Capacitors C_1 and C_2 isolate the load R_L and input signal V_{sn} from the dc biasing circuit, consisting of R_G, R_D, and R_S.

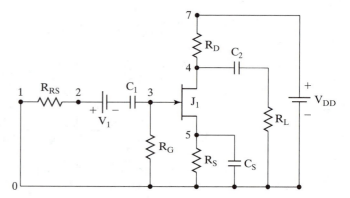

FIGURE 9.14

Equivalent circuit for transient analysis bias calculation.

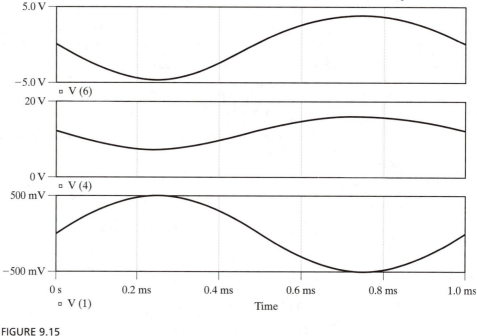

Example 9.3   N-Channel JFET amplifier

Temperature: 27.0

FIGURE 9.15

Transient responses for Example 9.3.

Example 9.4:   Frequency response of a JFET amplifier

If the JFET in Fig. 9.11 is replaced by the subcircuit model of Figure 9.16, plot the frequency response of the output voltage. The frequency is varied from 10 Hz to 100 MHz with a decade increment and 10 points per decade.

Solution   The PSpice schematic is shown in Fig. 9.17(a). The JFET model is shown in Fig. 9.17(b) and the setup for ac analysis in Fig. 9.17(c). The listing of the circuit file follows.

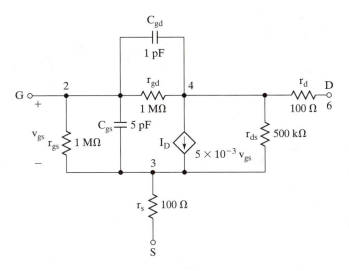

FIGURE 9.16

Subcircuit model for JFET of Example 9.4.

Example 9.4   An *n*-channel JFET amplifier

▲ ```
  .OPTIONS NOPAGE NOECHO
  * Input voltage has 0.5-V peak for frequency response.
  VIN 1 0 AC 0.5V
  VDD 7 0 DC 20V
  * Dummy voltage source of 0V
  VI 8 2 DC 0V
  VX 6 9 DC 0V
```
▲▲ ```
  RRS  1  8   50
  RG   3  0   0.5MEG
  RD   7  4   3.5K
  RS   5  0   1.5K
  RL   9  0   20K
  C1   2  3   1UF
  C2   4  6   1UF
  CS   5  0   10UF
  *  Calling subcircuit for TRANS
  XQ1  4  3  5  TRANS
  *  Subcircuit definition for TRANS
  .SUBCKT  TRANS  6  2  5
  RD   4  6   100
  RS   3  5   100
  RGS  2  3   1MEG
  CGS  2  3   5PF
  RGD  2  4   1MEG
  CGD  2  4   1PF
  RDS  4  3   500K
  *  Voltage-controlled current source with a gain of 5E-3
  G1   4  3  2  3  5E-3
  .ENDS  TRANS
```

▲ ▲ ▲ * Ac analysis for 100 Hz to 100 MHz with a decade increment and
    * 10 points per decade
    .AC DEC 10 10HZ 100MEGHZ
    * Plot the results of the ac analysis for the magnitudes and phases of
    * output voltage and the magnitudes of input and load currents.
    .PLOT AC VM(6) VP(6)
    .PLOT AC IM(VI) IM(VX)
    .PROBE
.END

The frequency response for Example 9.4 is shown in Fig. 9.18. The .PLOT statement generates graphical plots in the output file. If the .PROBE command is included, there is no need for the .PLOT command.

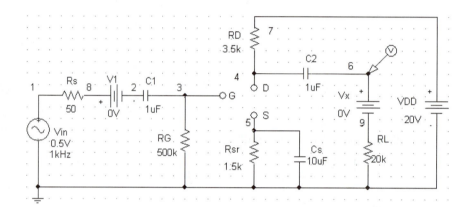

(a) Schematic

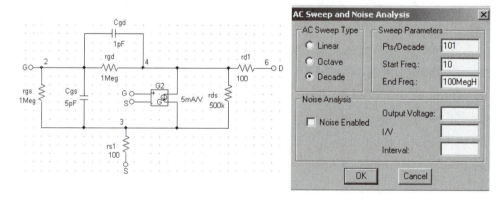

(b) JFET model

(c) Setup for ac analysis

FIGURE 9.17

PSpice schematic for Example 9.4.

Example 9.4   N-Channel JFET amplifier

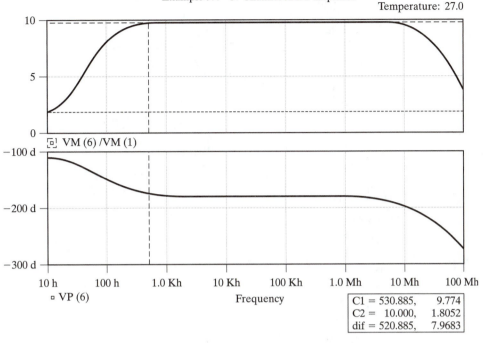

Temperature: 27.0

□ VM (6) /VM (1)

□ VP (6)

Frequency

C1 = 530.885,	9.774
C2 = 10.000,	1.8052
dif = 520.885,	7.9683

FIGURE 9.18

Frequency response for Example 9.4.

Notes: (a) The isolating capacitors C_1, C_2, and C_S set the low cut-off frequency, whereas the transistor capacitances C_{gs} and C_{gd} set the high cut-off frequency. A typical amplifier exhibits a band-pass characteristic.

(b) The elements and node numbers in the main circuit file and sub-circuit are independent of each other. For example C_2 in the main file is connected between nodes 4 and 6, whereas C_{gd} in the sub-circuit file is connected between nodes 2 and 4. They are independent.

Example 9.5:   Dc transfer characteristic of a JFET bootstrapped amplifier

A p-channel JFET bootstrapped amplifier is shown in Fig. 9.19. The output is taken from Node 5. Calculate and print the voltage gain, the input resistance, and the output resistance. The model parameters of the JFET are IS = 100E−14, RD = 10, RS = 10, BETA = 1E−3, and VTO = 5.

Solution   The PSpice schematic is shown in Fig. 9.20(a). The setup for finding the transfer characteristic is shown in Fig. 9.20(b). The listing of the circuit follows.

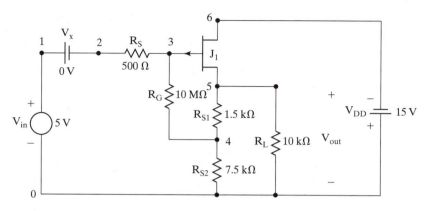

FIGURE 9.19

A *p*-channel JFET bootstrapped amplifier.

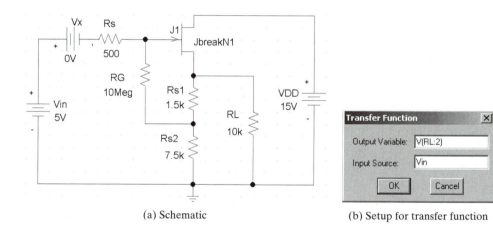

(a) Schematic

(b) Setup for transfer function

FIGURE 9.20

PSpice schematic for Example 9.5.

Example 9.5   Bootstrapped JFET amplifier

```
▲  VDD  0   6    15V
    *    Input voltage of 5 V dc
    VIN  1   0    DC   5V
    *    A dummy voltage source of 0 V
    VX   1   2    DC   0V
▲▲  RS   2   3    500
    RG   3   4    10MEG
    RS1  5   4    1.5K
    RS2  4   0    7.5K
    RL   5   0    10K
    *    p-channel JFET of model JMOD
```

```
JX   6   3   5   JMOD
*  Model statement for p-channel JFET
.MODEL JMOD PJF (IS=100E-14 RD=10 RS=10 BETA=1E-3 VTO=5)
▲▲▲ * Transfer function analysis between the output and input voltages
   .TF V(5) VIN
.END
```

The results of the transfer function analysis are

```
****      SMALL-SIGNAL CHARACTERISTICS
V(5)/VIN = -9.257E-01
INPUT RESISTANCE AT VIN =    4.375E+07
OUTPUT RESISTANCE AT V(5)  =   3.521E+02
   JOB CONCLUDED
   TOTAL  JOB TIME            9.50
```

Note: The bootstrapped amplifier increases the input resistance from $R_{in} = R_G = 10$ MΩ without bootstrap to $R_{in} = 43.75$ MΩ.

9.5   METAL OXIDE SILICON FIELD-EFFECT TRANSISTORS

The PSpice model of an *n*-channel MOSFET [5, 10, 12] is shown in Fig. 9.21. The small-signal model and the static (or dc) model generated by PSpice are shown in Figs. 9.22

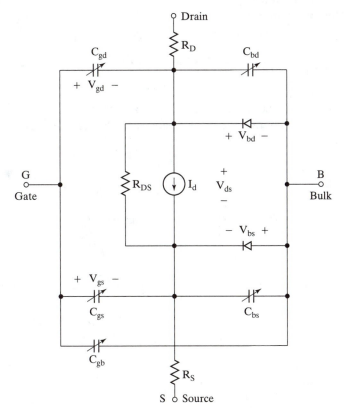

FIGURE 9.21

PSpice *n*-channel MOSFET model.

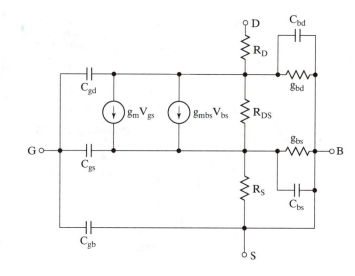

FIGURE 9.22

Small-signal n-channel MOSFET model.

and 9.23, respectively. The model parameters for a MOSFET device and the default values assigned by PSpice are given in Table 9.2. The model equations of MOSFETs that are used by PSpice are described in Schichman and Hodges [10], Vladimirescu and Liu [12], and the *PSpice Manual* [7].

The model statement for n-channel MOSFETs has the general form

```
.MODEL  MNAME  NMOS  (P1=A1  P2=A2  P3=A3 . . . PN=AN)
```

and the statement for p-channel MOSFETs has the form

```
.MODEL  MNAME  PMOS  (P1=A1  P2=A2  P3=A3 . . . PN=AN)
```

where MNAME is the model name; it can begin with any character, and its word size is normally limited to eight characters. NMOS and PMOS are the type symbols of

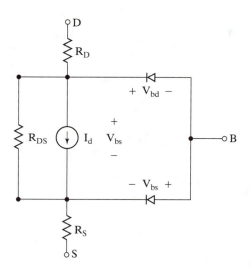

FIGURE 9.23

Static n-channel MOSFET model.

TABLE 9.2   Model Parameters of MOSFETs

Name	Model Parameters	Units	Default	Typical
LEVEL	Model type (1, 2, or 3)		1	
L	Channel length	meters	DEFL	
W	Channel width	meters	DEFW	
LD	Lateral diffusion length	meters	0	
WD	Lateral diffusion width	meters	0	
VTO	Zero-bias threshold voltage	Volts	0	0.1
KP	Transconductance	Amps/Volts^2	2E–5	2.5E–5
GAMMA	Bulk threshold parameter	Volts$^{1/2}$	0	0.35
PHI	Surface potential	Volts	0.6	0.65
LAMBDA	Channel-length modulation (LEVEL = 1 or 2)	Volts^{-1}	0	0.02
RD	Drain ohmic resistance	Ohms	0	10
RS	Source ohmic resistance	Ohms	0	10
RG	Gate ohmic resistance	Ohms	0	1
RB	Bulk ohmic resistance	Ohms	0	1
RDS	Drain-source shunt resistance	Ohms	∞	
RSH	Drain-source diffusion sheet resistance	Ohms/square	0	20
IS	Bulk p–n saturation current	Amps	1E–14	1E–15
JS	Bulk p–n saturation current/area	Amps/meters^2	0	1E–8
PB	Bulk p–n potential	Volts	0.8	0.75
CBD	Bulk-drain zero-bias p–n capacitance	Farads	0	5PF
CBS	Bulk-source zero-bias p–n capacitance	Farads	0	2PF
CJ	Bulk p–n zero-bias bottom capacitance/length	Farads/meters^2	0	
CJSW	Bulk p–n zero-bias perimeter capacitance/length	Farads/meters	0	
MJ	Bulk p–n bottom grading coefficient		0.5	
MJSW	Bulk p–n sidewall grading coefficient		0.33	
FC	Bulk p–n forward-bias capacitance coefficient		0.5	
CGSO	Gate-source overlap capacitance/ channel width	Farads/meters	0	
CGDO	Gate-drain overlap capacitance/ channel width	Farads/meters	0	
CGBO	Gate-bulk overlap capacitance/ channel length	Farads/meters	0	
NSUB	Substrate doping density	1/centimeter^3	0	
NSS	Surface-state density	1/centimeter^2	0	
NFS	Fast surface-state density	1/centimeter^2	0	
TOX	Oxide thickness	meters	∞	
TPG	Gate material type: +1 = opposite of substrate, −1 = same as substrate, 0 = aluminum		+1	
XJ	Metallurgical junction depth	meters	0	
UO	Surface mobility	centimeters^2/Volts · seconds	600	
UCRIT	Mobility degradation critical field (LEVEL = 2)	Volts/centimeter	1E4	
UEXP	Mobility degradation exponent (LEVEL = 2)		0	
UTRA	(Not used) Mobility degradation transverse field coefficient			

(Continued)

TABLE 9.2   *(Continued)*

Name	Model Parameters	Units	Default	Typical
VMAX	Maximum drift velocity	meters/second	0	
NEFF	Channel charge coefficient (LEVEL = 2)		1	
XQC	Fraction of channel charge attributed to drain		1	
DELTA	Width effect on threshold		0	
THETA	Mobility modulation (LEVEL = 3)	Volts^{-1}	0	
ETA	Static feedback (LEVEL = 3)		0	
KAPPA	Saturation field factor (LEVEL = 3)		0.2	
KF	Flicker noise coefficient		0	1E–26
AF	Flicker noise exponent		1	1.2

n-channel and *p*-channel MOSFETs, respectively. P1, P2, … and A1, A2, … are the parameters and their values, respectively.

In Table 9.2, *L* and *W* are the channel length and width, respectively. AD and AS are the drain and source diffusion areas. *L* is decreased by twice LD to get the effective channel length. Similarly, *W* is decreased by twice WD to get the effective channel width. *L* and *W* can be specified for the device, the model, or in the .OPTION statement. PSpice sets priority in selecting their values. The value specified for the device supersedes that for the model, which supersedes that in the .OPTION statement.

AD and AS are the drain and source diffusion areas. PD and PS are the drain and source diffusion perimeters. The drain *p–n*-saturation current can be specified either by IS in an absolute value, or by JS, which is multiplied by AD and AS. The zero-bias depletion capacitances can be specified (1) by CBD and CBS in absolute values; (2) by CJ, which is multiplied by AD and AS; or (3) by CJSW, which is multiplied by PD and PS.

Contact and bulk resistances are included in series with the drain, source, gate, and bulk (substrate). The MOSFET is modeled as an intrinsic device. RDS is a shunt resistance in parallel with the drain-source channel. These ohmic resistances can be specified in absolute values of RD, RS, RG, and RB. Alternatively, one could specify these resistances by RSH, which is multiplied by NRD, NRS, NRG, and NRB, respectively. NRD, NRS, NRG, and NRB are the relative resistivities of the drain, source, gate, and substrate in squares.

PD, PS, NRG, and NRB default to 0. NRD and NRS default to 1. Defaults for *L*, *W*, AD, and AS may be set in the .OPTIONS statement. The default value of AD or AS is 0. The default value of *L* or *W* is 100 μm.

The dc characteristics are defined by parameters VTO, KP, LAMBDA, PHI, and GAMMA, which are computed by PSpice by using the fabrication-process parameters NSUB, TOX, NSS, NFS, TPG, and so on. The values of VTO, KP, LAMDA, PHI, and GAMMA, which are specified in the model statement, supersede the values calculated by PSpice based on fabrication-process parameters. *VTO is positive for enhancement type n-channel MOSFETs and for depletion type p-channel MOSFETs. VTO is negative for enhancement type p-channel MOSFETs and for depletion type n-channel MOSFETs.*

PSpice incorporates three MOSFET device models. The LEVEL parameter selects among different models for the intrinsic MOSFET. If LEVEL = 1, the Schichman–Hodges model [10] is used. If LEVEL = 2, an advanced version of the Schichman–Hodges model, which is a geometry-based analytical model and incorporates extensive second-order effects [12], is used. If LEVEL = 3, a modified version of the Schichman–Hodges model, which is a semiempirical short-channel model [12], is used.

The LEVEL 1 model, which employs fewer fitting parameters, gives approximate results. However, it is useful for a quick and rough estimate of the circuit performances and it is normally adequate for the analysis of basic electronic circuits. The LEVEL 2 model, which can take into consideration various parameters, requires a great amount of CPU time for the calculations and could cause convergence problems. The LEVEL 3 model introduces a smaller error as compared to the LEVEL 2 model, and the CPU time is also approximately 25% less. The LEVEL 3 model is designed for MOSFETs with short channels.

The symbol for a metal-oxide silicon field-effect transistor (MOSFET) is *M*. The name of a MOSFET must start with *M* and takes the general form

```
M⟨name⟩   ND   NG   NS   NB   MNAME
+         [L=⟨value⟩]  [W=⟨value⟩]
+         [AD=⟨value⟩]  [AS=⟨value⟩]
+         [PD=⟨value⟩]  [PS=⟨value⟩]
+         [NRD=⟨value⟩]  [NRS=⟨value⟩]
+         [NRG=⟨value⟩]  [NRB=⟨value⟩]
```

where ND, NG, NS, and NB are the drain, gate, source, and bulk (or substrate) nodes, respectively. MNAME is the model name, and it can begin with any character; its word size is normally limited to eight characters. Positive current is the current that flows into a terminal. That is, the current flows from the drain node through the device to the source node for an *n*-channel MOSFET.

Some MOSFET Statements

```
M1    4   2   7    0 MMOD  L=10U  W=20U
.MODEL   MMOD   NMOS
M13  15   3   0    0  IRF150
.MODEL IRF150   NMOS (LEVEL=3 TOX=.10U L=3.0U LD=.5U W=2.0 WD=0
+XJ=1.2U
+        NSUB=4E14 IS=2.1E-14 RB=0 RD=.01 RS=.03 RDS=1E6 VTO=3.25
+        U0=550 THETA=.1 ETA=0 VMAX=1E6 CBS=1P CBD=4000P PB=.7 MJ=.5
+        RG=4.9 CGSO=1690P CGDO=365P CGBO=1P)
M2A   0   2  20   20  IRF9130
.MODEL IRF9130  PMOS (LEVEL=3 TOX=.1U L=3.0U LD=.5U W=1.3 WD=0
+XJ=1.2U
+        NSUB=4E14 IS=2.1E-14 RB=0 RD=.03 RS=.2 RDS=5E5 VTO=-3.7
+        U0=600 THETA=.1 ETA=0 VMAX=1E6 CBS=1P CBD=2000P PB=.7 MJ=.5
+        RG=5 CGSO=520P CGDO=180P CGBO=1P)
MA    0   2  15   15   PMOD  L=20U W=20U AD=100U AS=200U PD=50U
+                            PS=50U NRD=10 NRS=20 NRG=10
.MODEL PMOD PMOS
```

The PSpice schematic is shown in Fig. 9.24(a). The PSpice specifies the model parameters of a known MOSFET (e.g., IRF150) as shown in Fig. 9.24(b). If the specific MOSFET is not available in the PSpice library, the user can specify the model parameters of an Mbreak MOSFET.

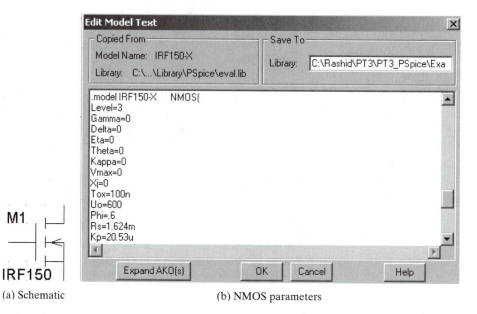

(a) Schematic   (b) NMOS parameters

FIGURE 9.24

Schematic and model parameters of MOSFETs.

9.6   MOSFET PARAMETERS

The data sheet for the n-channel MOSFET of type IRF150 is shown in Fig. 9.25. The library file EVAL.LIB of the student version of PSpice supports models for the n-type MOSFET of type IRF150 and the p-type MOSFET of type IRF9140. As an example, we shall generate approximate values of some parameters [6, 9] from the data sheet of IRF150.

From the data sheet, we get $I_{DSS} = 250 \ \mu A$ at $V_{GS} = 0 \ V$ and $V_{DS} = 100 \ V$. $V_{Th} = 2$ to $4 \ V$. The geometric mean, $V_{Th} = \text{VTO} = \sqrt{2 \times 4} = 2.83 \ V$. The constant K_p can be found from

$$I_D = K_p(V_{GS} - V_{Th})^2. \tag{9.3}$$

For $I_D = I_{DSS} = 250 \ \mu A$ and $V_{Th} = 2.83 \ V$, Eq. (9.3) gives $K_p = 250 \ \mu A/2.83^2 = 31.2 \ \mu A/V^2$. K_p is related to channel length L and channel width W by

$$K_p = \frac{\mu_a C_o}{2} \left(\frac{W}{L}\right) \tag{9.4}$$

where C_o is the capacitance per unit area of the oxide layer, a typical value for a power MOSFET being $3.5 \times 10^{-11} \ F/cm^2$ at a thickness of $0.1 \ nm$ (assumed), and μ_a is the surface mobility of electrons, $600 \ cm^2/(V \cdot s)$.

The ratio W/L can be found from Eq. (9.4):

$$\frac{W}{L} = \frac{2K_p}{\mu_a C_o} = \frac{2 \times 31.2 \times 10^{-6}}{600 \times 3.5 \times 10^{-11}} = 3000$$

Data Sheet No. PD-9.305E

INTERNATIONAL RECTIFIER I⊘R

HEXFET® TRANSISTORS IRF150

IRF151

N-Channel IRF152

IRF153

100 Volt, 0.055 Ohm HEXFET

The HEXFET® technology is the key to International Rectifier's advanced line of power MOSFET transistors. The efficient geometry and unique processing of the HEXFET design achieve very low on-state resistance combined with high transconductance and great device ruggedness.

The HEXFET transistors also feature all of the well established advantages of MOSFETs such as voltage control, freedom from second breakdown, very fast switching, ease of paralleling, and temperature stability of the electrical parameters.

They are well suited for applications such as switching power supplies, motor controls, inverters, choppers, audio amplifiers, and high energy pulse circuits.

Features:

- Fast Switching
- Low Drive Current
- Ease of Paralleling
- No Second Breakdown
- Excellent Temperature Stability

Product Summary

Part Number	V_{DS}	$R_{DS(on)}$	I_D
IRF150	100V	0.055Ω	40A
IRF151	60V	0.055Ω	40A
IRF152	100V	0.08Ω	33A
IRF153	60V	0.08Ω	33A

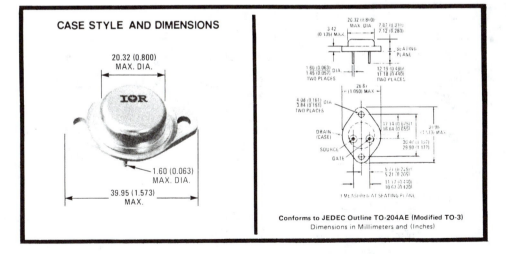

CASE STYLE AND DIMENSIONS

Conforms to JEDEC Outline TO-204AE (Modified TO-3)
Dimensions in Millimeters and (Inches)

FIGURE 9.25

Data sheet for MOSFET type IRF150 (Courtesy of International Rectifier).

IRF150, IRF151, IRF152, IRF153 Devices

Absolute Maximum Ratings

	Parameter	IRF150	IRF151	IRF152	IRF153	Units
V_{DS}	Drain - Source Voltage ①	100	60	100	60	V
V_{DGR}	Drain - Gate Voltage (R_{GS} = 20 kΩ) ①	100	60	100	60	V
I_D @ T_C = 25°C	Continuous Drain Current	40	40	33	33	A
I_D @ T_C = 100°C	Continuous Drain Current	25	25	20	20	A
I_{DM}	Pulsed Drain Current ③	160	160	132	132	A
V_{GS}	Gate - Source Voltage	± 20				V
P_D @ T_C = 25°C	Max. Power Dissipation	150	(See Fig. 14)			W
	Linear Derating Factor	1.2	(See Fig. 14)			W/K
I_{LM}	Inductive Current, Clamped	(See Fig. 15 and 16) L = 100µH				A
		160	160	132	132	
T_J T_{stg}	Operating Junction and Storage Temperature Range	−55 to 150				°C
	Lead Temperature	300 (0.063 in. (1.6mm) from case for 10s)				°C

Electrical Characteristics @ T_C = 25°C (Unless Otherwise Specified)

	Parameter	Type	Min.	Typ.	Max.	Units	Test Conditions	
BV_{DSS}	Drain - Source Breakdown Voltage	IRF150 IRF152	100	–	–	V	V_{GS} = 0V	
		IRF151 IRF153	60	–	–	V	I_D = 250µA	
$V_{GS(th)}$	Gate Threshold Voltage	ALL	2.0	–	4.0	V	V_{DS} = V_{GS}, I_D = 250µA	
I_{GSS}	Gate-Source Leakage Forward	ALL	–	–	100	nA	V_{GS} = 20V	
I_{GSS}	Gate-Source Leakage Reverse	ALL	–	–	-100	nA	V_{GS} = -20V	
I_{DSS}	Zero Gate Voltage Drain Current	ALL	–	–	250	µA	V_{DS} = Max. Rating, V_{GS} = 0V	
			–	–	1000	µA	V_{DS} = Max. Rating x 0.8, V_{GS} = 0V, T_C = 125°C	
$I_{D(on)}$	On-State Drain Current ②	IRF150 IRF151	40	–	–	A	V_{DS} > $I_{D(on)}$ x $R_{DS(on)}$ max., V_{GS} = 10V	
		IRF152 IRF153	33	–	–	A		
$R_{DS(on)}$	Static Drain-Source On-State Resistance ②	IRF150 IRF151	–	0.045	0.055	Ω	V_{GS} = 10V, I_D = 20A	
		IRF152 IRF153	–	0.06	0.08	Ω		
g_{fs}	Forward Transconductance ②	ALL	9.0	11	–	S (℧)	V_{DS} > $I_{D(on)}$ x $R_{DS(on)}$ max., I_D = 20A	
C_{iss}	Input Capacitance	ALL	–	2000	3000	pF	V_{GS} = 0V, V_{DS} = 25V, f = 1.0 MHz	
C_{oss}	Output Capacitance	ALL	–	1000	1500	pF	See Fig. 10	
C_{rss}	Reverse Transfer Capacitance	ALL	–	350	500	pF		
$t_{d(on)}$	Turn-On Delay Time	ALL	–	–	35	ns	V_{DD} = 24V, I_D = 20A, Z_o = 4.7Ω	
t_r	Rise Time	ALL	–	–	100	ns	See Figure 17.	
$t_{d(off)}$	Turn-Off Delay Time	ALL	–	–	125	ns	(MOSFET switching times are essentially	
t_f	Fall Time	ALL	–	–	100	ns	independent of operating temperature.)	
Q_g	Total Gate Charge (Gate-Source Plus Gate-Drain)	ALL	–	63	120	nC	V_{GS} = 10V, I_D = 50A, V_{DS} = 0.8 Max. Rating. See Fig. 18 for test circuit. (Gate charge is essentially	
Q_{gs}	Gate-Source Charge	ALL	–	27	–	nC	independent of operating temperature.)	
Q_{gd}	Gate-Drain ("Miller") Charge	ALL	–	36	–	nC		
L_D	Internal Drain Inductance	ALL	–	5.0	–	nH	Measured between the contact screw on header that is closer to source and gate pins and center of die.	Modified MOSFET symbol showing the internal device inductances.
L_S	Internal Source Inductance	ALL	–	12.5	–	nH	Measured from the source pin, 6 mm (0.25 in.) from header and source bonding pad.	

Thermal Resistance

R_{thJC}	Junction-to-Case	ALL	–	–	0.83	K/W	
R_{thCS}	Case-to-Sink	ALL		0.1	–	K/W	Mounting surface flat, smooth, and greased.
R_{thJA}	Junction-to-Ambient	ALL	–	–	30	K/W	Free Air Operation

FIGURE 9.25

(*Continued*)

IRF150, IRF151, IRF152, IRF153 Devices

Source-Drain Diode Ratings and Characteristics

I_S	Continuous Source Current (Body Diode)	IRF150 IRF151	—	--	40	A	Modified MOSFET symbol showing the integral reverse P-N junction rectifier. 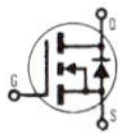
		IRF152 IRF153	-	-	33	A	
I_{SM}	Pulse Source Current (Body Diode) ③	IRF150 IRF151	—	—	160	A	
		IRF152 IRF153	—	—	132	A	
V_{SD}	Diode Forward Voltage ②	IRF150 IRF151	—	—	2.5	V	$T_C = 25°C$, $I_S = 40A$, $V_{GS} = 0V$
		IRF152 IRF153	—	—	2.3	V	$T_C = 25°C$, $I_S = 33A$, $V_{GS} = 0V$
t_{rr}	Reverse Recovery Time	ALL	—	600	-	ns	$T_J = 150°C$, $I_F = 40A$, $dI_F/dt = 100A/\mu s$
Q_{RR}	Reverse Recovered Charge	ALL	—	3.3	-	μC	$T_J = 150°C$, $I_F = 40A$, $dI_F/dt = 100A/\mu s$
t_{on}	Forward Turn-on Time	ALL					Intrinsic turn-on time is negligible. Turn-on speed is substantially controlled by $L_S + L_D$.

① $T_J = 25°C$ to $150°C$.   ② Pulse Test: Pulse width ≤ 300μs, Duty Cycle ≤ 2%.   ③ Repetitive Rating: Pulse width limited by max. junction temperature. See Transient Thermal Impedance Curve (Fig. 5).

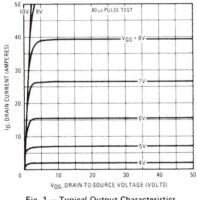

Fig. 1 – Typical Output Characteristics

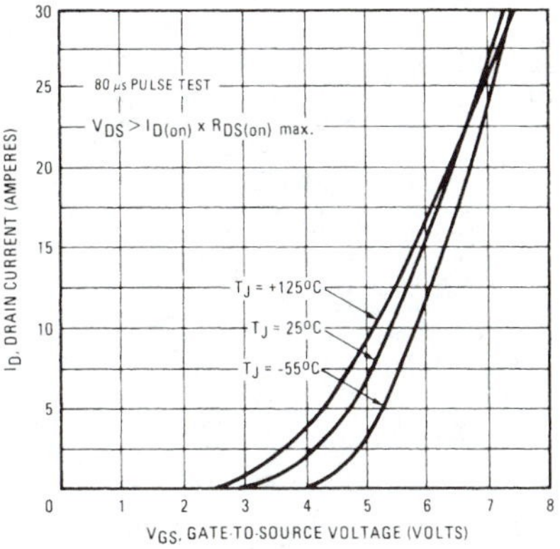

Fig. 2 – Typical Transfer Characteristics

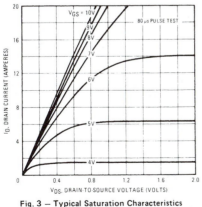

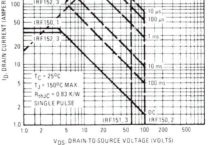

Fig. 3 – Typical Saturation Characteristics

Fig. 4 – Maximum Safe Operating Area

FIGURE 9.25

(Continued)

IRF150, IRF151, IRF152, IRF153 Devices

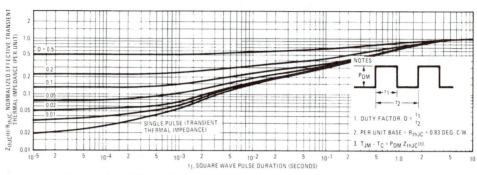

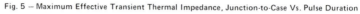

Fig. 5 — Maximum Effective Transient Thermal Impedance, Junction-to-Case Vs. Pulse Duration

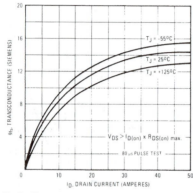

Fig. 6 — Typical Transconductance Vs. Drain Current

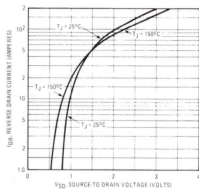

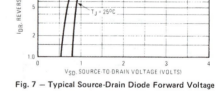

Fig. 7 — Typical Source-Drain Diode Forward Voltage

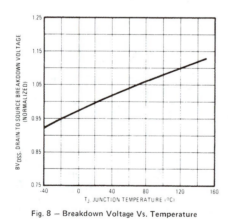

Fig. 8 — Breakdown Voltage Vs. Temperature

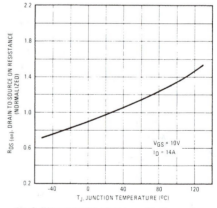

Fig. 9 — Normalized On-Resistance Vs. Temperature

FIGURE 9.25

(*Continued*)

IRF150, IRF151, IRF152, IRF153 Devices

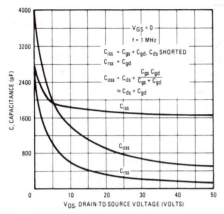

Fig. 10 — Typical Capacitance Vs. Drain-to-Source Voltage

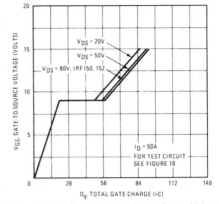

Fig. 11 — Typical Gate Charge Vs. Gate-to-Source Voltage

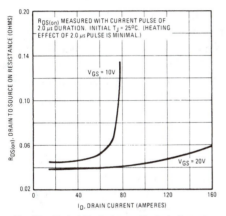

Fig. 12 — Typical On-Resistance Vs. Drain Current

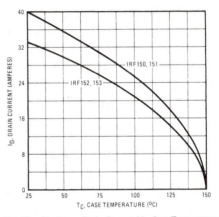

Fig. 13 — Maximum Drain Current Vs. Case Temperature

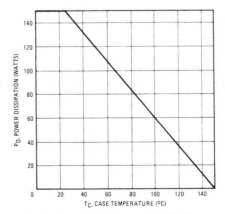

Fig. 14 — Power Vs. Temperature Derating Curve

FIGURE 9.25

(*Continued*)

Let $L = 1$ nm and $W = 3$ μm. $C_{rss} = 350 - 500$ pF at $V_{GS} = 0$ and $V_{DS} = 25$ V. The geometric mean, $C_{rss} = C_{gd} = \sqrt{350 \times 500} = 418.3$ pF at $V_{DG} = 25$ V.

For a MOSFET, the values of C_{gs} and C_{gd} remain relatively constant with changing V_{GS} or V_{DS}. They are determined mainly by the thickness and type of the insulating oxide. Although, the curves of the capacitances versus drain-source voltage show some variations, we will assume constant capacitances. Thus, $C_{gdo} = 418.3$ pF. $C_{iss} = 2000$ to 3000 pF. The geometric mean, $C_{iss} = \sqrt{2000 \times 3000} = 2450$ pF. Since C_{iss} is measured at $V_{GS} = 0$ V, $C_{gs} = C_{gso}$. That is,

$$C_{iss} = C_{gso} + C_{gd}$$

which gives $C_{gso} = C_{iss} - C_{sd} = 2450 - 418.3 = 2032$ pF $= 2.032$ nF. Thus, the PSpice model statement for MOSFET IRF150 is

```
.MODEL IRF150 NMOS (VTO=2.83 KP=31.2U L=1N W=3U CGDO=0.418N CGSO=2.032N)
```

The model can be used to plot the characteristics of the MOSFET. It may be necessary to modify the parameter values to conform with the actual characteristics. It should be noted that the parameters would differ from those given in the PSpice library, because their values are dependent on the constants used in derivations. Students are encouraged to run a circuit file with the PSpice library model and compare the results obtained with the preceding model statement.

9.7   EXAMPLES OF MOSFET AMPLIFIERS

The large number of parameters involved is an indication of the complexity of modeling a MOSFET. An accurate modeling requires a SPICE library file of a MOSFET. The parameters that are determined from the data sheet in Section 9.6 are the approximate values only. If a model parameter is not available, its typical value as indicated in Table 9.2 should be used. The following examples illustrate the PSpice simulation of MOSFET circuits.

Example 9.6:   Frequency response of a MOSFET amplifier with shunt-series feedback

An n-channel enhancement-type MOSFET amplifier with series-shunt feedback is shown in Fig. 9.26. Plot the magnitude of output voltage. The frequency is varied from 10 Hz to 100 MHz in decade steps with 10 points per decade. The peak input voltage is 100 mV. The model parameters of the MOSFET are VTO = 1, KP = 6.5E−3, CBD = 5PF, CBS = 2PF, RD = 5, RS = 2, RB = 0, RG = 0, RDS = 1MEG, CGSO = 1PF, CGDO = 1PF, and CGBO = 1PF. Print the details of the bias and operating points.

Solution   The PSpice schematic is shown in Fig. 9.27(a) and the setup for ac analysis in Fig. 9.27(b). The parameters of model MbreakN1 are specified by a model statement in file "RASH_MODEL.LIB." The listing of the circuit file follows.

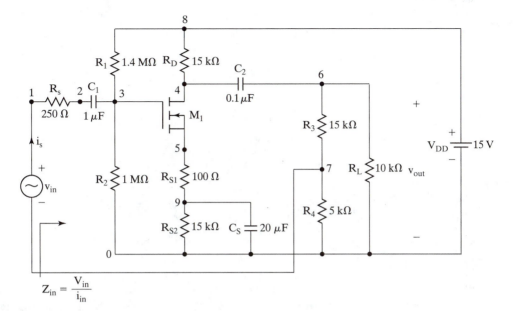

FIGURE 9.26

A MOSFET feedback amplifier.

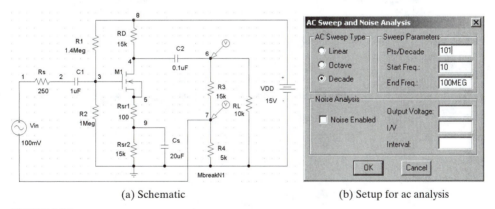

(a) Schematic   (b) Setup for ac analysis

FIGURE 9.27

PSpice schematic for Example 9.6.

Example 9.6   A MOSFET feedback amplifier

```
▲ *      Input voltage of a 100-mV peak for frequency response
    VIN   1   7   AC   100mV
    VDD   8   0   15V
▲▲ RS    1   2   250
    C1    2   3   1UF
    R1    8   3   1.4MEG
    R2    3   0   1MEG
    RD    8   4   15K
    RS1   5   9   100
    RS2   9   0   15K
```

```
          CS   9  0  20UF
          C2   4  6  0.1UF
          R3   6  7  15K
          R4   7  0  5K
          RL   6  0  10K
          *    MOSFET M1 with model MQ is connected to 4 (drain), 3 (gate), 5
          *    (source) and 5 (substrate).
          M1 4 3 5 5  MQ
          *    Model for MQ
          .MODEL MQ NMOS (VTO=1 KP=6.5E-3 CBD-5PF CBS=2PF RD=5 RS=2 RB=0
          +    RG=0 RDS=1MEG CGS0=1PF CGD0=1PF CGB0=1PF)
▲ ▲ ▲     *    Ac analysis for 10 Hz to 100 MHz with a decade increment and 10
          *    points per decade
          .AC  DEC 10  10HZ    100MEGHZ
          *    Plot the results of ac analysis: voltage at node 6.
          .PLOT  AC  VM(6)
          *    Print the details of the dc operating point.
          .OP
          .PROBE
   .END
```

The details of the bias and operating points are given next:

****   SMALL-SIGNAL BIAS SOLUTION   TEMPERATURE = 27.000 DEG C

NODE	VOLTAGE	NODE	VOLTAGE	NODE	VOLTAGE	NODE	VOLTAGE
(1)	0.0000	(2)	0.0000	(3)	6.2500	(4)	10.1000
(5)	4.9323	(6)	0.0000	(7)	0.0000	(8)	15.0000
(9)	4.8997						

```
    VOLTAGE SOURCE CURRENTS
    NAME          CURRENT
    VIN           0.000E+00
    VDD          -3.329E-04
    TOTAL POWER DISSIPATION  4.99E-03  WATTS
```

****   OPERATING POINT INFORMATION   TEMPERATURE = 27.000 DEG C
**** MOSFETS

```
NAME          M1
MODEL         MQ
ID            3.32E-04
VGS           1.32E+00
VDS           5.17E+00
VBS           0.00E+00
VTH           1.00E+00
VDSAT         3.17E-01
GM            2.06E-03
GDS           1.00E-06
GMB           0.00E+00
CBD           1.83E-12
CBS           2.00E-12
CGSOV         1.00E-16
CGDOV         1.00E-16
CGBOV         1.00E-16
CGS           0.00E+00
CGD           0.00E+00
CGB           0.00E+00
          JOB CONCLUDED
          TOTAL JOB TIME         17.41
```

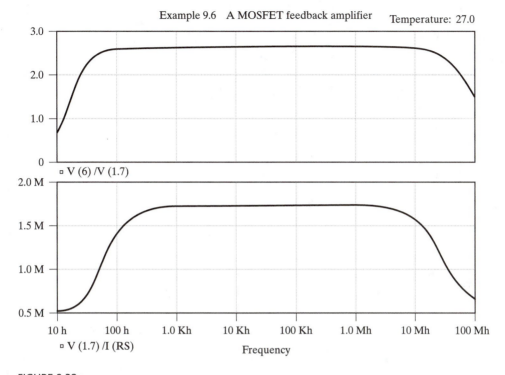

FIGURE 9.28

Frequency response for Example 9.6.

The frequency response for Example 9.6 is shown in Fig. 9.28. If the .PROBE command is included, there is no need for the .PLOT command.

Notes: The low cut-off frequency is set by the coupling and by-pass capacitances, whereas the high cut-off frequency is set by the internal transistor parameters. The feedback reduces the overall gain and increases the bandwidth. However, the gain-bandwidth product remains the same as that without feedback. That is, $A_{no\text{-}feedback} \times BW_{no\text{-}feedback} = A_{feedback} \times BW_{feedback}$.

Example 9.7:   Output impedance plot of a MOSFET amplifier with shunt-series feedback

For Fig. 9.26, plot the magnitude response of output impedance. The frequency is varied from 10 Hz to 100 MHz in decade steps with 10 points per decade.

Solution   The output impedance of the MOSFET feedback amplifier in Fig. 9.26 can be determined by short-circuiting the input source and connecting a test current source between terminals 0 and 6, as shown in Fig. 9.29. Let the peak value of the test current be 1 mA. The voltage at Node 6 is a measure of the output impedance: $Z_{out} = V(6)/1 \text{ mA} = V(6)\text{k}\Omega$. The PSpice schematic is shown in Fig. 9.30(a) and the setup for ac analysis in Fig. 9.30(b). The input source is set to zero.

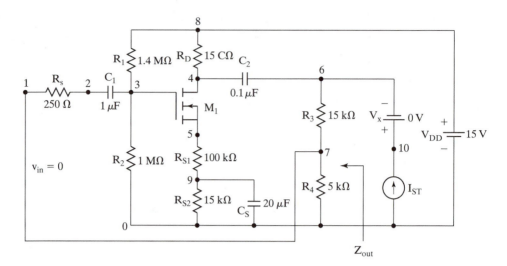

FIGURE 9.29

Equivalent circuit for output impedance calculation (Example 9.7).

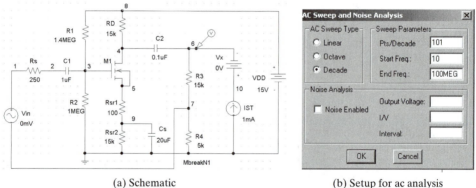

(a) Schematic (b) Setup for ac analysis

FIGURE 9.30

PSpice schematic for Example 9.7.

The listing of the circuit file follows.

Example 9.7   Output impedance of a MOSFET feedback amplifier

```
▲  *  Input source VIN is shorted.
   VIN  1  7  AC  0V
   *  Test current of a 1-mA peak for frequency response.
   IST  0  10  AC  1MA
   *  A dummy source of 0 V dc
   VX   10  6  DC  0V
   VDD  8   0  15V
▲▲ RS   1   2  250
   C1   2   3  1UF
   R1   8   3  1.4MEG
```

```
      R2   3   0   1MEG
      RD   8   4   15K
      RS1  5   9   100
      RS2  9   0   15K
      CS   9   0   20UF
      C2   4   6   0.1UF
      R3   6   7   15K
      R4   7   0   5K
      *      M1 with model MQ, whose substrate is connected to node 5
      M1 4  3   5   5   MQ
      *      Model for n-channel MOSFET with model name MQ
      .MODEL MQ NMOS (VTO=1 KP=6.5E-3 CBD=5PF CBS=2PF RD=5 RS=2 RB=0
      +    RG=0 RDS=1MEG CGSO=1PF CGDO=1PF  CGBO=1PF)
▲▲▲  *      Ac analysis for 10 Hz to 100 MHz with a decade increment and 10
      *     points per decade
      .AC   DEC   10   10HZ   10MEGHZ
      *     Plot the results of the ac analysis: voltage at node 6.
      .PLOT   AC   VM(6)
      .PROBE
.END
```

The frequency response of the output impedance for Example 9.7 is shown in Fig. 9.31. If the .PROBE command is included, there is no need for the .PLOT command.

Note: The output impedance remains constant during the mid-frequency range (e.g., 1 kHz to 10 MHz).

Example 9.7   Output impedance of a MOSFET feedback amplifier

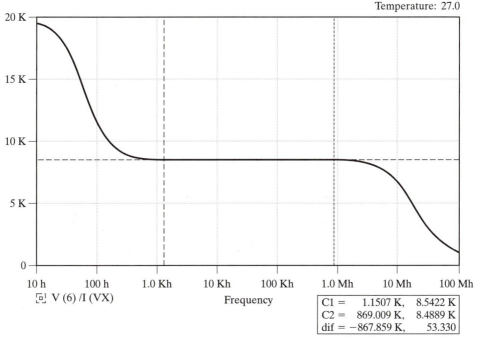

FIGURE 9.31

Output impedance for Example 9.7.

Example 9.8: Transient response and dc transfer characteristic of a CMOS inverter

A CMOS inverter circuit is shown in Fig. 9.32(a). The output is taken from Node 3. The input voltage is shown in Fig. 9.32(b). Plot the transient response of the output voltage from 0 to 80 μs in steps of 2 μs. If the input voltage is 5 V, calculate the voltage gain, the input resistance, and the output resistance. Print the small-signal parameters of the MOS and NMOS. The model parameters of the PMOS are $L = 1U$, $W = 20U$, VTO = -2, KP = 4.5E–4, CBD = 5PF, CBS = 2PF, RD = 5, RS = 2, RB = 0, RG = 0, RDS = 1MEG, CGSO = 1PF, CGDO = 1PF, and CGBO = 1PF. The model parameters of the NMOS are $L = 1U$, $W = 5U$, VTO = 2, KP = 4.5E–5, CBD = 5PF, CBS = 2PF, RD = 5, RS = 2, RB = 0, RG = 0, RDS = 1MEG, CGSO = IPF, CGDO = 1PF, and CGBO = 1PF.

Solution The PSpice schematic is shown in Fig. 9.33(a). The setup for finding the transfer characteristic is shown in Fig. 9.33(b). The listing of the circuit file follows.

Example 9.8 A CMOS inverter

```
▲ VDD   2   0   5V
    *   The input voltage is 5 V for dc analysis and pulse waveform for
    *   transient analysis.
    VIN   1   0   DC   5V   PULSE (0   5V   0   1NS   1NS   20US   40US)
▲▲ RL    3   0   100K
    *    PMOS with model PMOD
    M1    3   1   2   2   PMOD   L=1U   W=20U
    .MODEL PMOD PMOS (VTO=−2 KP=4.5E−4 CBD=5PF CBS=2PF RD=5 RS=2 RB=0
    +   RG=0 RDS=1MEG CGSO=1PF CGDO=1PF   CGBO=1PF)
    M2    3   1   0   0   NMOD   L=1U   W=5U
    *    NMOS with model NMOD
    .MODEL NMOD NMOS (VTO=2 KP=4.5E−5 CBD=5PF CBS=2PF RD=5 RS=2 RB=0
    +   RG=0 RDS=1MEG CGSO=1PF CGDO=1PF CGBO=1PF)
    *   Transient analysis from 0 to 80 μs in steps of 1 μs
▲▲▲ .TRAN   1US   80US
        *   Transfer-function analysis
        .TF   V(3)   VIN
        *   Print details of operating points.
        .OP
        .PLOT TRAN   V(3)   V(1)
        .PROBE
    .END
```

```
    ****        OPERATING POINT INFORMATION              TEMPERATURE = 27.000 DEG C
    **** MOSFETS
    NAME            M1              M2
    MODEL           MQ              NMOD
    ID           −5.00E−06       1.53E−11
    VGS           0.00E+00       5.00E+00
    VDS          −5.00E+00       2.27E−08
    VBS           0.00E+00       0.00E+00
    VTH          −2.00E+00       2.00E+00
    VDSAT         0.00E−00       3.00E+00
    GM            0.00E−00       5.09E−12
    GDS           1.00E−06       6.76E−04
    GMB           0.00E+00       0.00E+00
    CBD           1.86E−12       5.00E−12
    CBS           2.00E−12       2.00E−12
    CGSOV         2.00E−17       5.00E−18
    CGDOV         2.00E−17       5.00E−18
```

CGBOV	1.00E−18	1.00E−18
CGS	0.00E+00	0.00E+00
CGD	0.00E+00	0.00E+00
CGB	0.00E+00	0.00E+00

```
****      SMALL-SIGNAL CHARACTERISTICS
     V(3)/VIN = −2.392E−3
     INPUT RESISTANCE AT VIN = 1.000E+20
     OUTPUT RESISTANCE AT V(3) = 1.462E+03
        JOB CONCLUDED
        TOTAL JOB TIME          48.33
```

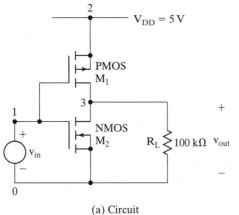

(a) Circuit

(b) Input voltage

FIGURE 9.32

A CMOS inverter (Example 9.8).

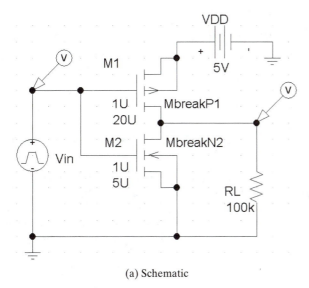

(a) Schematic

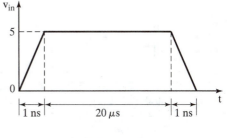

(b) Setup for transfer function

FIGURE 9.33

PSpice schematic for Example 9.8.

Example 9.8   A CMOS inverter

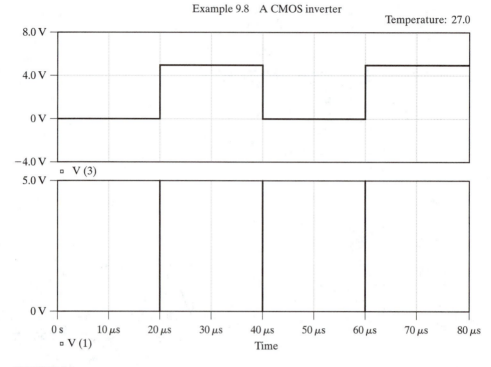

FIGURE 9.34

Frequency response for Example 9.8.

The frequency response for Example 9.8 is shown in Fig. 9.34. If the .PROBE command is included, there is no need for the .PLOT command. The .OPTIONS statement is essential to avoid convergence problems, and it is often necessary to adjust the .OPTIONS parameters (e.g., ABSTOL, RELTOL, and VNTOL) in order to let the simulation converge.

Note: The transition time from on-to-off or off-to-on is very short.

Example 9.9:   Finding the output current and resistance of an MOSFET active current source

The circuit diagram of a MOSFET current source is shown in Fig. 9.35. Use (a) the transfer function analysis to find the output current I_o and the output resistance R_o, (b) the dc biasing currents, and (b) the parametric sweep to find the output current I_o for $\lambda = 1\,k\Omega, 1\,G\Omega$. The MOS transistor has a length of 0.6 μm at minimum and can be expanded by integer increments of 0.3 μm. The minimum width is 0.9 μm and can be expanded by integer increments of 0.3 μm. Attributes of the NMOS are: L = 6 μm, W = 300 μm, AD = 720 μm, AS = 720 μm, PD = 302.4 μm, PS = 302.4 μm. Generally, AS = AD = (2.4 μm × W) and PS = PD = (2.4 μm + W).

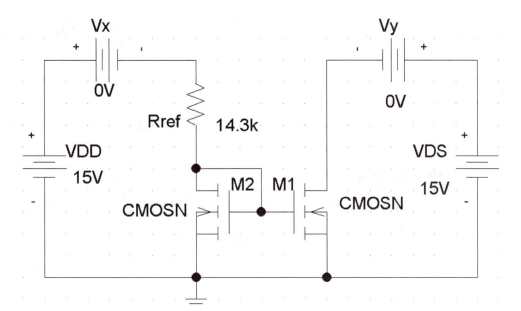

FIGURE 9.35

PSpice schematic for Example 9.9 [Ref. 8, Rashid].

Parameters for the CMOSN model of the NMOS are:

Model parameters are LEVEL=3   PHI=0.700000   TOX=9.6000E−09
    XJ=0.200000U   TPG=1   VTO=0.6684 DELTA=1.0700E+00 LD=4.2030E−08
    KP=1.7748E−04   UO=493.4   THETA=1.8120E−01 RSH=1.6680E+01
    GAMMA=0.5382   NSUB=1.1290E+17   NFS=7.1500E+11 VMAX=2.7900E+05
    ETA=1.8690E−02   KAPPA=1.6100E−01 CGDO=4.0920E−10 CGSO=4.0920E−10
    CGBO=3.7765E−10   CJ=5.9000E−04 MJ=0.76700   CJSW=2.0000E−11
    MJSW=0.71000   PB=0.9900000 WD=1.83E−07 LAMBDA=0.02

Solution

(a) The setup for the transfer function anaylysis as shown in Fig. 9.36(a) gives the output resistance and the output current whose values are found from the output file as follows:

$$\text{I(V\_Vy)/V\_VDD} = 6.901\text{E-05} \quad I_O = 6.901 \times 10^{-5} \times V_{DD} = 6.901 \times 10^{-5} \times$$
$$15 = 1035.2 \ \mu A$$

```
INPUT RESISTANCE AT V_VDD = 1.459E+04
OUTPUT RESISTANCE AT I(V_Vy) = 2.625E+06  Ro = 2.625 MΩ
```

(b) From the output file, we get the currents through the voltage sources are as follows:

```
Through V_Vx      9.659E-04     (965.9 μA)
Through V_Vy      9.712E-04     (971.2 μA)
```

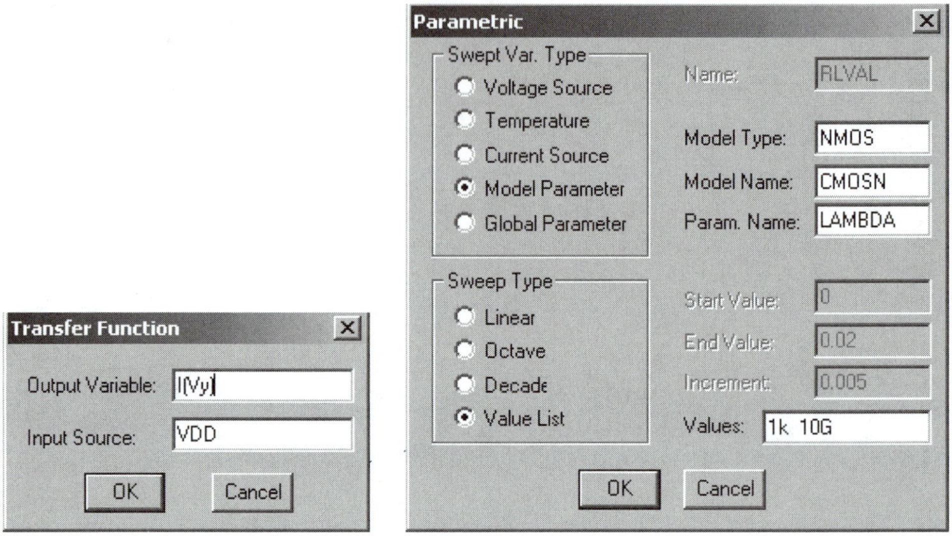

Parametric   ☒

Swept Var. Type
- ○ Voltage Source
- ○ Temperature
- ○ Current Source
- ◉ Model Parameter
- ○ Global Parameter

Name: RLVAL

Model Type: NMOS

Model Name: CMOSN

Param. Name: LAMBDA

Sweep Type
- ○ Linear
- ○ Octave
- ○ Decade
- ◉ Value List

Start Value: 0

End Value: 0.02

Increment: 0.005

Values: 1k 10G

OK   Cancel

Transfer Function   ☒

Output Variable: I(Vy)

Input Source: VDD

OK   Cancel

(a) Transfer function   (b) Parametric-variation of λ

FIGURE 9.36

Analysis setup.

(c) The setup for the parametric function analysis as shown in Fig. 9.36(b) gives the output resistance and the output currents for different values of the transistor's Lamda λ. From the output file, we get as follows:

For $\lambda = 1\ k\Omega$
$$I(V\_Vy)/V\_VDD = 6.901E\text{-}05 \quad I_O = 6.901 \times 10^{-5} \times V_{DD} = 6.901 \times 10^{-5} \times 15 = 1035.15\ \mu A$$

```
INPUT RESISTANCE AT V_VDD = 1.459E+04
OUTPUT RESISTANCE AT I(V_Vy) = 2.625E+06 R₀ = 2.625 MΩ
```

For $\lambda = 1\ G\Omega$
$$I(V\_Vy)/V\_VDD = 6.901E\text{-}05 \quad I_O = 6.901 \times 10^{-5} \times V_{DD} = 6.901 \times 10^{-5} \times 15 = 1035.15\ \mu A$$

```
INPUT RESISTANCE AT V_VDD = 1.459E+04
OUTPUT RESISTANCE AT I(V_Vy) = 2.625E+06 R₀ = 2.625 MΩ
```

Notes: (a) L = 6 μm, W = 72 μm, V_{TO} = 0.6684 V, and K_p = 1.7748 $\times$ 10^{-4} A/V^2. Thus, the effective K_p for the values of L and W is $K_{pe} = K_p W/L = 1.7748 \times 10^{-4} \times 72\,\mu/6\mu = 21.298 \times 10^{-4}$ A. Using the KVL on the input side, we get

$$V_{DD} = I_D R_{ref} + V_{Gs2} = K_{pe}\left(V_{GS2} - V_{TO}\right)^2 R_{ref} + V_{Gs2}$$

which gives an acceptable value of $V_{GS1} = V_{GS2} = 1.338$ V and $I_{D2} = I_O = K_{pe}(V_{GS2} - V_{TO})^2 =$ 955.4 μA (compared to 965.9 μA by PSpice).

(b) The effect of the transistor's Lambda λ on the output resistance and the output current are not significant. However, this example illustrates how to apply the paramertic variation of a model parameter on an output qualtity.

9.8   GALLIUM ARSENIDE MESFETS

The PSpice model of an *n*-channel GaAsFET (gallium arsenide FET) is shown in Fig. 9.37 [2, 11]. The small-signal model, which is generated by PSpice, is shown in Fig. 9.38. The model parameters for a GaAsFET device and the default values assigned by PSpice are listed in Table 9.3. The model equations of GaAsFETs that are used by PSpice are described in Curtice [5], Sussman–Fort, et al. [6], and the *PSpice Manual* [7].

The model statement of *n*-channel GaAsFETs has the general form

.MODEL   BNAME   GASFET (P1=A1   P2=A2   P3=A3 . . . PN=AN)

where GASFET is the type symbol of *n*-channel GaAsFETs. BNAME is the model name. It can begin with any character and its word size is normally limited to eight

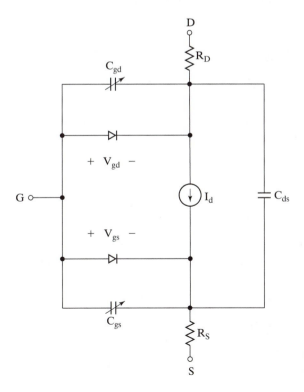

FIGURE 9.37

PSpice *n*-channel GaAsFET model.

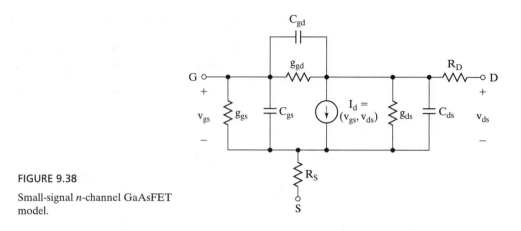

FIGURE 9.38

Small-signal *n*-channel GaAsFET
model.

characters, P1, P2, ... and A1, A2, ... are the parameters and their values, respectively.

RD, RS, and RG represent the contact and bulk resistances per unit area of the drain, source, and gate, respectively. The GaAsFET is modeled as an intrinsic device. The area value, which is the relative device area, is specified in the .MODEL statement and changes the actual resistance values. The default value of the area is 1.

The symbol for a gallium arsenide MESFET (GaAs MESFET or GaAsFET) is *B*. The name of a GaAs MESFET must start with *B*, and it takes the general form

B⟨name⟩   ND   NG   NS   BNAME   [(area) value]

TABLE 9.3   Model Parameters of GaAs MESFETs

Name	Area	Model Parameters	Units	Default	Typical
VTO		Threshold voltage	Volts	−2.5	−2.0
ALPHA		Tan *h* constant	Volts^{-1}	2.0	1.5
BETA		Transconductance coefficient	Amps/Volts^2	0.1	25U
LAMBDA		Channel-length modulation	Volts	0	1E−10
RG	*	Gate ohmic resistance	Ohms	0	1
RD	*	Drain ohmic resistance	Ohms	0	1
RS	*	Source ohmic resistance	Ohms	0	1
IS		Gate *p–n* saturation current	Amps	1E−14	
M		Gate *p–n* grading coefficient		0.5	
N		Gate *p–n* emission coefficient		1	
VBI		Threshold voltage	Volts	1	0.5
CGD		Gate-drain zero-bias *p–n* capacitance	Farads	0	1FF
CGS		Gate-source zero-bias *p–n* capacitance	Farads	0	6FF
CDS		Drain-source capacitance	Farads	0	0.3FF
TAU		Transit time	seconds	0	10PS
FC		Forward-bias depletion capacitance coefficient		0.5	
VTOTC		VTO temperature coefficient	Volts/°C	0	
BETATCE		BETA exponent temperature coefficient	%/°C	0	
KF		Flicker noise coefficient		0	
AF		Flicker noise exponent		1	

where ND, NG, and NS are the drain, gate, and source nodes, respectively. BNAME, which is the model name, can begin with any character, and its word size is normally limited to eight characters. Positive current flows into a terminal.

Some GaAs MESFET Statements

```
BIX  2  5  7  NMOD
.MODEL NMOD GASFET
BIM 15  1  0  GMOD
.MODEL GMOD GASFET (VTO=-2.5 BETA=60U VBI=0.5 ALPHA=1.5 TAU=10PS)
B5  7  9  3    NMOM 1.5
.MODEL MNOM GASFET (VTO=-2.5 BETA=32U VBI=0.5 ALPHA=1.5)
```

Example 9.10: Transient response and dc transfer characteristic of a GaAsFET inverter

A GaAsFET inverter with active load is shown in Fig. 9.39(a). The input voltage is a pulse wave-form, as shown in Fig. 9.39(b). Plot the transient response of the output voltage for a time dura-tion of 240 ps in steps of 2 ps. Plot the dc transfer characteristic if the input voltage is varied from -2.5 V to 1 V in steps of 0.1 V. The model parameters of the GaAsFET are VTO $= -2$, BETA $=$ 60U, VBI $= 0.5$, ALPHA $= 1.5$, and TAU $=$ 10PS, and those of B2 are VTO $= -2$, BETA $= 3$U, VBI $= 0.5$, and ALPHA $= 1.5$. Calculate the dc voltage gain, the input resis-tance, and the output resistance. Print the small-signal parameters for the dc analysis.

Solution   The PSpice schematic is shown in Fig. 9.40(a). A specific GaAsFET is not available in the PSpice library, the user can specify the model parameters of a Bbreak GaASFET as shown in Fig. 9.40(b). The setup for transient analysis is shown in Fig. 9.40(c) and for transfer function analysis is shown in Fig. 9.40(d). The listing of the circuit file follows.

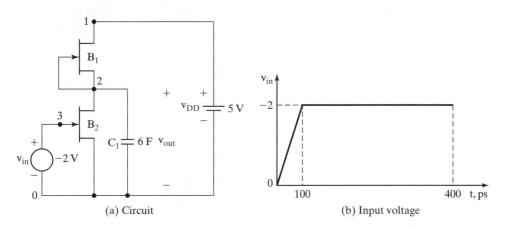

(a) Circuit   (b) Input voltage

FIGURE 9.39

GaAsFET inverter with active load.

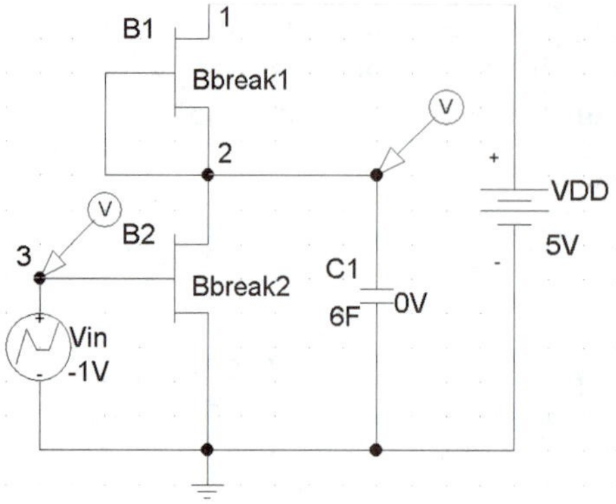

(a) Schematic

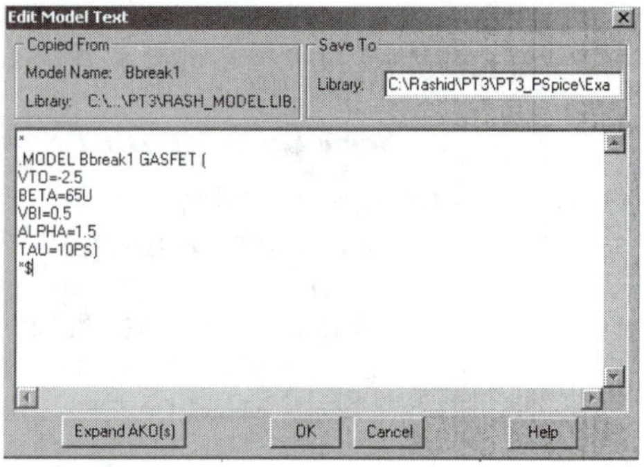

(b) GaAsFET parameters

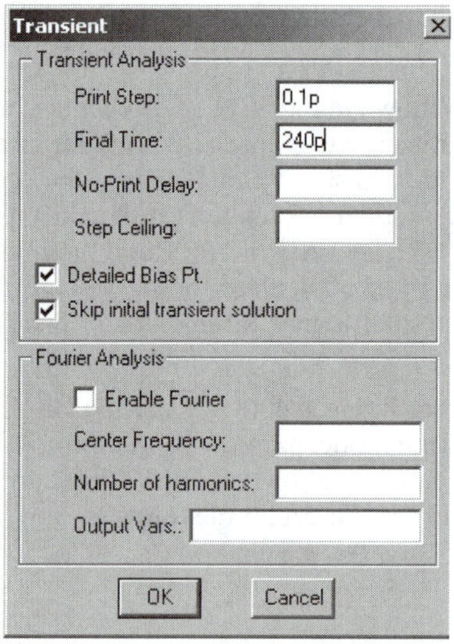

(c) Setup for transient analysis

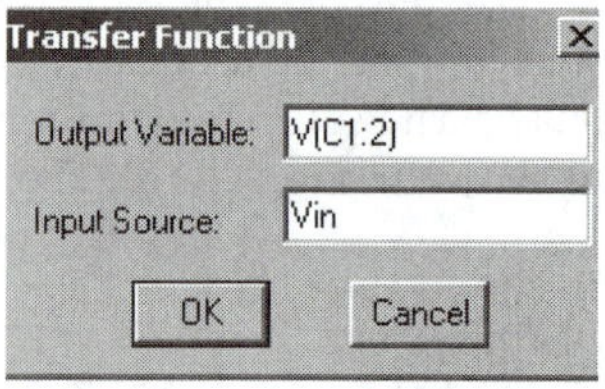

(d) Setup for transfer function

FIGURE 9.40

PSpice schematic for Example 9.10.

Example 9.10   A GaAsFET inverter with active load

```
▲ VDD    1  0  5V
   *      Pulsed input voltage
   VIN    3  0  DC  −2V  PWL (0 0 100PS −2V 1NS −2V)
▲▲ *      GaAsFET, which is connected to 1 (drain), 2 (gate) and
   *      2 (source), has a model of GF1.
   B1     1  2  2  GF1
   B2     2  3  0  GF2
   C1     2  0  6F  IC=0V
   *      Model for GF1
```

```
         .MODEL GF1 GASFET (VTO=-2.5 BETA=65U VBI=0.5 ALPHA=1.5 TAU=10PS)
         .MODEL GF2 GASFET (VTO=-2.5 BETA=32.5U VBI=0.5 ALPHA=1.5)
▲▲▲  *    Transient analysis for 0 to 240 ps with 2-ps increment
         .TRAN  2PS  240PS  UIC
         .DC VIN -2.5  1  0.1
         *    Plot the results of transient analysis.
         .PLOT TRAN  V(3)  V(2)
         .PLOT  DC  V(2)
         *    Dc transfer characteristics
         .TF   V(2)  VIN
         * Small-signal parameters for dc analysis
         .OP
         .PROBE
   .END
```

```
****      SMALL-SIGNAL BIAS SOLUTION           TEMPERATURE = 27.000 DEG C
NODE      VOLTAGE      NODE      VOLTAGE    NODE     VOLTAGE   NODE     VOLTAGE
(  1)     5.0000     (   2)     4.8787    (   3)   -1.0000   (  4)    -1.0000
       VOLTAGE SOURCE CURRENTS
       NAME            CURRENT
       VDD            -7.313E-05
       VIN             6.899E-12
       TOTAL POWER DISSIPATION    3.66E-04 WATTS
```

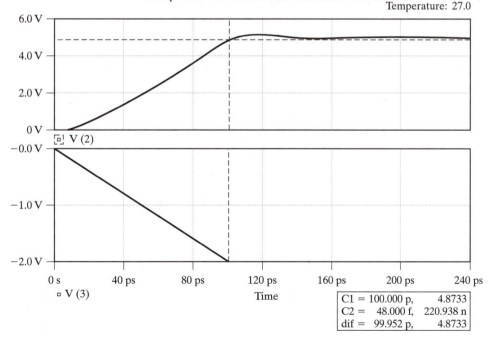

Example 9.10   A GaAsFET inverter with active load

Temperature: 27.0

FIGURE 9.41

Transient response for Example 9.10.

```
  ****    OPERATING POINT INFORMATION              TEMPERATURE = 27.000 DEG C
  **** GASFETS
  NAME        B1         B2
  MODEL       GF1        GF2
  ID          7.31E-05   7.31E-05
  VGS         0.00E+00  -1.00E+00
  VDS         1.21E-01   4.88E+00
  GM          5.85E-05   9.75E-05
  GDS         5.90E-04   1.93E-10
  CGS         0.00E+00   0.00E+00
  CGD         0.00E+00   0.00E+00
  CDS         0.00E+00   0.00E+00

  ****    SMALL-SIGNAL CHARACTERISTICS
     V(2)/VIN = -1.654E-01
     INPUT RESISTANCE AT VIN = 4.618E+11
     OUTPUT RESISTANCE AT V(2) = 1.696E+03
```

The transient response and the dc transfer characteristics for Example 9.9 are shown in Fig. 9.41. If the .PROBE command is included, there is no need for the .PLOT command.

Note: The transition time for the GaAsFET inverter from on-to-off or off-to-on is very short in the order of ps.

SUMMARY

The model statements for FETs can be summarized as follows:

```
B⟨name⟩ ND  NG  NS  BNAME  [⟨area⟩ value]
.MODEL  BNAME   GASFET (P1=V1 P2=V2 P3=V3 . . . PN=VN)
J⟨name⟩ ND  NG  NS  JNAME  [⟨area⟩ value]
.MODEL  JNAME   NJF (P1=V1 P2=V2 P3=V3 . . . PN=VN)
.MODEL  JNAME   PJF (P1=V1 P2=V2 P3=V3 . . . PN=VN)
N⟨name⟩ ND  NG  NS  NB  MNAME
+          [L=⟨value⟩] [W=⟨value⟩]
+          [AD=⟨value⟩] [AS=⟨value⟩]
+          [PD=⟨value⟩] [PS=⟨value⟩]
+          [NRD=⟨value⟩] [NRS=⟨value⟩]
+          [NRG=⟨value⟩] [NRB=⟨value⟩]
.MODEL  MNAME   NMOS (P1=V1 P2=V2 P3=V3 . . . PN=VN)
.MODEL  MNAME   PMOS (P1=V1 P2=V2 P3=V3 . . . PN=VN)
```

REFERENCES

[1] David J. Comer and Donald T. Comer, "*Teaching MOS Integrated Circuit Amplifier Design to Undergraduates.*" *IEEE Transactions on Education*, Vol. 44, No. 3, August 2001, pp. 232–238.

[2] W. R. Curtice, "A MESFET model for use in the design of GaAs integrated circuits." *IEEE Transactions on Microwave Theory and Techniques*, Vol. MTT-23, May 1980, pp. 448–456.

[3] L. M. Dang, "A simple current model for short channel IGFET and its application to circuit simulation." *IEEE Journal of Solid-State Circuits*, Vol. SC-14, No. 2, 1979, pp. 358–367.

[4] Paul R. Gray, Paul J. Hurst, Stephen H. Lewis, and Robert G. Meyer, *Electronic Devices and Circuits: Discrete and Integrated*, 4th ed. New York: John Wiley & Sons, 2001.

[5] J. F. Meyer, "MOS models and circuit simulation." *RCA Review*, Vol. 32, March 1971, pp. 42–63.

[6]   S. Natarajan, "An effective approach to obtain model parameters for BJTs and FETs from data books." *IEEE Transactions on Education*, Vol. 35, No. 2, 1992, pp. 164–169.

[7]   *PSpice Manual*. Irvine, CA: MicroSim Corporation, 1992.

[8]   M. H. Rashid, *Microelectronic Circuits: Analysis and Design*. Boston: PWS Publishing, 1999, Chapter 13.

[9]   M. H. Rashid, *SPICE for Power Electronics and Electric Power*. Englewood Cliffs, New Jersey: Prentice Hall, 1993.

[10]   H. Schichman and D. A. Hodges, "Modeling and simulation of insulated gate field effect transistor switching circuits." *IEEE Journal of Solid-State Circuits*, Vol. SC-3, September 1968, pp. 285–289.

[11]   S. E. Sussman-Fort, S. Narasimhan, and K. Mayaram, "A complete GaAs MESFET computer model for SPICE." *IEEE Transactions on Microwave Theory and Techniques*, Vol. MTT-32, April. 1984, pp. 471–473.

[12]   A. Vladimirescu and Sally Liu, *The Simulation of MOS Integrated Circuits Using SPICE2*. Memorandum no. M80/7, February 1980, University of California, Berkeley.

PROBLEMS

9.1   For the amplifier circuit in Fig. 9.11, calculate and plot the frequency responses of the output voltage and the input current. The frequency is varied from 10 Hz to 10 MHz in decade steps with 10 points per decade.

9.2   A shunt-shunt feedback is applied to the amplifier circuit in Fig. 9.11. This is shown in Fig. P9.2. Calculate and print the frequency responses of the output voltage and the input current. The frequency is varied from 10 Hz to 10 MHz in decade steps with 10 points per decade. The model parameters are $IS = 1N$, $VTO = -4$, $BETA = 0.5M$, $CGDO = 5.85P$, $CGSO = 3.49P$, and $LAMBDA = 2.395E - 3$.

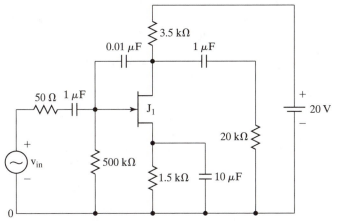

FIGURE P9.2

9.3   Repeat Example 9.5 if the transistor is an *n*-channel JFET. The model parameters are $IS = 100E - 14$, $RD = 10$, $RS = 10$, $BETA = 1E - 3$, and $VTO = -2$. Assume that $R_{S2} = 0$.

9.4   For the *n*-channel enhancement-type MOSFET in Fig. P9.4, plot the output characteristics if V_{DS} is varied from 0 to 15 V in steps of 0.1 V and V_{GS} is varied from 0 to 6 V in steps of 1 V. The model parameters are $L = 10U$, $W = 20U$, $VTO = 2.5$, $KP = 6.5E - 3$, $RD = 5$, $RS = 2$, $RB = 0$, $RG = 0$, and $RDS = 1MEG$.

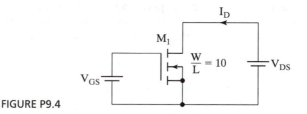

FIGURE P9.4

9.5 For Problem 9.4, plot the input characteristics if V_{GS} is varied from 0 to 6 V in steps of 0.1 V and V_{DS} = 15 V.

9.6 An inverter circuit is shown in Fig. P9.6. For the input voltage as shown in Fig. 9.32(b), plot the transient response of the output voltage from 0 to 80 μs in steps of 2 μs. If the input voltage is 5 V dc, calculate the voltage gain, the input resistance, and the output resistance. Print the small-signal parameters of the PMOS. The model parameters of the PMOS are VTO = −2.5, KP = 4.5E − 3, CBD = 5PF, CBS = 2PF, CGSO = 1PF, CGDO = 1PF, and CGBO = 1PF.

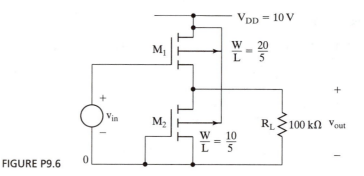

FIGURE P9.6

9.7 For the NMOS AND-logic circuit in Fig. P9.7, plot the transient response of the output voltage from 0 to 100 μs in steps of 1 μs. The model parameters of the p-channel depletion-type MOSFETs are VTO = 2, KP = 4.5E − 3, CBD = 5PF, CBS = 2PF, RD = 5, RS = 2, RB = 0, RG = 0, RDS = 1MEG, CGSO = 1PF, CGDO = 1PF, and CGBO = 1PF.

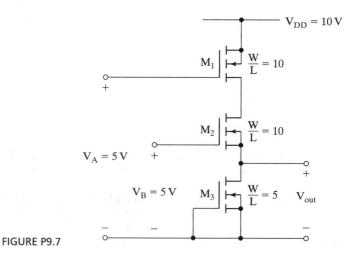

FIGURE P9.7

9.8 For the NMOS NAND-logic gate circuit in Fig. P9.8, plot the transient response of the output voltage from 0 to 100 μs in steps of 1 μs. The model parameters of the PMOS are VTO = −2.5, KP = 4.5E − 3,   CBD = 5PF, CBS = 2PF, RD = 5, RS = 2,   RB = 0, RG = 0, RDS = 1MEG, CGSO = 1PF, CGDO = 1PF, and CGBO = 1PF. The model parameters of the NMOS are VTO = 2.5, KP = 4.5E − 3, CBD = 5PF, CBS = 2PF, RD = 5, RS = 2, RB = 0, RG = 0, RDS = 1MEG, CGSO = 1PF, CGDO = 1PF, and CCBO = 1PF.

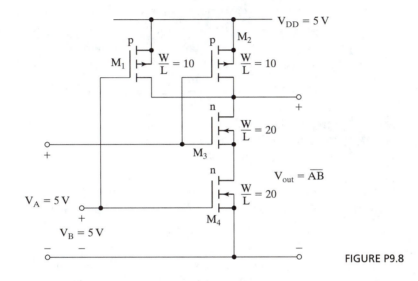

FIGURE P9.8

9.9 A MOSFET amplifier with active load is shown in Fig. P9.9. Plot the magnitudes of the output voltage and the input current. The frequency is varied from 10 Hz to 100 MHz with a decade increment and 10 points per decade. The peak input voltage is 200 mV. The model parameters of the NMOS are VTO = 2.5, KP = 4.5E − 2, CBD = 5PF, CBS = 2 PF, RD = 5, RS = 2, RB = 0, RG = 0, RDS = 1MEG, CGSO = 1PF, CGDO = 1 PF, and CGBO = 1PF. Print the details of the bias point and the small-signal parameters of the NMOS.

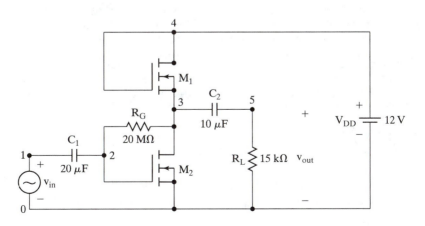

FIGURE P9.9

9.10 Use PSpice to perform a Monte Carlo analysis for six runs and for the frequency of Problem 9.2. The model parameter is $R = 1$ for resistors. The lot deviation for all resistances is $\pm15\%$. The transistor parameter having uniform deviations is

$$VTO = -4 \text{ V} \pm 1.5 \text{ V}.$$

(a) The greatest difference from the nominal run is to be printed.
(b) The maximum value of the output voltage is to be printed.
(c) The minimum value of the output voltage is to be printed.

9.11 Use PSpice to perform the worst-case analysis for Problem 9.10.
9.12 Use PSpice to perform a Monte Carlo analysis for six runs and for the transient response of Problem 9.6. The transistor parameter having uniform deviations is

$$VTO = -2.5 \text{ V} \pm 1.2 \text{ V}.$$

(a) The greatest difference from the nominal run is to be printed.
(b) The maximum value of the output voltage is to be printed.
(c) The minimum value of the output voltage is to be printed.

9.13 Use PSpice to perform the worst-case analysis for Problem 9.12.
9.14 Use PSpice to perform a Monte Carlo analysis for five runs and for the transient response of Problem 9.7. The transistor parameter having uniform deviations is

$$VTO = 2 \text{ V} \pm 1.5 \text{ V}.$$

(a) The greatest difference from the nominal run is to be printed.
(b) The maximum value of the output voltage is to be printed.
(c) The minimum value of the output voltage is to be printed.

9.15 Use PSpice to perform the worst-case analysis for Problem 9.14.
9.16 Use PSpice to perform a Monte Carlo analysis for five runs and for the transient response of Problem 9.8. The transistor parameter having uniform deviations is

$$VTO = -2.5 \text{ V} \pm 1.5 \text{ V}.$$

(a) The greatest difference from the nominal run is to be printed.
(b) The maximum value of the output voltage is to be printed.
(c) The minimum value of the output voltage is to be printed.

9.17 Use PSpice to perform the worst-case analysis for Problem 9.16.
9.18 Use PSpice to perform a Monte Carlo analysis for five runs and for the frequency response of Problem 9.9. The transistor parameter having uniform deviations is

$$VTO = 2.5 \text{ V} \pm 1.5 \text{ V}.$$

(a) The greatest difference from the nominal run is to be printed.
(b) The maximum value of the output voltage is to be printed.
(c) The minimum value of the output voltage is to be printed.

9.19 Use PSpice to perform the worst-case analysis for Problem 9.18.

9.20 Use PSpice to find the worst-case minimum and maximum values of the voltage gain $A_v = V_{out}/V_{in}$, the input resistance R_{in}, and the output resistance R_{out} for the JFET amplifier shown in Fig. 9.11.

9.21 Use PSpice to perform the sensitivity analysis of the JFET amplifier in Fig. 9.11 for Example 9.3. Which element has the highest sensitivity with respect to the output voltage?

9.22 Use PSpice to perform the sensitivity analysis of the MOSFET feedback amplifier in Fig. 9.26 for Example 9.6. Which element has the highest sensitivity with respect to the output voltage?

CHAPTER 10

Op-Amp Circuits

After completing this chapter, students should be able to

- Model SPICE op-amps
- Derive the parameters of linear and nonlinear models
- Perform the dc analysis, the transient analysis, and the ac analysis of op-amp circuits.

10.1   INTRODUCTION

An **op-amp** may be modeled as a linear amplifier to simplify the design and analysis of op-amp circuits. The linear models give reasonable results, especially for determining the approximate design values of op-amp circuits. However, the simulation of the actual behavior of op-amps is required in many applications to obtain accurate responses for the circuits. PSpice does not have any model for op-amps. However, an op-amp can be simulated from the circuit arrangement of the particular type of op-amp. The μA741 type of op-amp consists of 24 transistors, and it is beyond the capability of the student (or demo) version of PSpice. A macromodel, which is a simplified version of the op-amp and requires only two transistors, is quite accurate for many applications and can be simulated as a subcircuit or library file. Some manufacturers often supply macromodels of their op-amps [5]. In the absence of a complex op-amp model, the characteristics of op-amp circuits may be determined approximately by one of the following models:

       Dc linear model
       Ac linear model
       Nonlinear macromodel

10.2   DC LINEAR MODELS

An op-amp may be modeled as a voltage-controlled voltage source, as shown in Fig. 10.1(a). The input resistance is high, typically 2 MΩ, and the output resistance is very

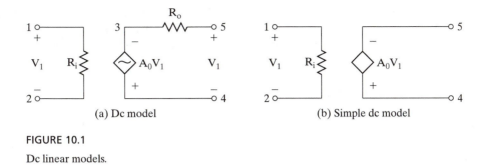

FIGURE 10.1

Dc linear models.

low, typically 75 Ω. For an ideal op-amp, the model in Fig. 10.1(a) can be reduced to that of Fig. 10.1(b). These models do not take into account the saturation effect and slew rate, which do exist in actual op-amps. The gain is also assumed to be independent of the frequency, but the gain of actual practical op-amps falls with the frequency. These simple models are normally suitable for dc or low-frequency applications.

10.3 AC LINEAR MODEL

The frequency response of an op-amp can be approximated by a single break frequency, as shown in Fig. 10.2(a). This characteristic can be modeled by the circuit of Fig. 10.2(b). This is a high-frequency model of op-amps. If an op-amp has more than one break frequency, it can be represented by using as many capacitors as the number of breaks. R_i is the input resistance and R_o is the output resistance.

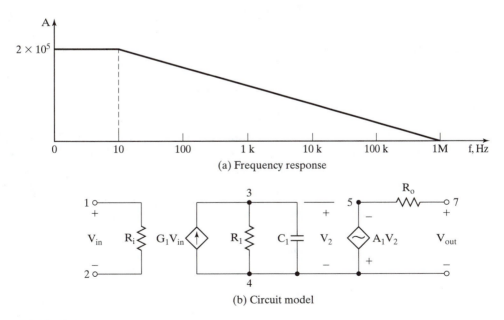

FIGURE 10.2

Ac linear model with a single break frequency.

The dependent sources of the op-amp model in Fig. 10.2(b) have a common node. Without this, PSpice will give an error message because there is no dc path from the nodes of the dependent current source. The common node could be either with the input stage or with the output stage. This model does not take into account the saturation effect and is suitable only if the op-amp operates within the linear region.

The output voltage can be expressed as

$$V_{out} = -A_1 V_2 = \frac{-G_1 R_1 A_1}{1 + R_1 C_1 s} = \frac{-A_0 V_{in}}{1 + R_1 C_1 s}$$

Substituting $s = j2\pi f$ yields

$$V_{out} = \frac{-A_0 V_{in}}{1 + j2\pi f R_1 C_1} = \frac{-A_0 V_{in}}{1 + jf/f_b}$$

where

$$f_b = 1/(2\pi R_1 C_1) \text{ is called the } break \ frequency, \text{ in hertz,}$$

and

$$A_0 = G_1 R_1 A_1 = \text{the } large\text{-}signal \ (\text{or } dc) \ gain \text{ of the op-amp.}$$

Thus, the open-loop voltage gain is

$$A(f) = \frac{V_{out}}{V_{in}} = -\frac{A_0}{1 + jf/f_b}$$

For μA741 op-amps, $f_b = 10$ Hz, $A_0 = 2 \times 10^5$, $R_i = 2$ MΩ, and $R_o = 75$ Ω. Letting $R_1 = 10$ kΩ, and $G_1 = 1$, $C_1 = 1/(2\pi \times 10 \times 10 \times 10^3) = 1.5619$ μF and $A_1 = A_0/(G_1 R_1) = 2 \times 10^5/(1 \times 10^4) = 20$. Alternatively, letting $C_1 = 1$ μF gives $R_1 = 1/(2\pi f_b C_1) = 1/(2\pi \times 10 \times 1 \ \mu F) = 15.915$ kΩ.

10.4   NONLINEAR MACROMODEL

The circuit arrangement of the **op-amp macromodel** is shown in Fig. 10.3 [1,2,5]. The macromodel can be used as a subcircuit with the .SUBCKT command. However, if an op-amp is used in various circuits, it is convenient to have the macromodel as a library file, namely, EVAL.LIB, and it is not required to type the statements of the macromodel in every circuit where the macromodel is employed. The library file EVAL.LIB that comes with the student version of PSpice has macromodels for op-amps, comparators, diodes, MOSFETs, BJTs, and SCRs. The macromodels for the linear operational amplifier of type LM324, the linear operational amplifier of type μA741, and the voltage comparator of type LM111 are included in the EVAL.LIB file. The professional version of PSpice supports library files for many devices.

The macromodel of the μA741 op-amp is simulated at room temperature. The library file EVAL.LIB contains the op-amp macromodel model as a subcircuit definition μA741 with a set of .MODEL statements. This op-amp model contains nominal (not worst-case) devices and does not consider the effects of temperature.

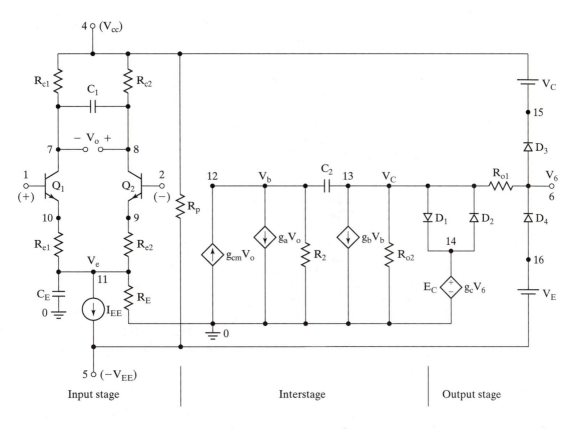

FIGURE 10.3

Circuit diagram of op-amp macromodel.

The listing of the library file EVAL.LIB follows:

```
*  Library file "EVAL.LIB" for UA741 op-amp
*  connections:    noninverting input
*                   :   inverting input
*                   :   :
*                   :   :       positive power supply
*                   :   :           negative power supply
*                   :   :           :   :       output
*                   :   :           :   :   :
.SUBCKT UA741     1   2       4   5   6
*                 Vi+ Vi−  Vp+ Vp− Vout
Q1     7    1   10  UA741QA
Q2     8    2   9   UA741QB
RC1    4    7   5.305165D+03
RC2    4    8   5.305165D+03
C1     7    8   5.459553D−12
RE1    10   11  2.151297D+03
RE2    9    11  2.151297D+03
```

```
IEE    11   5    1.666000D-05
CE     11   0    3.000000D-12
RE     11   0    1.200480D+07
GCM    0    12   11   3     5.960753D-09
GA     12   0    8    7     1.884955D-04
R2     12   0    1.000000D+05
C2     12   13   3.000000D-11
GB     13   0    12   0     2.357851D+02
RO2    13   0    4.500000D+01
D1     13   14   UA741DA
D2     14   13   UA741DA
EC     14   0    6    3    1.0
RO1    13   6    3.000000D+01
D3     6    15   UA741DB
VC     4    15   2.803238D+00
D4     16   6    UA741DB
VE     16   5    2.803238D+00
RP     4    5    18.16D+03
*   Models for diodes and transistors
.MODEL UA741DA D (IS=9.762287D-11)
.MODEL UA741DB D (IS=8.000000D-16)
.MODEL UA741QA NPN  (IS=8.000000D-16 BF=9.166667D+01)
.MODEL UA741QB NPN  (IS=8.309478D-16 BF=1.178571D+02)
*   End of library file
*   End of subcircuit definition
.ENDS
```

Note: The macromodel in Fig. 10.3 consists of 25 elements with 17 nodes, whereas the simple model in Fig. 10.1(a) consists of 3 elements with 5 nodes and the linear ac model in Fig. 10.2(b) requires 6 elements with 6 nodes. As a result, an op-amp circuit with only one macronode may exceed the node limit of 65, and the simulation will be terminated.

10.5   EXAMPLES OF OP-AMP CIRCUITS

Example 10.1: DC transfer characteristic of an inverting amplifier circuit with a simple op-amp model

An inverting amplifier is shown in Fig. 10.4. The output is taken from node 5. Calculate and print the voltage gain, the input resistance, and the output resistance. The op-amp, which is modeled by the circuit in Fig. 10.1(a), has $A_0 = 2 \times 10^5$, $R_i = 2$ MΩ, and $R_o = 75$ Ω.

Solution   The PSpice schematic is shown in Fig. 10.5(a). The setup for finding the transfer characteristic is shown in Fig. 10.5(b). The listing of the subcircuit file follows.

Example 10.1   Inverting amplifier

```
▲ *   Input voltage is 1.5 V dc.
  VIN  1   0   DC   1.5V
▲▲ R1    2   3   10K
```

```
R2    4  0  6.67K
RF    3  5  20K
*  Calling subcircuit OPAMP
XA1  2  1  2  0  OPAMP
XA2  3  4  5  0  OPAMP
*  Subcircuit definition for OPAMP
.SUBCKT OPAMP 1  2  5  4
RI   1  2  2MEG
RO   3  5  75
*  Voltage-controlled voltage source with a gain of 2E+5. The polarity of
*  the output voltage is taken into account by changing the location of
*  the controlling nodes.
EA  3  4  2  1  2E+5
*  End of subcircuit definition
.ENDS    OPAMP
*  Transfer-function analysis calculates and prints the dc gain,
*  the input resistance, and the output resistance.
.TF  V(5)  VIN
.END
```

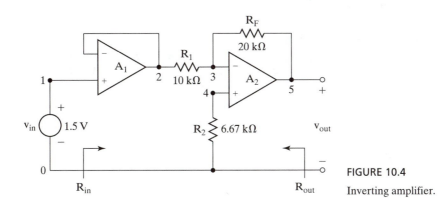

FIGURE 10.4

Inverting amplifier.

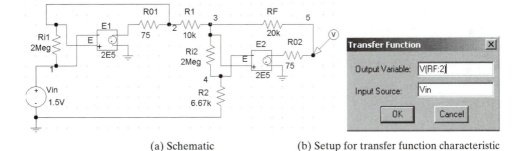

(a) Schematic

(b) Setup for transfer function characteristic

FIGURE 10.5

PSpice schematic for Example 10.1.

The results of the transfer function analysis by the .TF command are as follows:

```
****     SMALL-SIGNAL BIAS SOLUTION                    TEMPERATURE = 27.000 DEG C
NODE       VOLTAGE      NODE     VOLTAGE     NODE    VOLTAGE        NODE      VOLTAGE
(   1)     1.5000     (   2)     1.5000    (   3)    15.11E-06    (   4)    50.21E-09
(   5)    -2.9999     (XA1.3)    1.5112    (XA2.3)   -3.0112
     VOLTAGE SOURCE CURRENTS
     NAME              CURRENT
     VIN             -3.778E-12
     TOTAL POWER DISSIPATION   5.67E-12 WATTS

****      SMALL-SIGNAL CHARACTERISTICS
      V(5)/VIN = -2.000E+00
      INPUT RESISTANCE AT VIN = 3.970E+11
      OUTPUT RESISTANCE AT V(5) = 1.132E-03
          JOB CONCLUDED
          TOTAL JOB TIME          2.42
```

Note: The unity-gain input-buffer stage yields a very high input resistance (e.g., R_{in} = 0.397 GΩ).

Example 10.2: Transient response of an integrator with a linear ac op-amp model

An integrator circuit is shown in Fig. 10.6(a). For the input voltage as shown in Fig. 10.6(b), plot the transient response of the output voltage for a duration of 0 to 4 ms in steps of 50 μs. The op-amp that is modeled by the circuit in Fig. 10.2(b) has R_i = 2 MΩ, R_o = 75 Ω, C_1 = 1.5619 μF, R_1 = 10 kΩ, A_0 = 2 × 10^5, G_1 = 1, and A_1 = 20.

Solution   The PSpice schematic is shown in Fig. 10.7(a). The setup for transient analysis is shown in Fig. 10.7(b). The listing of the circuit file follows.

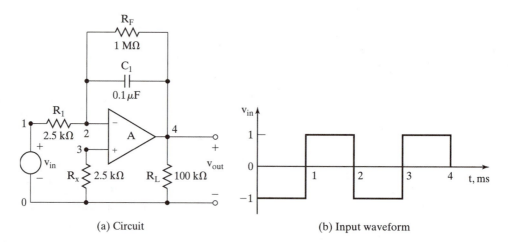

(a) Circuit   (b) Input waveform

FIGURE 10.6

Integrator circuit.

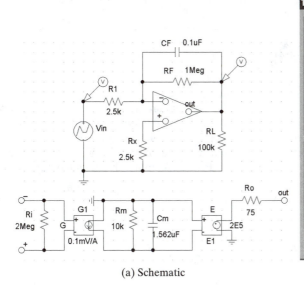

(a) Schematic

(b) Setup for transient analysis

FIGURE 10.7

PSpice schematic for Example 10.2.

Example 10.2 Integrator circuit

```
▲ *  The input voltage is represented by a piecewise linear waveform.
  *  To avoid convergence problems due to a rapid change, the input
  *  voltage is assumed to have a finite slope.
  VIN  1  0  PWL (0 0 1NS −1V 1MS −1V  1.0001MS IV 2MS 1V
  +     2.0001MS −1V 3MS −1V 3.0001MS 1V  4MS 1V)
▲▲ R1  1  2  2.5K
    RF  2  4  1MEG
    RX  3  0  2.5K
    RL  4  0  100K
    C1  2  4  0.1UF
    *  Calling subcircuit OPAMP
    XA1  2  3  4  0  OPAMP
    *  Subcircuit definition for OPAMP
    .SUBCKT OPAMP  1  2  7  4
    RI  1  2  2.0E6
    *  Voltage-controlled current source with a gain of 1
    GB  4  3  1  2  0.1M
    R1  3  4  10K
    C1  3  4  1.5619UF
    *  Voltage-controlled voltage source with a gain of 2E+5
    EA  4  5  3  4  2E+5
    RO  5  7  75
    *  End of subcircuit OPAMP
    .ENDS
```

```
▲▲▲ *  Transient analysis for 0 to 4 ms with 50-μs increment
       .TRAN  50US  4MS
       *  Plot the results of transient analysis
       .PLOT  TRAN  V(4)  V(1)
       .PLOT  AC  VM(4)  VP(4)
       .PROBE
.END
```

Note: The elements and node numbers in the main circuit file and subcircuit are independent of each other. For example, C_1 in the main file is connected between nodes 2 and 4, where as C_1 in the subcircuit file is connected between nodes 3 and 4. They are independent.

The transient response for Example 10.2 is shown in Fig. 10.8. The .PLOT statements generate graphical plots in the output file. If the .PROBE command is included, there is no need for the .PLOT commands.

Note: The output voltage, which is a ramp waveform for a square-wave input voltage, is given by

$$V_o(t) \approx -\frac{1}{R_1 C_1} \int V_{in}\, dt = -4 \times 10^{-3}\, t.$$

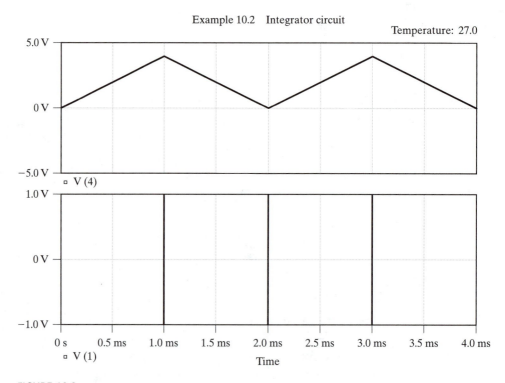

Example 10.2   Integrator circuit

Temperature: 27.0

FIGURE 10.8

Transient response for Example 10.2.

Example 10.3: Transient response of an integrator with a linear ac op-amp model

A practical differentiator circuit is shown in Fig. 10.9(a). For the input voltage as shown in Fig. 10.9(b), plot the transient response of the output voltage for a duration of 0 to 4 ms in steps of 10 μs. The op-amp, which is modeled by the circuit in Fig. 10.2(b), has $R_i = 2$ MΩ, $R_o = 75$ Ω, $C_1 = 1.5619$ μF, $R_1 = 10$ kΩ, $A_0 = 2 \times 10^5$, $G_1 = 1$, and $A_1 = 20$.

Solution   The PSpice schematic is shown in Fig. 10.10(a). The setup for transient analysis is shown in Fig. 10.10(b). The listing of the circuit file follows.

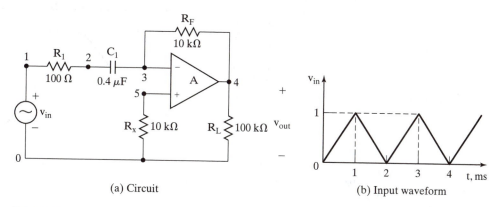

(a) Circuit   (b) Input waveform

FIGURE 10.9

Differentiator circuit.

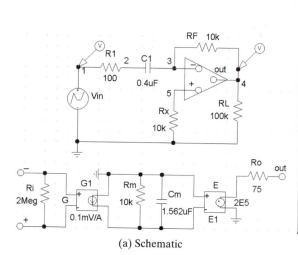

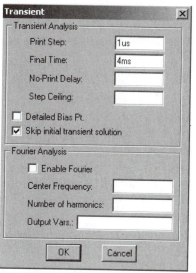

(a) Schematic   (b) Setup for transient analysis

FIGURE 10.10

PSpice schematic for Example 10.3.

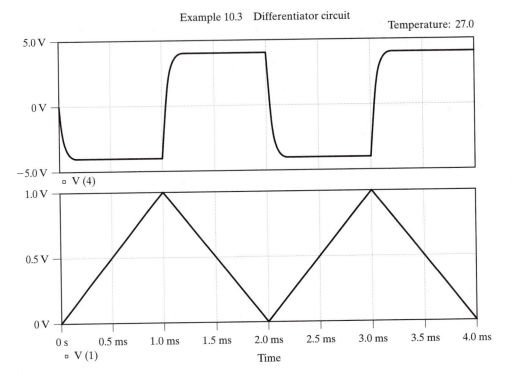

FIGURE 10.11

Transient response for Example 10.3.

Example 10.3 Differentiator circuit

```
▲ *  The maximum number of points is changed to 410. The default
   *  value is only 201.
   .OPTIONS  NOPAGE  NOECHO LIMPTS=410
   *  Input voltage is a piecewise linear waveform for transient analysis.
   VIN  1  0  PWL (0  0  1MS  1  2MS  0  3MS  1  4MS  0)
▲▲ R1  1  2  100
   RF  3  4  10K
   RX  5  0  10K
   RL  4  0  100K
   C1  2  3  0.4UF
   *  Calling op-amp OPAMP
   XA1  3  5  4  0  OPAMP
   *  Op-amp subcircuit definition
   .SUBCKT  OPAMP  1  2  7  4
   RI  1  2  2.0E6
   *  Voltage-controlled current source with a gain of 0.1M
   GB  4  3  1  2  0.1M
   R1  3  4  10K
   C1  3  4  1.5619UF
   *  Voltage-controlled voltage source with a gain of 2E+5
   EA  4  5  3  4  2E+5
```

```
         RO   5   7   75
         *   End of subcircuit OPAMP
         .ENDS  OPAMP
▲ ▲ ▲*   Transient analysis for 0 to 4 ms with 50 µs increment
         .TRAN  10US  4MS
         *  Plot the results of transient analysis at Node 4.
         .PLOT  TRAN  V(4)  V(1)
         .PROBE
  .END
```

The transient response for Example 10.3 is shown in Fig. 10.11. If the .PROBE command
is included, there is no need for the .PLOT command.

Note: The output voltage, which is a square-wave for a ramp-wave input voltage, is given by

$$v_o(t) \approx -R_1 C_1 \frac{dv_{in}}{dt} = \mp 4V.$$

Example 10.4: Frequency response of a band-pass filter

A filter circuit is shown in Fig. 10.12. Plot the frequency response of the output voltage. The frequency is varied from 10 Hz to 100 MHz with an increment of 1 decade and 10 points per decade.
For the op-amp modeled by the circuit in Fig. 10.2(b), $R_i = 2\ \text{M}\Omega$, $R_o = 75\ \Omega$, $C_1 = 1.5619\ \mu\text{F}$,
$R_1 = 10\ \text{k}\Omega$, $A_0 = 2 \times 10^5$, $G_1 = 1$, and $A_1 = 20$.

Solution The PSpice schematic is shown in Fig. 10.13(a). The setup for ac analysis is shown in
Fig. 10.13(b). The listing of the circuit file follows.

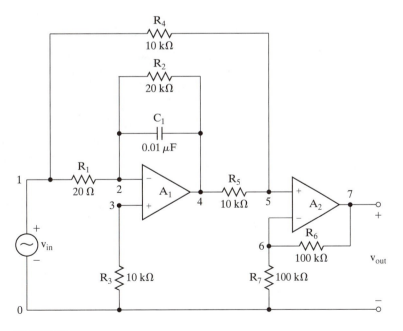

FIGURE 10.12

A filter circuit.

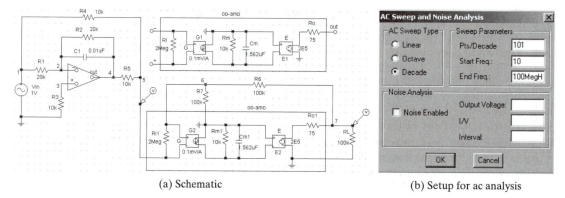

(a) Schematic   (b) Setup for ac analysis

FIGURE 10.13

PSpice schematic for Example 10.4.

Example 10.4   A filter circuit

▲ *   Input voltage is a 1-V peak for ac analysis or frequency response.
    VIN 1 0 AC 1
    R1   1 2 20K
    R2   2 4 20K
    R3   3 0 10K
    R4   1 5 10K
    R5   4 5 10K
    R6   6 7 100K
    R7   6 0 100K
    C1   2 4 0.01UF
▲▲ *   Subcircuit call for OPAMP
     XA1 2 3 4 0 OPAMP
     XA2 5 6 7 0 OPAMP
     *   Subcircuit definition for OPAMP
     .SUBCKT OPAMP 1 2 7 4
     RI 1 2 2.0E6
     *   Voltage-controlled current source with a gain of 0.1 M
     GB 4 3 1 2 0.1M
     R2 3 4 10K
     C1 3 4 1.5619UF
     *   Voltage-controlled voltage source of gain 2E+5
     EA 4 5 3 4 2E+5
     RO 5 7 75
     *   End of subcircuit definition
     .ENDS OPAMP
▲▲▲ *   AC analysis for 10 Hz to 100 MHz with a decade increment and
      *   10 points per decade
      .AC DEC 10 10HZ 100MEGHZ
      *   Plot the results of ac analysis
      .PLOT AC VM(7) VP(7)
      .PROBE
   .END

The frequency response for Example 10.4 is shown in Fig. 10.14. If the .PROBE command is included, there is no need for the .PLOT command.

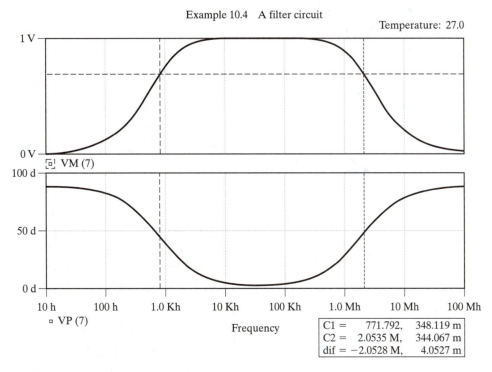

Example 10.4   A filter circuit

Temperature: 27.0

☐ VM (7)

☐ VP (7)

Frequency

C1 =	771.792,	348.119 m
C2 =	2.0535 M,	344.067 m
dif =	−2.0528 M,	4.0527 m

FIGURE 10.14

Frequency response for Example 10.4.

Note: The output of the integrator of the first stage is summed up with the input signal by the noninverting amplifier in the second stage. The integrating capacitor C_1 sets the low cut-off frequency. The filter exhibits a band-pass characteristic.

Example 10.5: Frequency response of a wide band-pass filter

A band-pass active filter is shown in Fig. 10.15. The op-amp can be modeled as a macromodel, as shown in Fig. 10.3. The description of the UA741 macromodel is listed in the library file EVAL.LIB. Plot the frequency response of the output voltage at Node 4 if the frequency is varied from 100 Hz to 1 MHz with an increment of 1 decade and 10 points per decade. The peak input voltage is 1 V.

Solution   The PSpice schematic is shown in Fig. 10.16(a). The setup for ac analysis is shown in Fig. 10.16(b). The listing of the circuit file follows.

Example 10.5   Band-pass active filter

```
▲ *  Input voltage of 1 V peak for frequency response
    VIN  1  0  AC  1
▲▲ R1  1  2  5K
    R2  3  4  1.5K
    R3  2  0  265K
    C1  2  4  0.01UF
    C2  2  3  0.01UF
    RL  4  0  15K
```

```
VCC   6   0   DC   12V
VEE   0   7   DC   12V
*  Subcircuit call for UA741
X1   0   3   6   7   4   UA741
*  Vi+ Vi- Vp+ Vp- Vout
*  Call library file EVAL.LIB
.LIB  EVAL.LIB
*  AC analysis for 100 Hz to 1 MHz with a decade increment and 10
*  points per decade
.AC  DEC  10  100HZ  1MEGHZ
*  Plot the results of the ac analysis: magnitude of voltage at node 4.
.PLOT  AC  VM(4)
.PROBE
.END
```

▲▲▲

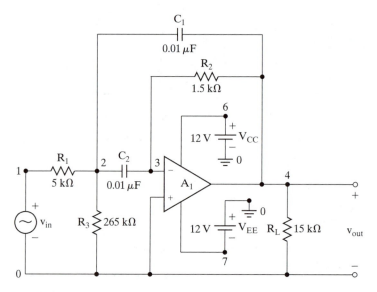

FIGURE 10.15

Band-pass active filter.

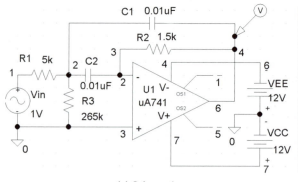

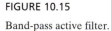

(a) Schematic

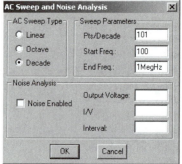

(b) Setup for ac analysis

FIGURE 10.16

PSpice schematic for Example 10.5.

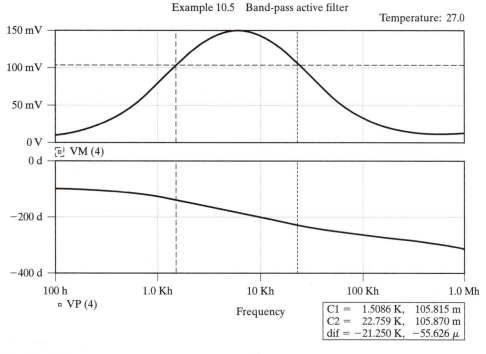

FIGURE 10.17

Frequency response for Example 10.5.

The frequency response for Example 10.5 is shown in Fig. 10.17. If the .PROBE command is included, there is no need for the .PLOT command.

Note: Figure 10.17 gives the bandwidth as 21.25 kHz. The peak gain of the narrow-width filter can be calculated from

$$K_p \approx -\frac{R_2}{2R_1} = \frac{1.5k}{2 \times 5k} = 0.15$$

and its center frequency can be found from

$$f_c = \frac{1}{2\pi\sqrt{C_1 R_1 C_2 R_2}} = \frac{1}{2\pi\sqrt{0.01^2 \times 10^{-6} \times 5 \times 1.5}} = 5811 \text{ Hz}$$

Example 10.6: Transient response of a free-running multivibrator

A free-running multivibrator circuit is shown in Fig. 10.18. Plot the transient response of the output voltage for a duration of 0 to 4 ms in steps of 20 μs. The op-amp can be modeled as a macromodel as shown in Fig. 10.3. The description of the UA741 macromodel is listed in library file EVAL.LIB. Assume that the initial voltage of the capacitor C_1 is −5 V.

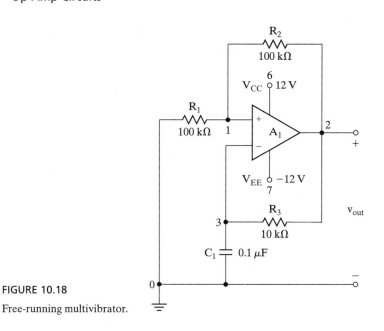

FIGURE 10.18

Free-running multivibrator.

Solution The PSpice schematic is shown in Fig. 10.19(a). The setup for transient response is shown in Fig. 10.19(b). The listing of the circuit file follows.

Example 10.6   Free-running multivibrator

```
▲   VCC  6  0  DC  12V
    VEE  0  7  DC  12V
▲▲  R1   1  0  100K
    R2   1  2  100K
    R3   2  3  10K
    C1   3  0  0.1UF  IC=-5V
    *  Subcircuit call for UA741
    XA1  1    3    6  7    2  UA741
    *    Vi+  Vi-  Vp+ Vp-  Vout
    *  Call library file EVAL.LIB
    .LIB  EVAL.LIB
▲▲▲ *   Transient analysis from 0 to 4 ms in steps of 20 μs
       .TRAN  10US  4MS  UIC
       .PROBE
   .END
```

The transient response for Example 10.6 is shown in Fig. 10.20.

Note: The threshold voltage at which the switching transition can be calculated from

$$V_{th} \approx \frac{R_1}{R_1 + R_2} V_{CC} = \frac{100k \times 12}{100k + 100k} = 6 \text{ V}$$

and the output frequency can be found from

$$f_o = \frac{1}{2C_1 R_3 \ln(1 + \dfrac{2R_1}{R_2})} = \frac{1}{2 \times 0.1 \times 10^{-3} \times 10 \times \ln 3} = 455.1 \text{ Hz}$$

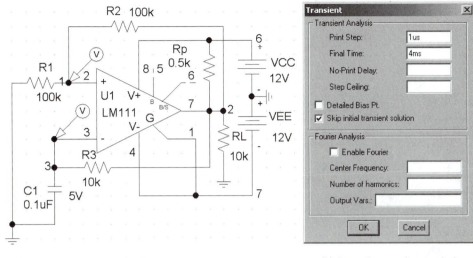

(a) Schematic

(b) Setup for transient analysis

FIGURE 10.19

PSpice schematic for Example 9.6.

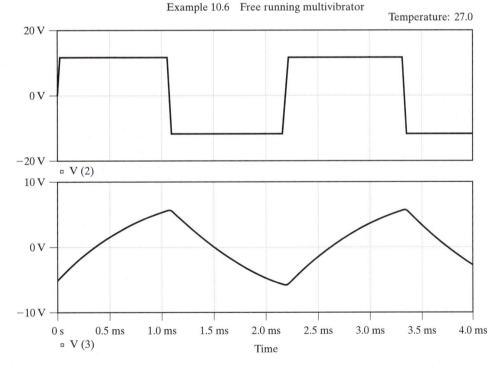

FIGURE 10.20

Transient response for Example 10.6.

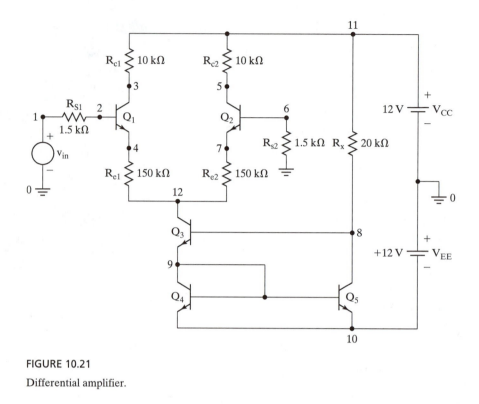

FIGURE 10.21

Differential amplifier.

Example 10.7: Transfer function analysis of a BJT differential amplifier

The circuit diagram of a differential amplifier with a transistor current source is shown in Fig. 10.21. Calculate the dc voltage gain, the input resistance, and the output resistance. The input voltage is 0.25 V(dc). The model parameters of the bipolar transistors are BF = 50, RB = 70, and RC = 40.

Solution   The PSpice schematic is shown in Fig. 10.22(a). A specific BJT is not available in the PSpice library, but the user can specify the model parameters of a Qbreak BJT. The setup for transfer function analysis is shown in Fig. 10.22(b). The parameters of model QbreakN6 are specified by a model statement in file 'RASH_MODEL.LIB.' The listing of the circuit file follows.

Example 10.7   Differential amplifier

```
▲  VCC   11   0    12V
   VEE   0    10   12V
   VIN   1    0    DC   0.25V
▲▲ RC1   11   3    10K
   RC2   11   5    10K
   RE1   4    12   150
   RE2   7    12   150
   RS1   1    2    1.5K
   RS2   6    0    1.5K
```

```
RX    11   8   20K
*   Model for NPN BJTs with model name QN
.MODEL QN NPN (BF=50 RB=70 RC=40)
Q1    3    2   4   QN
Q2    5    6   7   QN
Q3    12   8   9   QN
Q4    9    9   10  QN
Q5    8    9   10  QN
▲▲▲ *  DC transfer function analysis
       .TF  V(3,5)  VIN
.END
```

The results of the transfer-function analysis by the .TF commands are as follows:

```
****    SMALL-SIGNAL BIAS SOLUTION              TEMPERATURE = 27.000 DEG C
   NODE     VOLTAGE    NODE    VOLTAGE    NODE    VOLTAGE    NODE    VOLTAGE
 (   1)      .2500   (    2)    .2190   (   3)   1.6609   (   4)    -.5575
 (   5)    11.3460   (    6)   -.0020   (   7)   -.7057   (   8)  -10.4430
 (   9)   -11.2220   (   10) -12.0000   (  11)  12.0000   (  12)    -.7157
VOLTAGE SOURCE CURRENTS
NAME            CURRENT
VCC           -2.221E-03
VEE           -2.243E-03
VIN           -2.068E-05
TOTAL POWER DISSIPATION  5.36E-02 WATTS

****    SMALL-SIGNAL CHARACTERISTICS
        V(3,5)/VIN = -2.534E+01
        INPUT RESISTANCE AT VIN = 3.947E+04
        OUTPUT RESISTANCE AT V(3,5) = 2.000E+04
            JOB CONCLUDED
            TOTAL JOB TIME          4.01
```

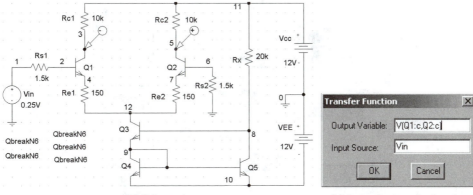

(a) Schematic   (b) Setup for transfer function

FIGURE 10.22

PSpice schematic for Example 10.6.

Note: If the output is taken from the collector of $Q1$ or $Q2$, the voltage gain will become half. That is, $A_d = -25.34/2 = -12.67$ V/V. The small-signal voltage gain can be found from

$$A_d = -\frac{\beta_f R_{C1}}{r_{\pi1} + (1 + \beta_f)R_{e1}}$$

where $r_{\pi1} = \dfrac{25.6mV}{I_{B1}}$.

Example 10.8: Finding the biasing currents and CMRR of a BJT differential amplifier

The circuit diagram of a BJT differential amplifier is shown in Fig. 10.23. (a) Find the dc biasing collector currents of the transistors. (b) Use the transfer function analysis to find the differential gain A_d, the common-mode gain A_c, and the CMRR. (c) Plot the frequency response of the differential gain and the common-mode gain. Use the PSpice model parameters of the transistor 2N2222.

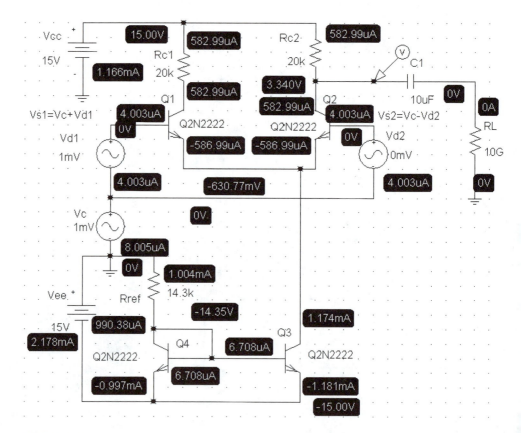

FIGURE 10.23

Differential-amplifier with differential and common-mode signals.

Solution

(a) All dc voltages and currents can be displayed by enabling 'Display Results on Schematic' from the analysis menu as shown in Fig. 10.24. Figure 10.23 shows $I_{C1} = I_{C2} = 582.99 \, \mu A$ and $I_o = 1.174 \, mA$.

Using the KVL on the input side, we get

$$V_{CC} = I_{ref} R_{ref} + V_{BE4}$$

which gives the input reference current I_{ref} through R_{ref} as

$$I_{ref} \approx I_o = \frac{V_{CC} - V_{BE4}}{Rref} = \frac{15 - 0.7}{14.3 \times 10^3} = 1mA \quad \text{(Compared to 1.174 } mA \text{ by PSpice)}$$

(b) For the differential gain, $V_{d1} = 1 \, mV$ and $V_c = 0 \, mV$. The setup for the differential mode transfer function is shown in Fig. 10.25(a). From the output file, we get the results of the differential mode as follows:

**** SMALL-SIGNAL CHARACTERISTICS

V($N_0003)/V_Vd1 = 1.951E+02 $A_d = 195.1$

INPUT RESISTANCE AT V_Vd1 = 1.448E+04

OUTPUT RESISTANCE AT V($N_0003) = 1.868E+04

For the common-mode gain, $V_{d1} = 0 \, mV$ and $V_c = 1 \, mV$. The setup for the common mode transfer function is shown in Fig. 10.25(b). From the output file, we get the results of the common mode as follows:

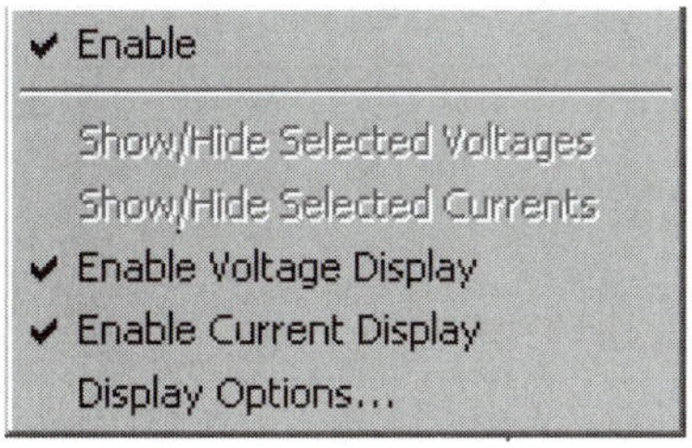

FIGURE 10.24

Enabling display results on schematic

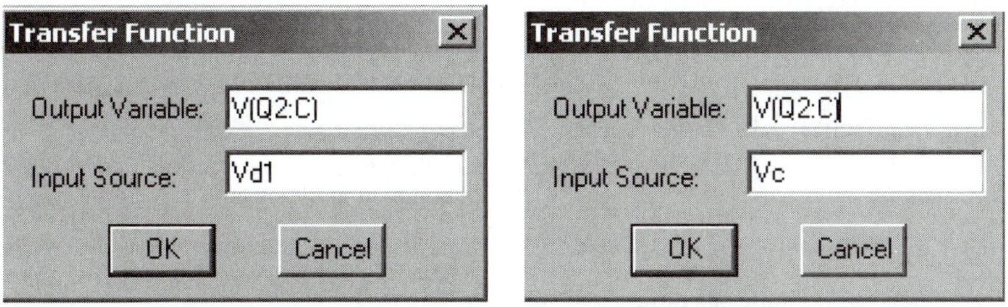

| (a) Differential mode | (b) Common mode |

FIGURE 10.25

Setup for transfer function analysis.

**** SMALL-SIGNAL CHARACTERISTICS

V($N_0003)/V_Vc = −1.318E-01 A_c = −0.1318

INPUT RESISTANCE AT V_Vc = 5.382E+06

OUTPUT RESISTANCE AT V($N_0003) = 1.868E+04

$$CMRR = \left| \frac{A_d}{A_c} \right|$$

$$= \left| \frac{195.1}{-0.1318} \right| = 1480.27 \; or \; 63.4 \; dB$$

(c) For the differential gain, $V_{d1} = 1$ mV and $V_c = 0$ mV. With a frequency sweep from 1 Hz to 10 MHz with 101 pints per decade gives the frequency response as shown in Fig. 10.26(a) which gives the low-frequency differential gain as $A_d = 195.1$.

For the common-mode gain, $V_{d1} = 0$ mV and $V_c = 1$ mV. The frequency response is shown in Fig. 10.26(b), which also gives the low-frequency common-mode gain as $A_c = 0.1318$.

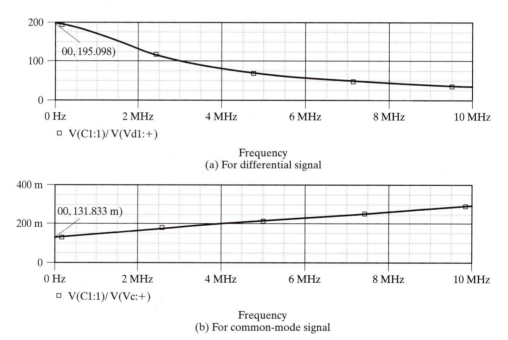

(a) For differential signal

(b) For common-mode signal

FIGURE 10.26

Frequency response of a differential-amplifier.

Example 10.9: Finding the CMRR of a BJT differential amplifier with active load

The circuit diagram of a BJT differential amplifier with active load is shown in Fig. 10.27.
(a) Find the dc biasing collector currents of the transistors. (b) Use the transfer function
analysis to find the differential gain A_d, the common-mode gain A_c, and the CMRR. Use
the PSpice model parameters of the transistor 2N2222 and 2N2907A.

Solution

(a) All dc voltages and currents are is displayed by enabling 'Display Results on Schematic' from the analysis menu as shown in Fig. 10.24. Figure 10.27 shows $I_{C1} = 584.98\ \mu A$, $I_{C2} = 581.91\ \mu A$, and $I_o = 1.174\ mA$.

(b) For the differential gain, $V_{d1} = 1\ mV$ and $V_c = 0\ mV$. The setup for the differential mode is shown in Fig. 10.25(a). From the output file, we get the results of the differential mode as follows:

$$**** \text{ SMALL-SIGNAL CHARACTERISTICS}$$

$$V(\$N\_0007)/V\_Vd1 = 1.928E+03 \qquad A_d = 1928$$

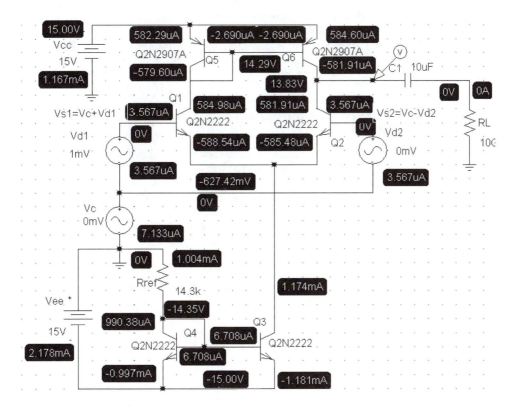

FIGURE 10.27

Differential-amplifier with active load.

INPUT RESISTANCE AT V_Vd1 = 3.756E+04

OUTPUT RESISTANCE AT V($N_0007) = 8.597E+04

For the common-mode gain, $V_{d1} = 0$ mV and $V_c = 1$ mV. The setup for the common mode is shown in Fig. 10.25(b). From the output file, we get the results of the common mode as follows:

**** SMALL-SIGNAL CHARACTERISTICS

V($N_0007)/V_Vc = −2.850E−03 $A_c = -2.850 \times 10^{-3}$

INPUT RESISTANCE AT V_Vc = 6.924E+06

OUTPUT RESISTANCE AT V($N_0007) = 8.597E+04

$$CMRR = \left| \frac{A_d}{A_c} \right|$$

$$= \left| \frac{1928}{-2.850 \times 10^{-3}} \right| = 676.49 \times 10^3 \ or \ 116.6 \ dB$$

Note: Replacing the load resistors R_{C1} and R_{C2} in the amplifier (as shown in Fig. 10.22) by active loads consisting of BJTs (as shown in Fig. 10.27) has increased the differential gain by almost 10 times and lowered the common-mode gain A_c significantly. As a result, there is a significant improvement in the CMRR.

Example 10.10: Finding the biasing currents and CMRR of a MOSFET differential amplifier

The circuit diagram of a MOSFET differential amplifier is shown in Fig. 10.28. (a) Find the dc biasing collector currents of the transistors. (b) Use the transfer function analysis to find the differential gain A_d, the common-mode gain A_c, and the CMRR. The MOS transistor has a length of 0.6 μm at minimum and can be expanded by integer increments of 0.3 μm. The minimum width is 0.9 μm and can be expanded by integer increments of 0.3um. Attributes of the NMOS are: L = 6 μm, W = 300 μm, AD = 720 μm, AS = 720 μm, PD = 302.4 μm, and PS = 302.4 μm. Generally, AS = AD = (2.4 μm × W) and PS = PD = (2.4 μm + W).
Parameters for the CMOSN model of the NMOS are:
Model parameters are LEVEL=3   PHI=0.700000   TOX=9.6000E−09
    XJ=0.200000U   TPG=1   VTO=0.6684 DELTA=1.0700E+00 LD=4.2030E−08
    KP=1.7748E−04   UO=493.4 THETA=1.8120E−01 RSH=1.6680E+01
    GAMMA=0.5382   NSUB=1.1290E+17   NFS=7.1500E+11   VMAX=2.7900E+05
    ETA=1.8690E−02   KAPPA=1.6100E−01 CGDO=4.0920E−10   CGSO=4.0920E−10
    CGBO=3.7765E−10 CJ=5.9000E−04 MJ=0.76700   CJSW=2.0000E−11
    MJSW=0.71000   PB=0.9900000 WD=1.83E−07 LAMBDA=0.02

Solution

(a) All dc voltages and currents can be displayed by enabling 'Display Results on Schematic' from the analysis menu as shown in Fig. 10.24. Figure 10.28 shows

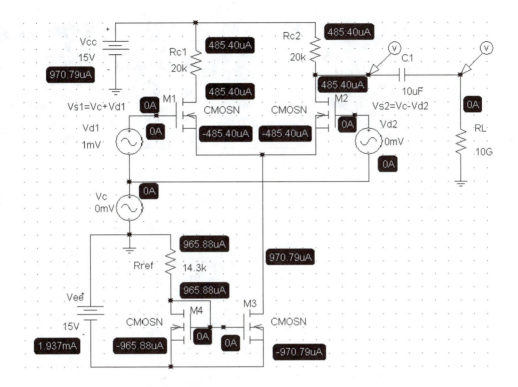

FIGURE 10.28

MOS differential-amplifier with differential and common-mode signals.

$I_{D1} = I_{D2} = 485.4\ \mu A$ and $I_o = 970.79\ \mu A$. $L = 6\ \mu m$, $W = 72\ \mu m$, $V_{TO} = 0.6684\ V$, and $K_p = 1.7748 \times 10^{-4}\ A/V^2$. Thus, the effective K_p for the values of L and W is $K_{pe} = K_p(W/L) = 1.7748 \times 10^{-4} \times (72\ \mu/6\ \mu) = 21.298 \times 10^{-4}\ A$.

Using the KVL on the input side, we get

$$V_{DD} = I_D R_{ref} + V_{Gs4} = K_{pe}(V_{Gs4} - V_{TO})^2 R_{ref} + V_{Gs4}$$

which gives an acceptable value of $V_{GS3} = V_{GS4} = 1.338\ V$ and $I_{ref} = I_{D4} = I_O = K_{pe}(V_{GS4} - V_{TO})^2 = 955.4\ \mu A$ (compared to 970.9 μA by PSpice).

(b)   For the differential gain, $V_{d1} = 1\ mV$ and $V_c = 0\ mV$. The setup for the differential mode transfer function is shown in Fig. 10.29(a). From the output file, we get the results of the differential mode as follows:

**** SMALL-SIGNAL CHARACTERISTICS

V($N_0001)/V_Vd1 = 2.547E+01 $A_d = 25.47$

INPUT RESISTANCE AT V_Vd1 = 1.000E+20

OUTPUT RESISTANCE AT V($N_0001) = 1.997E+04

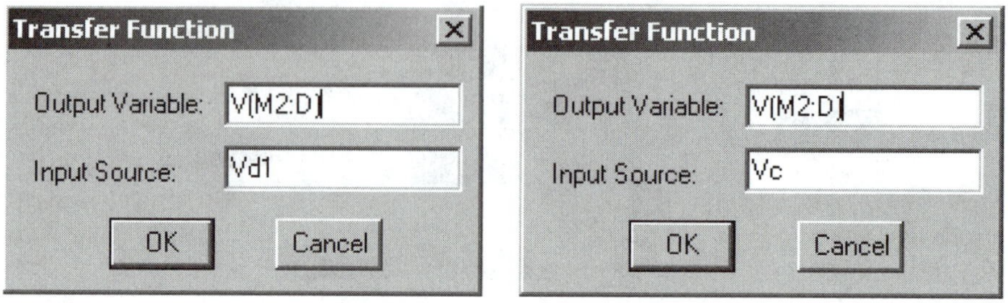

Transfer Function ☒	Transfer Function ☒
Output Variable: V[M2:D]	Output Variable: V[M2:D]
Input Source: Vd1	Input Source: Vc
OK Cancel	OK Cancel
(a) Differential mode	(b) Common mode

FIGURE 10.29

Setup for transfer function analysis.

For the common-mode gain, $V_{d1} = 0$ mV and $V_c = 1$ mV. The setup for the common mode transfer function is shown in Fig. 10.29(b). From the output file, we get the results of the common mode as follows:

**** SMALL-SIGNAL CHARACTERISTICS

V($N_0001)/V_Vc = −3.818E−03 $A_c = -3.818 \times 10^{-3}$

INPUT RESISTANCE AT V_Vc = 1.000E+20

OUTPUT RESISTANCE AT V($N_0001) = 1.997E+04

$$CMRR = \left| \frac{A_d}{A_c} \right|$$

$$= \left| \frac{25.47}{-3.818 \times 10^{-3}} \right| = 6671 \; or \; 76.484 \; dB$$

Example 10.11: Finding the CMRR of an MOS differential amplifier with active load

The circuit diagram of an MOS differential amplifier with active load is shown in Fig. 10.30. (a) Find the dc biasing collector currents of the transistors. (b) Use the transfer function analysis to find the differential gain A_d, the common-mode gain A_c, and the CMRR. (c) Complete a table of A_d, A_c and CMRR for L = 1 μm, 2 μm, 3 μm, 4 μm, 5 μm, and 6 μm. Use the PSpice model parameters of the NMOS in Example 10.10.

Parameters for the CMOSP model of the PMOS are:

LEVEL=3 PHI=0.700000 TOX=9.6000E−09 XJ=0.200000U TPG=−1
+ VTO=−0.9352 DELTA=1.2380E−02 LD=5.2440E−08 KP=4.4927E−05
+ UO=124.9 THETA=5.7490E−02 RSH=1.1660E+00 GAMMA=0.4551

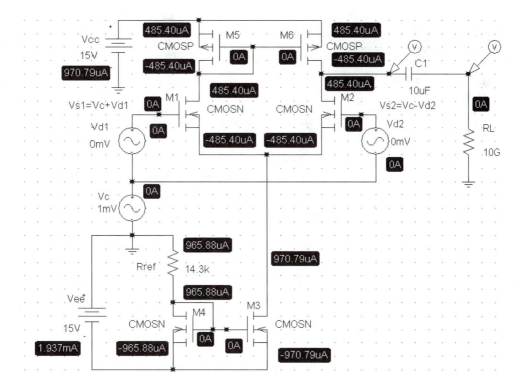

FIGURE 10.30

Differential-amplifier with active load.

+ NSUB=8.0710E+16 NFS=5.9080E+11 VMAX=2.2960E+05 ETA=2.1930E−02
+ KAPPA=9.3660E+00 CGDO=2.1260E−10 CGSO=2.1260E−10
+ CGBO=3.6890E−10 CJ=9.3400E−04 MJ=0.48300 CJSW=2.5100E−10
+ MJSW=0.21200 PB=0.930000 WD=1.75E−07 LAMBDA=0.02)

Solution

(a) All dc voltages and currents are is displayed by enabling 'Display Results on Schematic' from the analysis menu as shown in Fig. 10.24. Figure 10.30 shows $I_{D1} = I_{D2} = 485.4 \ \mu A$ and $I_o = 970.79 \ \mu A$.

(b) For the differential gain, $V_{d1} = 1$ mV and $V_c = 0$ mV. The setup for the differential mode is shown in Fig. 10.29(a). From the output file, we get the results of the differential mode as follows:

**** SMALL-SIGNAL CHARACTERISTICS

V($N_0007)/V_Vd1 = 1.130E+04 $A_d = 1.130 \times 10^4$

INPUT RESISTANCE AT V_Vd1 = 1.000E+20

OUTPUT RESISTANCE AT V($N_0007) = 4.414E+06

TABLE 10.1   Effect of MOS length on the voltage gains

L (μm)	1	2	3	4	5	6
A_d	103.2	857.3	2542	5017	8012	11300
A_c	-8.810×10^{-3}	-1.192×10^{-3}	-4.681×10^{-4}	-2.706×10^{-4}	-1.873×10^{-4}	-1.428×10^{-4}
CMRR $= \left| \dfrac{A_d}{A_c} \right|$	11.714×10^3	71.921×10^4	543.04×10^4	241.667×10^5	427.763×10^5	791.31×10^5

For the common-mode gain, $V_{d1} = 0$ mV and $V_c = 1$ mV. The setup for the common mode is shown in Fig. 10.29(b). From the output file, we get the results of the common mode as follows:

**** SMALL-SIGNAL CHARACTERISTICS

V(N_0007)/V_Vc = −1.428E−04 A_c = −1.428Ex-4

INPUT RESISTANCE AT V_Vc = 1.000E+20

OUTPUT RESISTANCE AT V(N_0007) = 4.414E+06

$$CMRR = \left| \frac{A_d}{A_c} \right|$$

$$= \left| \frac{1.130 \times 10^4}{-1.428 \times 0^{-4}} \right| = 79.13 \times 10^6 \; or \; 157.97 \; dB$$

Repeating the simulation for different values of L, we get the values in Table 10.1.

Note: The differential gain of an MOS amplifier can be varied by changing the MOS length L as illustrated through an example in Table 10.1.

REFERENCES

[1] G. Boyle, B. Cohn, D. Pederson, and J. Solomon, "Macromodeling of integrated circuit operational amplifiers." *IEEE Journal of Solid-State Circuits*, Vol. SC-9, No. 6, December 1974, pp. 353–364.

[2] I. Getreu, A. Hadiwidjaja, and J. Brinch, "An integrated-circuit comparator macromodel." *IEEE Journal of Solid-State Circuits*, Vol. SC-11, No. 6, December 1976, pp. 826–833.

[3] J. R. Hufault, *OP-AMP Network Design*. New York: John Wiley & Sons, 1986.

[4] F. W. Hughes, *OP-AMP Handbook*. Englewood Cliffs, New Jersey: Prentice Hall, 1986.

[5] *Linear Circuits—Operational Amplifier Macromodels*. Dellas, Texas: Texas Instruments, Inc., 1990.

[6] S. Progozy, "Novel applications of SPICE in engineering education." *IEEE Transactions on Education*, Vol. 32, No. 1, February 1990, pp. 35–38.

[7] M. H. Rashid, *Microelectronic Circuits: Analysis and Design*. Boston: PWS Publishing, 1999.

[8] C. F. Wojslow, *Operational Amplifiers*. New York: John Wiley & Sons, 1986.

PROBLEMS

10.1 Plot the frequency response of the integrator in Fig. 10.6 if the frequency is varied from 10 Hz to 100 kHz with a decade increment and 10 points per decade. The peak input voltage is 1 V.

10.2 Plot the frequency response of the differentiator in Fig. 10.9 if the frequency is varied from 10 Hz to 100 kHz with a decade increment and 10 points per decade. The peak input voltage is 1 V.

10.3 Repeat Example 10.2 if the macromodel of the op-amp in Fig. 10.3 is used. The supply voltages are $V_{CC} = 15$ V and $V_{EE} = -15$ V.

10.4 Repeat Example 10.3 if the macromodel of the op-amp in Fig. 10.3 is used. The supply voltages are $V_{CC} = 15$ V and $V_{EE} = -15$ V.

10.5 A full-wave precision rectifier is shown in Fig. P10.5. If the input voltage is $v_{in} = 0.1 \sin(2000\pi t)$, plot the transient response of the output voltage for a duration of 0

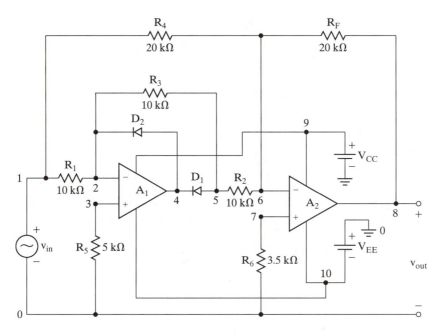

FIGURE P10.5

to 1 ms in steps of 10 μs. The op-amp can be modeled by the circuit of Fig. 10.2(b), and has $R_i = 2$ MΩ, $R_o = 75$ Ω, $C_1 = 1.5619$ μF, $R_1 = 10$ kΩ, $A_o = 2 \times 10^5$, $G_1 = 1$, and $A_1 = 20$. Use the default values for the diode model. The supply voltages are $V_{CC} = 12$ V and $V_{EE} = 12$ V.

10.6 For Fig. P10.5, plot the dc transfer characteristics. The input voltage is varied from -1 V to 1 V in steps of 0.01 V.

10.7 For Fig. P10.7, plot the dc transfer characteristics. The input voltage is varied from -10 V to 10 V in steps of 0.1 V. The op-amp can be modeled as a macromodel, as shown in Fig. 10.3. The description of the macromodel is listed in library file EVAL.LIB. Use the default values for the diode model.

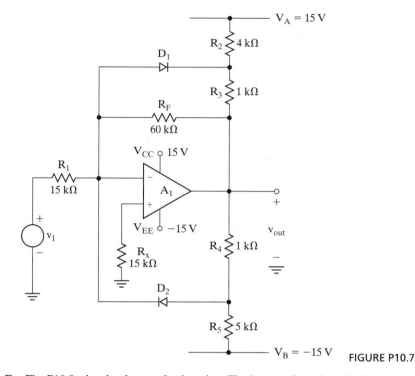

FIGURE P10.7

10.8 For Fig. P10.8, plot the dc transfer function. The input voltage is varied from -10 V to 10 V in steps of 0.1 V. The Zener voltages are $V_{Z1} = V_{Z2} = 6.3$ V. The op-amp can be modeled as a macromodel, as shown in Fig. 10.3. The description of the macromodel is listed in library file EVAL.LIB. The dc supply voltages of the op-amp are $V_{CC} = |V_{EE}| = 12$ V.

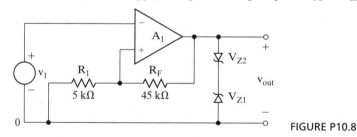

FIGURE P10.8

10.9 An integrator circuit is shown in Fig. P10.9(a). For the input voltage as shown in Fig. P10.9(b), calculate the slew rate of the amplifier by plotting the transient response of the output voltage

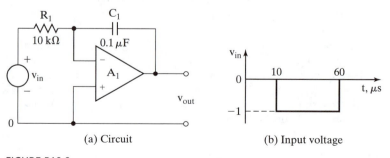

(a) Circuit

(b) Input voltage

FIGURE P10.9

for a duration of 0 to 200 µs in steps of 2 µs. For the op-amp modeled by the circuit in Fig. 10.2(b), $R_1 = 2$ MΩ, $R_0 = 75$ Ω, $C_1 = 1.5619$ µF, $R_1 = 10$ kΩ, and $A_0 = 2 \times 10^5$.

10.10 Repeat Problem 10.9 if the macromodel of the op-amp in Fig. 10.3 is used. The supply voltages are $V_{CC} = 12$ V and $V_{EE} = -12$ V.

10.11 A sine-wave oscillator is shown in Fig. P10.11. Plot the transient response of the output voltage for a duration of 0 to 2 ms in steps of 0.1 ms. The op-amp can be modeled by the

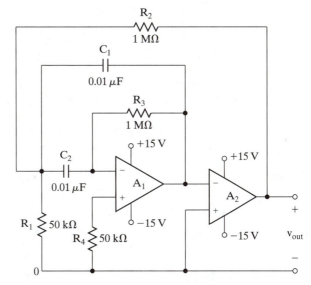

FIGURE P10.11

circuit of Fig. 10.2(b), and it has $R_i = 2$ MΩ, $R_o = 75$ Ω, $C_1 = 1.5619$ µF, $R_1 = 10$ kΩ, and $A_0 = 2 \times 10^5$.

10.12 For the gyrator in Fig. P10.12, plot the frequency response of the input impedance. The frequency is varied from 10 Hz to 10 MHz with a decade increment and 10 points per decade. For the op-amp modeled by the circuit in Fig. 10.2(b), $R_i = 2$ MΩ, $R_o = 75$ Ω, $C_1 = 1.5619$ µF, $R_1 = 10$ kΩ, and $A_0 = 2 \times 10^5$.

10.13 Use PSpice to perform a Monte Carlo analysis for five runs and for the dc analysis of Example 10.7. The output voltage is taken between Nodes 3 and 5. The model parameter is $R = 1$ for resistors. The lot deviation for all resistances is ±15%. The transistor parameter having uniform deviations is

$$BF = 50 \pm 20$$

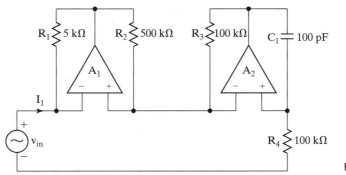

FIGURE P10.12

(a) The greatest difference of the output voltage from the nominal run is to be printed.

(b) The maximum value of the output voltage is to be printed.

(c) The minimum value of the output voltage is to be printed.

10.14 Use PSpice to perform the worst-case analysis for Problem 10.13.

10.15 Use PSpice to perform a Monte Carlo analysis for six runs and for the transient response of Problem 10.5. The model parameter is $R = 1$ for resistors. The lot deviation for all resistances is $\pm 20\%$.

(a) The greatest difference of the output from the nominal run is to be printed.

(b) The maximum value of the output voltage is to be printed.

(c) The minimum value of the output voltage is to be printed.

10.16 Use PSpice to perform the worst-case analysis for Problem 10.15.

10.17 Use PSpice to perform a Monte Carlo analysis for five runs and for the dc response of Problem 10.7. The model parameter is $R = 1$ for resistors. The lot deviation for all resistances is $\pm 15\%$. The diode parameters having uniform deviations are

$$V_{Z1} = V_{Z2} = 6.3 \text{ V} \pm 1.3 \text{ V}$$

(a) The greatest difference of the output voltage from the nominal run is to be printed.

(b) The maximum value of the output voltage is to be printed.

(c) The minimum value of the output voltage is to be printed.

10.18 Use PSpice to perform the worst-case analysis for Problem 10.17.

10.19 Use PSpice to perform a Monte Carlo analysis for five runs and for the dc response of Problem 10.8. The model parameter is $R = 1$ for resistors. The lot deviation for all resistances is $\pm 15\%$. The diode parameters having uniform deviations are

$$V_{Z1} = V_{Z2} = 6.3 \text{ V} \pm 1.3 \text{ V}$$

(a) The greatest difference of the output voltage from the nominal run is to be printed.

(b) The maximum value of the output voltage is to be printed.

(c) The minimum value of the output voltage is to be printed.

10.20 Use PSpice to perform the worst-case analysis for Problem 10.19.

10.21 Use PSpice to perform a Monte Carlo analysis for five runs and for the transient response of Problem 10.11. The model parameter is $R = 1$ for resistors and $C = 1$ for capacitors. The lot deviations for all resistances and capacitances are $\pm 15\%$.

(a) The greatest difference of the output voltage from the nominal run is to be printed.

(b) The maximum value of the output voltage is to be printed.

(c) The minimum value of the output voltage is to be printed.

10.22 Use PSpice to perform the worst-case analysis for Problem 10.21.

10.23 Use PSpice to plot the worst-case minimum and maximum values of the instantaneous output voltage $v_o(t)$ for Problem 10.5 if all resistances have uniform tolerances of $\pm 10\%$. Assume an operating temperature of 25°C.

10.24 Use PSpice to plot the worst-case minimum and maximum dc transfer characteristics for Problem 10.7 if all resistances have uniform tolerances of $\pm 10\%$. Assume an operating temperature of 25°C.

10.25 Use PSpice to plot the worst-case minimum and maximum values of the instantaneous output voltage $v_o(t)$ for Problem 10.11 if all resistances have uniform tolerances of $\pm 10\%$. Assume an operating temperature of 25°C.

Difficulties

After completing this chapter, students should be able to

- List the common types of simulation errors in SPICE and how to overcome these errors
- Handle convergence problems that are common in PSpice especially in circuits with rapidly switching voltages or currents
- Use the options setup and know its values.

11.1   INTRODUCTION

An input file may not run for various reasons, and it is necessary to know what to do when the program does not work. To run a program successfully requires the knowledge of what would not work, why not, and how to fix the problem. There can be many reasons why a program does not work; this chapter covers the problems that are commonly encountered and their solutions. The problems discussed could be due to one or more of the following causes:

Large circuits
Running multiple circuits
Large outputs
Long transient runs
Convergence
Analysis accuracy
Negative component values
Power-switching circuits
Floating nodes
Nodes with fewer than two connections
Voltage source and inductor loops.

11.2 LARGE CIRCUITS

The memory requirement (RAM) for the PSpice analysis depends on the size of a circuit file. The peak memory usage (MEMUSE) of a circuit can be found by including the ACCT option in the .OPTIONS statement and by looking at the MEMUSE number in the output file of that circuit. If the circuit does not fit into RAM, PSpice will show "Out of memory."

PSpice sends the results of an analysis (including those for the .PRINT and .PLOT statements) to an output file or one of the temporary files. It should be noted that the results are not sent to RAM.

For DOS, the total available memory can be checked by the command CHKDSK. The possible remedies for a memory problem would be the following:

1. To break the circuit file into pieces and run the pieces separately
2. To remove other resident software to release enough RAM.

11.3 RUNNING MULTIPLE CIRCUITS

A set of circuits may be placed into one input file and run as a single job. Each circuit should begin with a title statement and end with a .END command. It is important to note that there should not be any blank space or comment line between the .END statement of the preceding circuit and the title line of the following circuit.

PSpice will run all the circuits in the input file and then process each one in sequence. PSpice will store the results in a single output file, which will contain the outputs from each circuit in the order in which they appear in the input file. This feature is most suitable for running a set of large circuits overnight. For example, two circuits are combined into a single input file as follows.

Example 3.4   Transfer-function analysis

```
▲  VIN   1   0   DC    1V    ; Dc input voltage of 1 V
▲▲ R1    1   2   1K
   R2    2   0   20K
   RP    2   6   1.5K
   RE    3   0   250
   F1    4   3   VX     40  , Current-controlled current-source
   RO    4   3   100K
   RL    4   5   2K
   VX    6   3   DC    OV    ; Measures the current through $R_P$
   VY    5   0   DC    OV    ; Measures the current through $R_L$
▲▲▲ .TF  V(4)  VIN          ; Transfer-function analysis
 .END                       ; End of circuit file
```

Example 3.6   Dc sweep

```
▲  VIN   1   0    DC    1V   ; Dc input voltage of 1 V
▲▲ R1    1   2    1K
   R2    2   0    20K
   RP    2   6    1.5K
   RE    3   0    RMOD   250  ; Resistance with model RMOD
```

```
      .MODEL    RMOD    RES (R = 1.0)    ; Model statement for RE
      F1    4    3       VX    40   ; Current-controlled current-source
      RO    4    3       100K
      RL    4    5       2K
      VX    6    3       DC    OV    ; Measures the current through R_P
      VY    5    0       DC    OV    ; Measures the current through R_L
▲ ▲ ▲ *   Dc sweep for VIN from 0 to 1 V with 0.5 V increment.
      *   using the listed values of parameter R in model RMOD
      .DC  VIN   0   1.5    0.5   RES   RMOD (R) LIST    0.75  1.0  1.25
      .PRINT  DC  V(1)    V(4)      ; Prints a table in the output file
      .PROBE                        ; Graphical waveform analyzer
  .END                             ; End of circuit file
```

The PSpice schematic is shown in Fig. 11.1. One circuit is run for transfer function and the other is run for the dc sweeps with different values of R_{e1}.

11.4   LARGE OUTPUTS

A large output file will be generated if an input file is run with several circuits, for several temperatures, or with the sensitivity analysis. This will not be a problem with a hard disk. For a PC with floppy disks, the diskette may be filled with the output file. The best solutions for this problem are to

1. Direct the output to the printer instead of a file
2. Direct the output to an empty diskette in a second drive instead of the one containing PSPICE1.EXE by specifying that the PSpice programs are on drive A and the input and output files on drive B. (The command to run a circuit file would be A:PSPICE B:EX2-1.CIR B:EX2-1.OUT.)

Note: Direct the output to the hard drive if possible.

11.5   LONG TRANSIENT RUNS

Long transient runs can be avoided by the following limit options:

1. LIMPTS limits the number of print steps in a run and has a default value of 0 (meaning no limit). LIMPTS can be specified as a positive value as high as 32,000. The number of print steps is the final analysis time divided by the print interval time (plus 1).

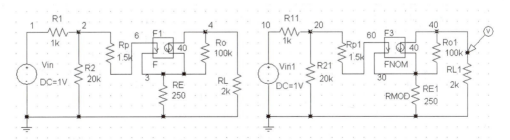

FIGURE 11.1

Combining two circuits in one PSpice schematic.

2. ITL5 is the number of total iterations in a run and has a default value of 5000. ITL5 can be set as high as 2×10^9. It is often convenient to turn it off by setting ITL5 = 0, which is the effect of setting ITL5 to infinity.

3. The user can limit data points to the Probe limit of 8000 by suppressing a part of the output at the beginning of the run with a third parameter on the .TRAN statement. For a transient analysis from 0 to 10 ms in steps of 10 μs that should print output only from 8 ms to 10 ms, the command would be .TRAN 10US 10MS 8MS.

LIMPTS and ITL5 limits can be set from the "Change Options Analysis" menu or typed in the .OPTIONS statement (see options in Fig. 6.5 and Tables 6.2 and 6.3 in Section 6.10 on page 169) as follows:

```
.OPTIONS  LIMPTS=8000  ITL5=0
```

11.6 CONVERGENCE PROBLEMS

PSpice uses iterative algorithms. These algorithms start with a set of node voltages, and each iteration calculates a new set, which is expected to be closer to a solution of Kirchhoff's voltage and current laws. That is, an initial guess is used, and the successive iterations are expected to converge to the solution. Convergence problems may occur in the following processes:

Dc sweep

Bias-point calculation

Transient analysis.

11.6.1 Dc Sweep

If a convergence problem occurs, the analysis fails and PSpice skips the remaining points of the dc sweep. The convergence problem often occurs in analyzing a circuit with regenerative feedback, such as one with Schmitt triggers. While calculating the hysteresis of such circuits, it is necessary to jump discontinuously from one solution to another at the crossover point, and the analysis fails.

A hysteresis characteristic can be obtained by using transient analysis with a piecewise linear (PWL) voltage source with a very slowly rising ramp. A very slow ramp will cause the input voltage to change slowly until the circuit switches, so that the hysteresis characteristics due to upward and downward switching can be calculated.

Example 11.1: Plotting the characteristic of a Schmitt-trigger circuit

An emitter-coupled Schmitt-trigger circuit is shown in Fig. 11.2(a). Plot the hysteresis characteristics of the circuit from the results of the transient analysis. The input voltage, which is varied slowly from 1 V to 3 V and from 3 V to 1 V, is as shown in Fig. 11.2(b). The model parameters of the transistors are IS = 1E − 16, BF = 50, BR = 0.1, RB = 50, RC = 10, TF =0.12 NS, TR = 5 NS, CJE = 0.4 PF, PE = 0.8, ME = 0.4, CJC = 0.5 PF, PC = 0.8, MC = 0.333, CCS = 1PF, and VA = 50. Print the job statistical summary of the circuit.

Solution The input voltage is varied very slowly from 1 V to 3 V and from 3 V to 1 V, as shown in Fig. 11.2(b).

The PSpice schematic is shown in Fig. 11.3(a). The ACCT option is selected from the Options menu as shown in Fig. 11.3(b). The listing of the circuit file follows.

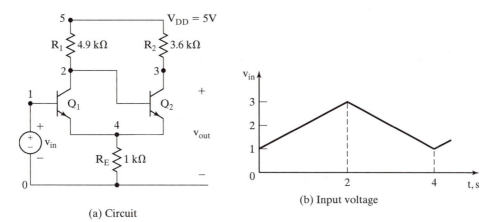

(a) Circuit

(b) Input voltage

FIGURE 11.2

Schmitt-trigger circuit.

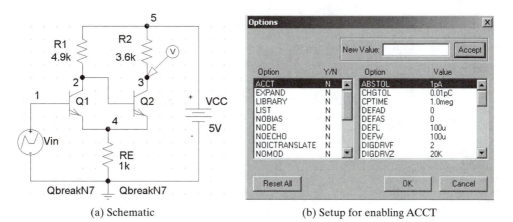

(a) Schematic

(b) Setup for enabling ACCT

FIGURE 11.3

PSpice schematic for Example 11.1.

Example 11.1 Emitter-coupled trigger circuit

```
*   Dc supply voltage of 5 V
▲ VDD 5  0  DC  5
   *  PWL waveform for transient analysis
   VIN  1  0  PWL (0  1V  2  3V  4  1V)
▲▲ R1  5  2  4.9K
   R2  5  3  3.6K
   RE  4  0  1K
   *   Q1 and Q2 with model QM
   Q1  2  1  4  QM
   Q2  3  2  4  QM
   *  Model parameters for QM
   .MODEL  QM NPN (IS=1E-16 BF=50 BR=0.1 RB=50 RC=10 TF=0.12NS TR=5NS
```

```
    +   CJE=0.4PF PE=0.8 ME=0.4 CJC=0.5PF PC=0.8 MC=0.333 CCS=1PF VA=50)
▲ ▲ ▲  * Transient analysis from 0 to 4 s in steps of 0.01 s
        .TRAN 0.01 4
        *  Printing the accounts summary
        .OPTIONS  ACCT
        .PROBE
 .END
```

The job statistical summary obtained from the output file is as follows:

**** JOB STATISTICS SUMMARY

NUNODS	NCNODS	NUMNOD	NUMEL	DIODES	BJTS	JFETS	MEETS	GASFETS
6	6	10	7	0	2	0	0	0
NSTOP	NTTAR	NTTBR	NTTOV	IFILL	IOPS	PERSPA		
12	37	39	13	2	71	72.917		
NUMTTP	NUMRTP	NUMNIT	MEMUSE					
285	55	1293	8906					

	SECONDS	ITERATIONS
MATRIX SOLUTION	20.94	5
MATRIX LOAD	59.29	
READIN	1.54	
SETUP	.05	
DC SWEEP	0.00	0
BIAS POINT	5.54	77
AC and NOISE	0.00	0
TRANSIENT ANALYSIS	122.81	1293
OUTPUT	0.00	
TOTAL JOB TIME	124.90	

The hysteresis characteristics for Example 11.1 are shown in Fig. 11.4. The default x-axis for a transient analysis is time. The x-axis setting as shown in Fig. 11.4 is changed from the x-axis settings of the plot menu in Probe.

11.6.2   Bias Point

If the node voltage(s) of a circuit changes very rapidly, PSpice may not find a stable bias point and the calculation will fail. This generally occurs in an oscillator circuit consisting of transistors and capacitors. The failure of the bias-point calculation prevents other analyses (e.g., ac analysis, sensitivity, etc.), and the PSpice simulation will stop.

The problems in calculating the bias point can be minimized by giving PSpice initial guesses for node voltages, so that it starts out much closer to the solution. The initial guesses for node voltages can be assigned by the .NODESET statement (e.g., .NODESET $V(1) = 2.5$ V). It requires a little judgment in assigning appropriate node voltages. Without any initial guesses, PSpice may not find the bias point of a circuit; but with a carefully selected initial node voltage, it will find the bias point very promptly. In the PSpice schematic, the node voltage can be set at a specific voltage as shown in Fig. 11.5(a) within the library special.slb as shown in Fig. 11.5(b).

A convergence problem in the bias-point calculation does not generally occur. This is because PSpice contains an algorithm for automatically scaling the power supplies if it is having trouble finding a solution. This algorithm first tries with the full-power supplies. If there is no convergence, and the program cannot find the bias point, then it cuts the power supplies to one-quarter strength and tries again. If there is still

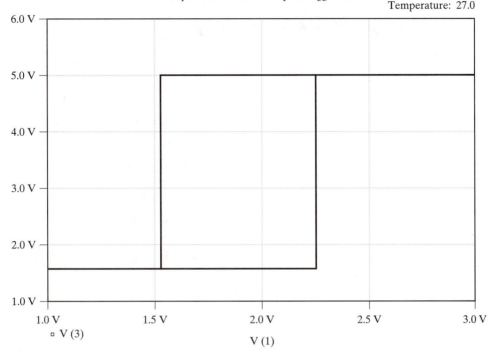

FIGURE 11.4

Hysteresis characteristics for Example 11.1.

no convergence, then the supplies are cut by another factor of four, to one-sixteenth strength, and so on.

The program will definitely find a solution for some values of the supplies scaled back to 0 V. Once it finds a solution, it then slowly adjusts the power supplies until it finds the bias point at full strength. While this algorithm is in effect, a message such as

```
Power supplies cut back to 25%
```

(or some other percentage) appears on the screen.

11.6.3 Transient Analysis

In case of failure due to a convergence problem, the transient analysis skips the remaining time. The remedies available for transient analysis follow:

1. Change the relative accuracy, RELTOL, from 0.001 to 0.01.
2. Set the iteration limits at any point during transient analysis by the ITL4 option. Setting ITL4 = 50 (by the statement .OPTIONS ITL4 = 50) will allow 50 iterations at each point. More iteration points require a longer simulation time. This is not recommended for circuits that do not have a convergence problem in transient analysis.

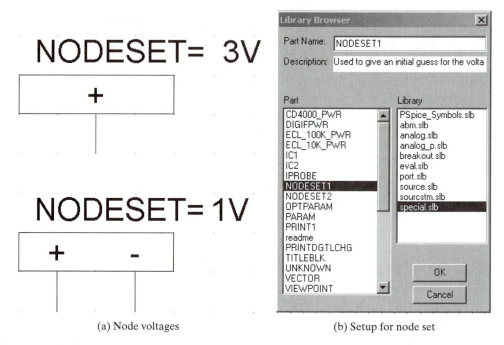

(a) Node voltages   (b) Setup for node set

FIGURE 11.5

Setup for node voltages.

11.7 ANALYSIS ACCURACY

The accuracy of PSpice's results is controlled by the following parameters in the .OPTIONS statement:

1. RELTOL controls the relative accuracy of all the voltage and currents that are calculated. The default value of RELTOL is 0.001 (0.1%), which is more accurate than necessary for many applications. The speed can be increased by setting RELTOL = 0.01 (1%), which would increase the average speed by a factor of 1.5.

2. VNTOL can limit the accuracy of all voltages to a finite value; the default value is 1 μV.

3. ABSTOL can limit the accuracy of all currents to a finite value; the default value is 1 μA.

4. CHGTOL can limit the accuracy of capacitor-charges/inductor-fluxes to a finite value.

RELTOL, VNTOL, ABSTOL, and CHGTOL limits can be set from the "Change Options Analysis" (see options in Fig. 6.5 and Tables 6.2 and 6.3 in Section 6.10) menu or typed in the .OPTIONS statement as follows:

```
.OPTIONS     ABSTOL = 5.00U     RELTOL = 0.01     VNTOL = 0.1
```

11.8   NEGATIVE COMPONENT VALUES

PSpice allows negative values for resistors, capacitors, and inductors. If a .NOISE analysis is performed on a circuit with negative values, the noise contribution will be calculated from the absolute values of resistors and will not generate negative noise. However, negative components may cause instabilities during transient analysis.

Example 11.2: Transfer function analysis of a circuit with negative resistors

A circuit with negative resistances is shown in Fig. 11.6. Calculate the voltage gain, the input resistance, and the output resistance.

Solution   The PSpice schematic with negative resistances is shown in Fig. 11.7. The listing of the circuit file follows.

Example 11.2   Circuit with negative components

```
▲  *  Dc input voltage of 1 V
   VIN   1  0   DC   1V
   *   Negative resistances
   R1    1  2   -40
   R2    2  0  -20
   R3    2  0   25
   *   Dc transfer-function analysis
   .TF   V(2)   VIN
.END
```

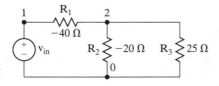

FIGURE 11.6

A circuit with negative resistances.

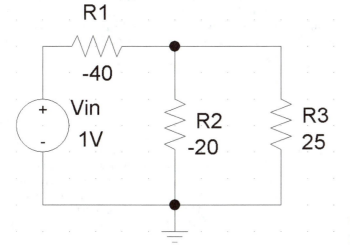

FIGURE 11.7

PSpice schematic for Example 11.3.

```
****        SMALL-SIGNAL BIAS SOLUTION        TEMPERATURE =   27.000 DEG C
NODE    VOLTAGE       NODE    VOLTAGE     NODE    VOLTAGE     NODE    VOLTAGE
(   1)    1.0000    (    2)    .7143
    VOLTAGE SOURCE CURRENTS
    NAME              CURRENT
    VIN             7.143E-03
    TOTAL POWER DISSIPATION  -7.14E-03 WATTS

****        SMALL-SIGNAL CHARACTERISTICS
    V(3) /VIN  -  7.143E-01
    INPUT  RESISTANCE AT VIN =  -1.400E+02
    OUTPUT  RESISTANCE AT V(2) =  -2.857E+01
       JOB CONCLUDED
       TOTAL JOB TIME              1.98
```

11.9 POWER-SWITCHING CIRCUITS

The switching period of a power-switching circuit may consist of many switching intervals of rapidly changing voltages and currents. The transient response of power-switching circuits may extend over many switching cycles. PSpice will try to keep the internal time step relatively short compared to the switching period, which may cause long transient runs. This problem can be solved by transforming the switching circuit into an equivalent circuit, which can represent a "quasi-steady state" of the actual circuit and can accurately model the actual circuit's response.

Example 11.3: Transient response of a single-phase full-bridge inverter with switches

A single-phase full-bridge resonant inverter is shown in Fig. 11.8. The transistors and diodes can be considered as switches whose on-state resistance is $10\ \text{m}\Omega$ and whose on-state voltage is $0.2\ \text{V}$. Plot the transient response of the capacitor voltage and the current through the load from 0 to 2 ms in steps of 10 μs. The output frequency of the inverter is $f_{\text{out}} = 4\ \text{kHz}$.

Solution When transistors Q_1 and Q_2 are turned on, the voltage applied to the load will be V_s, and the resonant oscillation will continue for the whole resonant period, first through Q_1 and Q_2 and then through diodes D_1 and D_2. When transistors Q_3 and Q_4 are turned on, the load voltage will be $-V_s$, and the oscillation will continue for another whole period, first through Q_3 and Q_4 and then through diodes D_3 and D_4. The resonant period of a series RLC circuit is approximately calculated as

$$\omega_r = \left(\frac{1}{LC} - \frac{R^2}{4L^2} \right)^{1/2}$$

For $L = L_1 = 50\ \mu\text{H}$, $C = C_1 = 6\ \mu\text{F}$, and $R = R_1 + R_{1(\text{sat})} + R_{2(\text{sat})} = 0.5 + 0.1 + 0.1 = 0.52\ \Omega$, $\omega_r = 57572.2\ \text{rad/s}$, and $f_r = \omega_r/2\pi = 9162.9\ \text{Hz}$. The resonant period is $T_r = 1/f_r = 1/9162.9 = 109.1\ \mu\text{s}$. The period of the output voltage is $T_{\text{out}} = 1/f_{\text{out}} = 1/4000 = 250\ \mu\text{s}$.

The switching action of the inverter can be represented by two voltage-controlled switches, as shown in Fig. 11.9(a). The switches are controlled by the voltages, as shown in Fig. 11.9(b). The on-time of switches, which should be approximately equal to the resonant period of the output voltage, is assumed to be 112 μs. The switch S_2 is delayed by 115 μs to take overlap into account. The model parameters of the switches are RON = 0.01, ROFF = 10E + 6, VON = 0.001, and VOFF = 0.0.

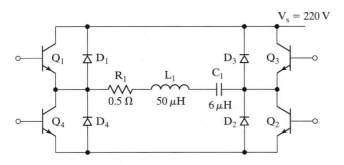

FIGURE 11.8

Single-phase full-bridge resonant inverter.

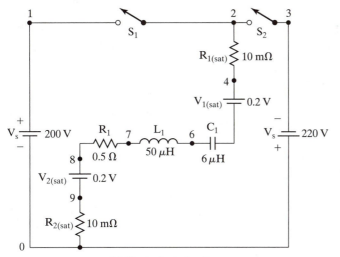

(a) Equivalent circuit

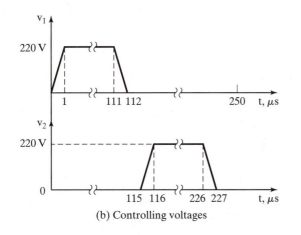

(b) Controlling voltages

FIGURE 11.9

Equivalent circuit for Fig. 11.4.

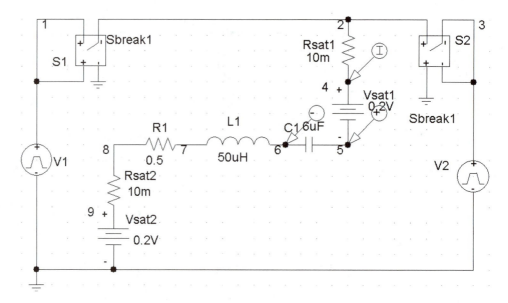

FIGURE 11.10

PSpice schematic for Example 11.3.

The PSpice schematic with voltage-controlled switches is shown in Fig. 11.10. The listing of the circuit file follows.

Example 11.3   Full-bridge resonant inverter

```
▲ *    The controlling voltage for switch S1:
   V1   1   0   PULSE (0   220V   0   1US   1US   110US   250US)
   *    The controlling voltage for switch S2 with a delay time of 115 μs
   V2   3   0   PULSE (0   −220V   115US   1US   1US   110US   250US)
   *    Voltage-controlled switches with model SMOD
▲▲ S1   1   2   1   0   SMOD
   S2   2   3   0   3   SMOD
   *   Switch model parameters for SMOD
   MODEL   SMOD   VSWITCH (RON=0.01   ROFF=10E+6   VON=0.001   VOFF=0.0)
   RSAT1   2   4   10M
   VSAT1   4   5   DC   0.2V
   RSAT2   9   0   10M
   VSAT2   8   9   DC   0.2V
   *   Assuming an initial capacitor voltage of −250 V to reduce settling time
   C1   5   6   6UF IC=−250V
   L1   6   7   50UH
   R1   7   8   0.5
   *   Switch model parameters for SMOD
   MODEL   SMOD   VSWITCH (RON=0.01   ROFF=10E+6   VON=0.001   VOFF=0.0)
▲▲▲ *   Transient analysis with UIC condition
    .TRAN   2US   500US   UIC
    .PROBE
 .END
```

The transient response for Example 11.3 is shown in Fig. 11.11.

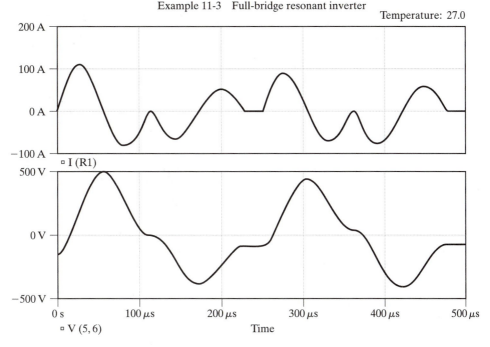

FIGURE 11.11

Transient responses for Example 11.3.

11.10   FLOATING NODES

PSpice does not allow floating nodes. A floating node is present if there is no dc path from a particular node to the ground. In general, PSpice requires that there must be a dc path from every node to the ground. Resistors, inductors, diodes, and transistors provide dc paths. If the circuit file contains a floating node, PSpice will indicate a read-in error on the screen, and the output file will contain a message indicating the node number. For example, if node 25 is floating, the output file will contain the following message:

```
ERROR: Node 25 is floating
```

Floating nodes can occur in many circuits, as shown in Fig. 11.12. Node 4 in Fig. 11.12(a) is floating because it does not have a dc path. The floating-node condition can be avoided by connecting node 4 to node 0, as shown by dotted lines (or by connecting node 3 to node 2). A similar situation can occur in voltage-controlled and current-controlled sources, as shown in Fig. 11.12(b) and 11.12(c). The model of an op-amp shown in Fig. 11.12(d) has five floating nodes. Nodes 0, 3, and 5 can be connected (or, alternatively, nodes 1, 2, and 4 may be connected) to avoid floating nodes.

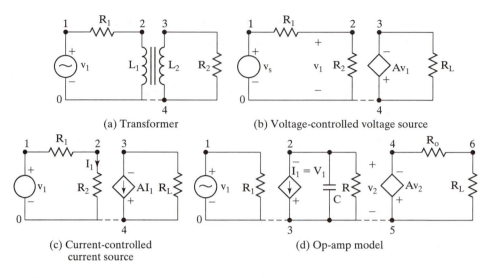

FIGURE 11.12

Typical circuits with floating nodes.

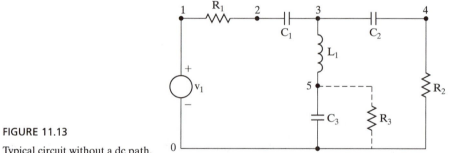

FIGURE 11.13

Typical circuit without a dc path.

Capacitors can also cause floating nodes, because there is no dc path between the two sides of a capacitor. Let us consider the circuit of Fig. 11.13. Nodes 3 and 5 do not have dc paths; however, dc paths can be provided by connecting a very large resistance R_3 (say, 100 MΩ) across capacitor C_3 as shown by the dotted lines.

Example 11.4: Providing dc paths to a passive filter

A passive filter is shown in Fig. 11.14. The output is taken from node 9. Plot the magnitude and phase of the output voltage separately against the frequency. The frequency should be varied from 100 Hz to 10 kHz in steps of 1 decade and 10 points per decade.

Solution The nodes between C_1 and C_3, C_3 and C_5, and C_5 and C_7 do not have dc paths to the ground. Therefore, the circuit cannot be analyzed without connecting resistors R_3, R_4, and R_5, as

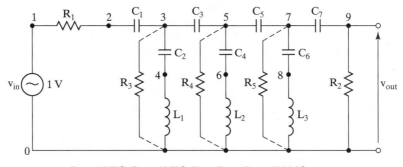

$R_1 = 10 \text{ K}\Omega, R_2 = 10 \text{ K}\Omega, R_3 = R_4 = R_5 = 200 \text{ M}\Omega$
$C_1 = 7 \text{ nF}, C_2 = 70 \text{ nF}, C_3 = 6 \text{ nF}, C_4 = 22 \text{ nF}, C_5 = 7.5 \text{ nF}$
$C_6 = 12 \text{ nF}, C_7 = 10.5 \text{ nF}, L_1 = 1.5 \text{ mH}$
$L_2 = 1.75 \text{ mH}, L_3 = 2.5 \text{ mH}$

FIGURE 11.14

A passive filter.

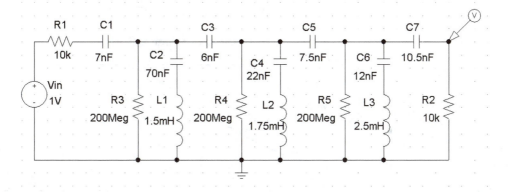

FIGURE 11.15

PSpice schematic for Example 11.4.

shown in Fig. 11.14 by dotted lines. If the values of these resistances are very high, say 200 MΩ, their influence on the ac analysis would be negligible.

The PSpice schematic of a passive filter with dc paths is shown in Fig. 11.15. The listing of the circuit file follows.

Example 11.4 A passive filter

```
▲ *    Input voltage is 1 V peak for ac analysis or frequency response.
   VIN  1  0  AC  1
▲▲ R1  1  2  10K
   R2  9  0  10K
   *   Resistances R3, R4, and R5 are connected to provide dc paths
```

```
R3  3  0  200MEG
R4  5  0  200MEG
R5  7  0  200MEG
C1  2  3  7NF
C2  3  4  70NF
C3  3  5  6NF
C4  5  6  22NF
C5  5  7  7.5NF
C6  7  8  12NF
C7  7  9  10.5NF
L1  4  0  1.5MH
L2  6  0  1.75MH
L3  8  0  2.5MH
```
▲▲▲ `* Ac analysis for 100 Hz to 10 kHz with a decade increment and`
`   * 10 points per decade`
`   .AC  DEC  10 100  10KHZ`
`   * Plot the results of ac analysis for the magnitude of voltage`
`   * at node 9.`
`   .PLOT  AC  VM(9)  VP(9)`
`   .PLOT  AG  VP(9)`
`   .PROBE`
` .END`

The frequency response for Example 11.4 is shown in Fig. 11.16.

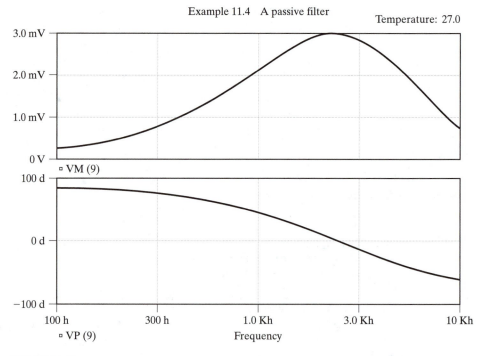

Example 11.4   A passive filter

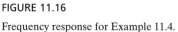

FIGURE 11.16

Frequency response for Example 11.4.

11.11 NODES WITH LESS THAN TWO CONNECTIONS

PSpice requires that each node must be connected to at least two other nodes. Otherwise, PSpice will give an error message similar to the following:

    ERROR: Less than two connections at node 10

This means that node 10 must have at least one more connection. A typical situation is shown in Fig. 11.17(a), where Node 3 has only one connection. This problem can be solved by short-circuiting resistance R_2, as shown by the dotted lines.

An error message may be indicated in the output file for a circuit with voltage-controlled sources as shown in Fig. 11.17(b). The input to the voltage-controlled source will not be considered to have connections during the check by PSpice. This is because the input draws no current and it has infinite impedance. A very high resistance (say, $R_i = 10\text{ G}\Omega$) may be connected from the input to the ground, as shown by the dotted lines.

11.12 VOLTAGE SOURCE AND INDUCTOR LOOPS

If there is a voltage source E with zero resistance ($R = 0$), PSpice will try to divide E by 0, which is impossible. It should be noted that a zero value for voltage source E will not cause problems. PSpice always looks for zero-resistance loops, and if there are any, PSpice indicates a read-in error on the screen. For example, if loop voltage V15 is involved, the output file will contain the following message:

    ERROR: Voltage loop involving V15

The zero-resistance components are independent voltage sources (V), inductors (L), voltage-controlled voltage sources (E), and current-controlled voltage sources (H). Typical circuits with such loops are shown in Fig. 11.18.

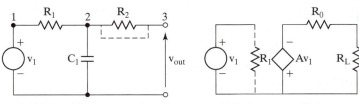

| (a) Node with one connection | (b) Voltage-controlled source |

FIGURE 11.17

Typical circuits with less than two connections at a node.

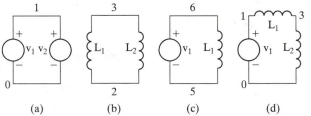

| (a) | (b) | (c) | (d) |

FIGURE 11.18

Typical circuits with zero-resistance loops.

The problem of a zero-resistance loop can be solved by adding a very small series resistance to at least one component in the loop. The resistor's value should be small enough so that it does not disturb the operation of the circuit, but it should not be less than $1\,\mu\Omega$.

11.13   RUNNING A PSpice FILE ON SPICE

PSpice should give the same results as SPICE-2G (or SPICE). However, there could be small differences due to different methods used for handling convergence problems. If a circuit file is developed for running on PSpice, it may be necessary to make some changes before running it on SPICE, because the following features are *not* available in SPICE:

1. Extended output variables for .PRINT and .PLOT statements are not available. SPICE allows only voltages between nodes $V(x)$ or $V(x, y)$ and currents through voltage sources.
2. Group delay is not available.
3. The gallium-arsenide model is not available.
4. The nonlinear magnetic (transformer) model is not available.
5. Voltage-controlled and current-controlled switches are not available.
6. The temperature coefficients for capacitors and inductors and exponential temperature coefficients for resistors are not available.
7. The model parameters RG, RDS, L, W, and WD are not available in the MOS-FET's .MODEL statement.
8. The sweep variable is limited to an independent current or voltage source.
9. Sweeping of model parameters or temperature is not allowed.
10. The .LIB and .INCLUDE statements are not available.
11. The current description of the input file must be typed in uppercase rather than in lowercase.

11.14   RUNNING A SPICE FILE ON PSpice

If a circuit file is developed for running on SPICE-2G (or SPICE), it may be necessary to make some changes before running it on PSpice, because the following features are *not* available in PSpice:

1. .DISTO (small-signal distortion) analysis and distortion output variables (HD2, DIM3, etc.) are not available. The .DISTO analysis should be replaced by a .TRAN analysis and a .FOUR analysis.
2. The IN = option in the .WIDTH statement is not available.
3. Temperature coefficients for resistors must be put into a .MODEL statement instead of in the resistor statement.
4. The voltage coefficients for capacitors and the current coefficients for inductors are placed in the .MODEL statements instead of in their own statements.

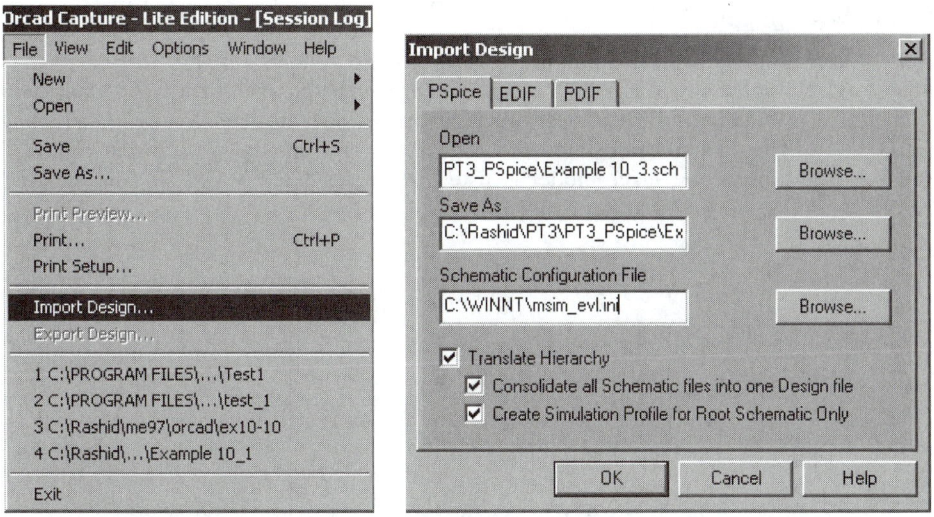

(a) OrCAD Capture (b) Importing PSpice schematic to OrCAD Capture

FIGURE 11.19

OrCAD Capture for importing PSpice schematic files.

11.15   RUNNING AN EARLIER VERSION OF SCHEMATICS

PSpice Schematics is being continuously improved with new versions. However, all versions have the same platform and the schematic files have the extension of $*.slb$. Thus, the newer version can run schematic files drawn in earlier versions.

OrCAD, Inc., owns the Microsim PSpice. The OrCAD platform is slightly different from the PSpice Schematics platform, and the file extension is $*.OPJ$. The PSpice schematic files can be imported to OrCAD Capture as shown in Fig. 11.19(a). It will require identifying the location of the Schematic Configuration file: $C:\WINNT\msim\_evl.ini$ as shown in Fig. 11.19(b). OrCAD's PSpice AD Lite Edition can run all PSpice files with extensions $*.CIR$.

REFERENCES

[1]   Wolfram Blume, "Computer circuit simulation." *BYTE*, Vol. 11, No. 7, July 1986, p. 165.

[2]   *OrCAD 9.2 Demo*. San Jose, California: Cadence Design Systems, Inc., 2001. http://www.cadencepcb.com/products/downloads/OrCADdemo/default.asp

[3]   *PSpice 9.1 Student Version*. San Jose, California: Cadence Design Systems, Inc., 2001. http://www.cadencepcb.com/products/downloads/PSpicestudent/default.asp

[4]   *PSpice Design Community*. San Jose, California: Cadence Design Systems, Inc., 2001. http://www.PSpice.com

[5]   *PSpice Manual*. Irvine, California: MicroSim Corporation, 1992.

[6]   M. H. Rashid, *SPICE for Power Electronics and Electric Power*. Englewood Cliffs, New Jersey: Prentice Hall, 1993.

PROBLEMS

11.1 For the inverter circuit in Fig. P9.6, plot the hysteresis characteristics.

11.2 For the circuit in Fig. P11.2, plot the hysteresis characteristics from the results of the transient analysis. The input voltage is varied slowly from -4 V to 4 V and from 4 V to -4 V. The op-amp can be modeled as a UA741 macromodel, as shown in Fig. 10.3. The description of the macromodel is listed in library file EVAL.LIB. The supply voltages are $V_{CC} = 12$ V and $V_{EE} = -12$ V.

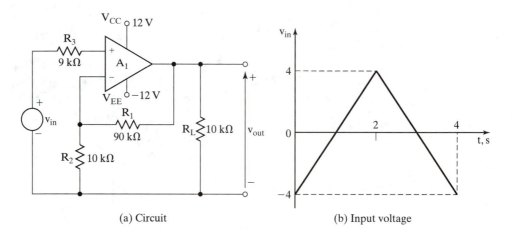

(a) Circuit   (b) Input voltage

FIGURE P11.2

APPENDIX A

Drawing in PSpice Schematics

A.1 LEARNING OBJECTIVES

The objectives of this appendix are to

- Familiarize students with the electronic circuit simulation software PSpice schematics
- Explain how to draw electrical circuits by PSpice schematics.

A.2 INSTALLING THE SOFTWARE

To install PSpice schematics, complete the following steps:

1. Place the Schematics CD-ROM in the CD drive.
2. From Windows, enter the File Manager, and click the left mouse button (**CLICKL**) on the CD drive.
3. Left click on *setup.exe*, and then select **File**, **Run**, **OK**.
4. Click **OK** to select *Install Schematics and PSpice A/D*.
5. Click **OK** to select default *C:OrCAD Dem*.
6. Select **Yes** to create *Design Center* icons as shown in Fig. A.1.
7. Click the left mouse button once on the Design Center icon; the Design Center window will open.

A.3 OVERVIEW

The Design Center software package has three major interactive programs: *Schematics*, *PSpice*, and *Probe*. *Schematics* is a powerful program that lets us build circuits by drawing them within a window on the monitor. *PSpice A/D* allows us to analyze the circuit

FIGURE A.1

Icons for PSpice schematics and PSpice A/D.　(a) Schematics　(b) PSpice A/D

FIGURE A.2

General layout of PSpice schematics.

created by *Schematics*, and generate voltage and current solutions. *Probe* is a graphic postprocessor that lets us display plots of voltages, currents, impedance, and power.

The general layout of the Design Center is shown in Fig. A.2. At the top of the display you will find 11 menus. However, **File**, **Edit**, **Draw**, **View**, and **Analysis** are most frequently used. For any help, check with the **Help** menu.

A.3.1 Mouse

The mouse follows an *object–action* sequence. First you select an object, and then you perform an action, as illustrated in the following exercise:

- Click the mouse on a menu title so that it stays open. Then, click on the command you want.
- A single click on the left mouse button (**CLICKL**) selects an item.
- A double click on the left mouse button (**DCLICKL**) *performs an action* such as to end a mode or edit a selection.
- To drag a selected item, *left click, hold down, and move the mouse* (**CLICKLH**). Release the left button when placed.
- To end a mode or edit a selection, *right click once* (**CLICKR**).
- To repeat an action, *double click the right* mouse button (**DCLICKR**).

The following table lists the actions and functions of the mouse in PSpice:

Button	Action	Function
Left	Single click	Select an item
	Double click	End a mode
	Double click on a selected object	Edit a selection
	Shift + single click	Extend a selection
Right	Single click	Abort the mode
	Double click	Repeat an action

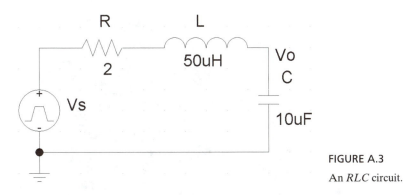

FIGURE A.3

An *RLC* circuit.

A.4 THE CIRCUIT ANALYSIS PROCESS

As an example, we will draw and analyze the pulse response of an *RLC* circuit as shown in Fig. A.3. The steps to draw and analyze a circuit are as follows:

1. Draw the circuit under *Schematics*.
2. Select the mode of analysis under *Schematics*.
3. Simulate the circuit under *PSpice*.
4. Display the results under *Probe*.

A.5 DRAWING THE CIRCUIT

The steps to draw a circuit are as follows:

1. Get components from the Get New Part command.
2. Place them on the drawing board.
3. Wire the components together.
4. Level components.
5. Set component values or models.
6. View the schematic.
7. Save the schematic.

A.5.1 Getting and Placing Components

Let us start by placing a pulse source, a resistor, an inductor, a capacitor, and a ground on the drawing board. From the schematic editor, you can place parts from the component libraries onto your schematic. Use the Get New Part command in Draw menu shown in Fig. A.4.

Choose the Get New Part command from the Draw Menu, and choose Browse to browse the list of libraries; or use this command to enter the name of a known part, shown in Fig. A.5:

- Choose **OK**, or double click after selecting the part. The chosen part becomes the "current part" and is ready to be placed on your schematic. (The cursor is replaced by the shape of the part.)

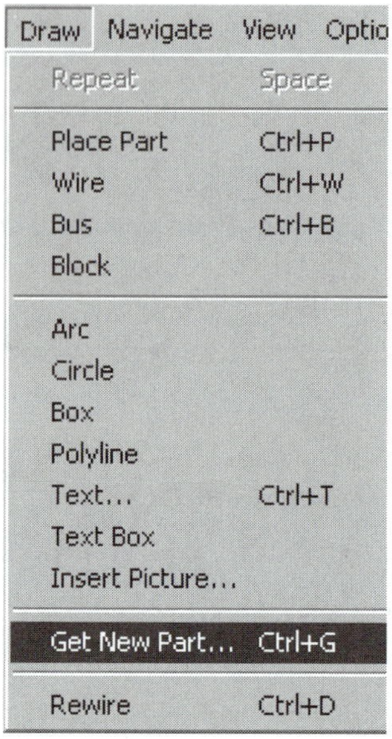

FIGURE A.4

Draw menu.

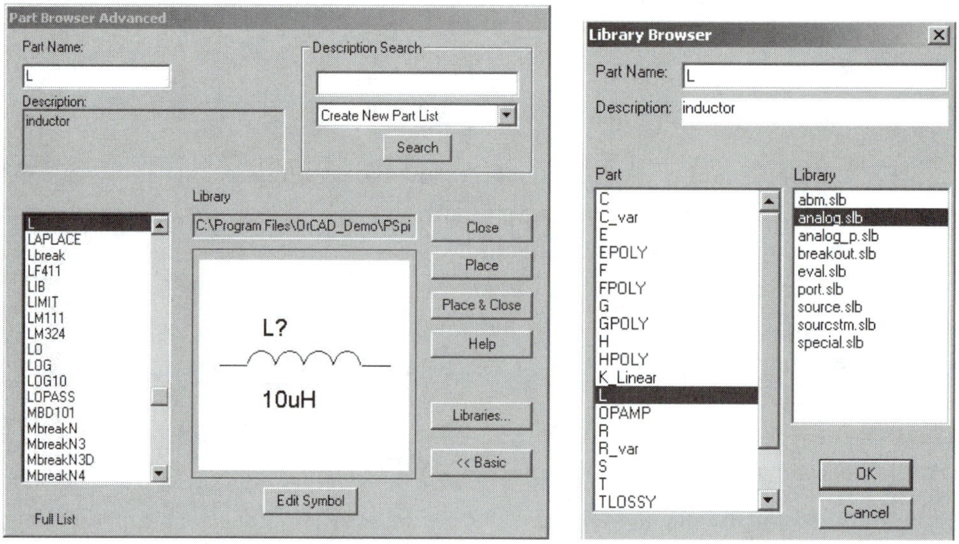

(a) From the part name

(b) From the part library

FIGURE A.5

Get new part.

- Left click to place an instance of the part; double click to place the part and end the mode; or right click to end the mode without placing the part. To drag, point to an object, press and hold the left mouse button, and move the mouse. When the object is where you want it, release the mouse button.

- To move a component, point to it, press and hold the left mouse button, and drag it to a new location.
- To remove a component, select it, and choose Delete from the **Edit** menu.

To demonstrate this technique, place a pulse source (VPULSE) from **source.slb** library, a resistor (R), an inductor (L), and a capacitor (C) from **analog.slb** library, and a ground symbol (AGND) from **port.slb** library. Arrange them as shown in Fig. A.6.

A.5.2 Rotating Components

Now rotate the capacitor so it can be wired neatly into the circuit. Each time you rotate a component, it turns clockwise 90 degrees. The Flip Menu is used to flip a selected object to produce a mirror-image of the object. Complete the following exercise:

- To rotate the capacitor (or other component), select it and choose Rotate from the **Edit** menu, as shown in Fig. A.7. If you select an area of your schematic, the area is rotated around the center of the selection box. This is done as shown in Fig. A.8.
- To flip a component, select it and choose Flip from the **Edit** menu.

Hint: If the Rotate command is dimmed, the capacitor isn't selected. Try again by pointing to it so the pointer becomes a hand, and then click the left mouse button.

- To deselect the selected capacitor (or other selected component), click it with the *right* mouse button or click an empty spot with the left mouse button.
- To drag or rotate two or more components at once, first select them by drawing a rectangle around the components, and then drag or rotate the rectangle.
- To draw a rectangle around components, point above and beside one of the components you want to select. Press and hold the left mouse button, and drag diagonally until the rest of the components are in the rectangle that appears.
- To deselect one of the selected components, click it with the *right* mouse button. To deselect everything on the marked rectangle, click an empty spot with the left mouse button.

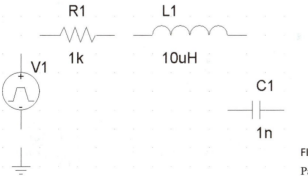

FIGURE A.6

Parts for an *RLC* circuit.

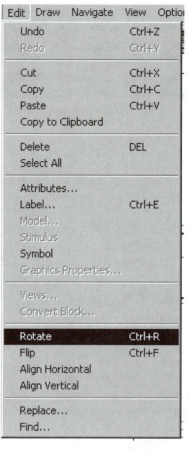

FIGURE A.7

Edit menu.

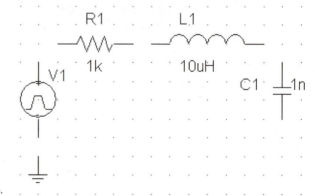

FIGURE A.8

After rotating component C1 in Fig. A.6.

A.5.3 Wiring Components

Once components are placed on the drawing board, you need to connect them. You can use the schematic editor to draw wires on your schematic and/or make vertices as follows:

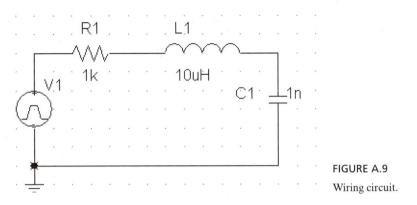

FIGURE A.9

Wiring circuit.

- To draw a wire, choose Wire from the **Draw** menu.
- When the cursor changes to a pencil shape, left click to start drawing.
- Move the mouse in any direction to extend the wire.
- Double click the left mouse button to end the wire and terminate the mode, or left click to form a vertex (corner) and continue drawing the wire.
- Wire the circuit as shown in Fig. A.9.

Hint: To bring back the last command used (for wire) (shown in the lower **right-hand** corner after *Cmd:*), **DCLICKR**.

A.5.4   Labeling a Component

You can also Place a label on selected wires, bus segments, or ports. Wire and bus segments or ports may display multiple labels; however, all labels for a segment will contain the same text. Each component in a circuit can be labeled. We will assign levels R, L, C, V_s, and V_o by completing the following steps:

1. To edit a label, select a wire, bus segment, or port for which to edit the label.
2. Choose Attributes from the **Edit** menu, or double-click on the label to bring up the dialog box.
3. Enter the text for the label, as shown in Fig. A.10(a).
4. Choose **OK**.
5. Change all levels R, L, C, V_s, and V_o as shown in Fig. A.10(b).

Hint: To move the level, **CLICKL** on it and move it to a desired location.

6. Retrieve text from **Text** on the **Draw** menu.

A.5.5   Adding Text

You can place text anywhere on your schematic and size it to suit your needs by completing the following steps:

1. To add text to your schematic, choose Text from the **Draw** menu.
2. In the dialog box, type in the desired text, as shown in Fig. A.11(a).

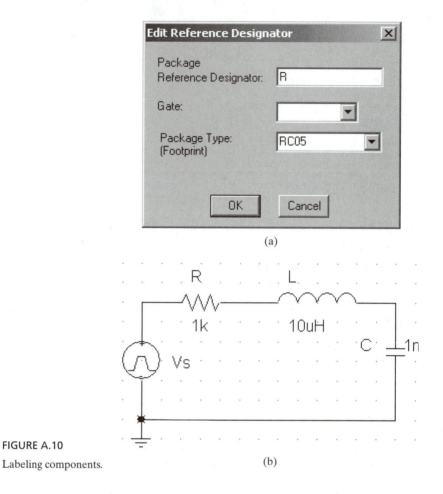

(a)

FIGURE A.10

Labeling components.

(b)

3. To change the font size, modify the Font Size shown in the dialog box to suit your needs.

4. Choose **OK**.

5. Move the text to the desired location on the schematic, and left click to place the text. Right click to exit text mode.

6. Place the output voltage, V_o, as shown in Fig. A.11(b).

A.5.6 Editing Attributes

An attribute of a schematic item consists of a name/value pair. Complete the following exercise:

1. To edit the attributes of a selected object(s), select the object(s) to edit.

2. Choose Attributes from the **Edit** menu or double-click on the attribute text to bring up the Edit Attributes dialog box directly as shown in Fig. A.12(a).

3. Select an individual attribute: A dialog box appears in which you can enter a new value for the attribute.

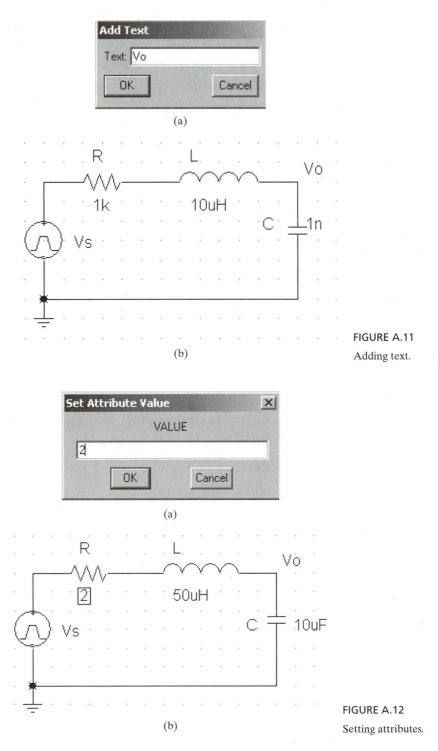

(a)

(b)

FIGURE A.11
Adding text.

(a)

(b)

FIGURE A.12
Setting attributes.

4. Select an entire part: A dialog box appears showing all of the attributes that may be edited for that part.
5. Set $R = 2 \, \Omega$, $L = 50 \, \mu H$, and $C = 10 \, \mu F$, as shown in Fig. A.12(b).

Hint: A quick way to set a component's value is to double click it.
Another Hint: Changes made to attributes in the schematic editor occur on the particular part instance only. Changes do not affect the underlying symbol in the library.

6. To change the value of an attribute, select it from the list. The name and value should appear in the edit fields at the top of the dialog.
7. Change the value in the value edit field and press the Save Attribute button.
8. To delete an attribute, select it from the list and press the Delete button.
9. To change whether the attribute name and/or value is shown on the schematic, select it and press the Change Display button.
10. To add a new attribute, type the name and value in the edit fields, and press Save Attribute. Set the pulse source Vs as shown in Fig. A.13(a). Double click it and then type v1 = 0V, v2 = 1V, TD (delay time) = 0, TR (rise time) = 1ns, TF (fall time) = 1 ns, PW (pulse width) = 0.5 ms, and PER (period) = 1 ms.

Hint: Protected attributes are marked with an '*' and cannot be changed in the Schematic Editor.

11. Choose OK to incorporate the change.

A.5.7 Viewing the Schematic

You can change the viewing scale on your schematic from the menu shown in Fig. A.14. The following are the options available:

Menu Item	Function
Fit	Resets the viewing scale so that all parts, wires, and text can be seen.
In	Allows you to view an area on the schematic at closer range (i.e., to magnify). After selecting this item, a crosshair appears. Move the crosshair to the area on which you want to zoom in.
Out	Changes the viewing scale so that you can view the schematic from a greater distance (i.e., to view more of the schematic on the screen). After selecting this item, a crosshair appears. Move the crosshair to define the center of the viewing area.
Area	Allows you to select a rectangular area on your schematic to be expanded to fill the screen. If you already have a selection box on your screen when you choose Area, the contents of the selection box will be expanded. Or you can drag the mouse to form a selection box around the portion of the schematic you want to have expanded. The items contained within the selection box will be expanded to fill the screen.
Entire Page	Allows you to view the entire schematic page at once.

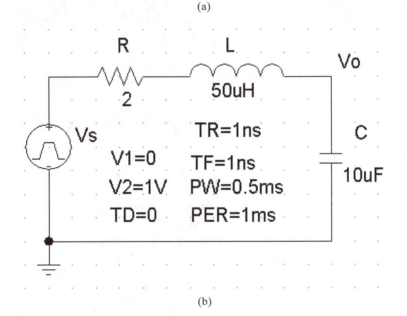

FIGURE A.13

Changing source attributes.

A.5.8   Saving The Circuit File

Your schematic files are automatically saved into the current directory, unless otherwise specified. To save an unnamed circuit or make a copy of a circuit, use the *Save As* command from the **File** menu, as shown in Fig. A.15:

1. To save your schematic file, choose **Save** from the **File** menu to save changes to the current file. If this is a new file, you will be prompted for a file name.

2. Type a file name in the text entry box. You do not need to type a file name extension. A circuit name must be a valid DOS file name. All schematic files are

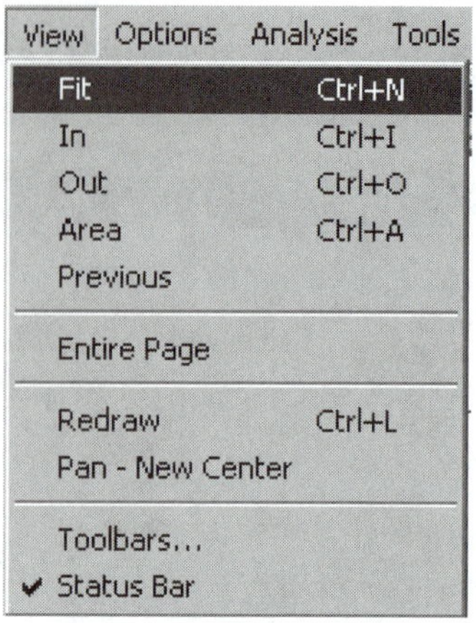

View Options Analysis Tools

Fit	Ctrl+N
In	Ctrl+I
Out	Ctrl+O
Area	Ctrl+A
Previous	
Entire Page	
Redraw	Ctrl+L
Pan - New Center	
Toolbars...	
✔ Status Bar	

FIGURE A.14

View menu.

File Edit Draw Navigate View Optio

New	
Open...	
Close	
Export...	
Save	Ctrl+S
Save As...	
Checkpoint...	
Print...	
Print Setup...	
Edit Library	
Symbolize...	
Reports...	
View Messages	F10
1 Example A_12.sch	
2 C:\Rashid\...\EXB-11.SCH	
3 C:\Rashid\...\EXB-10.SCH	
4 C:\Rashid\...\exa-27.sch	
Exit	

FIGURE A.15

File menu.

automatically given a file name extension SCH. For example, the file name FIGA_13 will be saved as FIG A_13.SCH

3. Choose **Save As** from the **File** menu to save changes to a different file name.

4. Choose OK.

A.6   COPYING AND CAPTURING SCHEMATICS

A.6.1   Copy to Clipboard

You can use this command from the **Edit** menu to copy an item or items from your schematic for use in another Windows program:

1. Use the mouse to select a rectangular area on the schematic to copy.
2. Choose **Copy to Clipboard** from the **Edit** menu.
3. Open the Windows program into which you want to paste the item(s).
4. Use the **Paste** command from the other Windows application to paste the item into the new file.

A.6.2   Cut, Copy or Paste

The **Cut** command deletes items from your schematic and places them in the Paste buffer. The **Copy** command copies items from your schematic to the **Paste** buffer. The **Paste** command places the contents of the Paste buffer on the schematic.

To cut or copy an item(s) on your schematic, complete the following steps:

1. Select the item(s) on your schematic to be cut or copied.
2. Choose **Cut** or **Copy** from the **Edit** menu.

The following steps allow you to paste a cut or copied item(s) onto your schematic:

1. After cutting or copying an item(s) with the **Cut** or **Copy** commands, choose **Paste** from the **Edit** menu.
2. Place the cursor at the desired location for the item that was cut or copied, and left click to place the item(s). Right click to exit the mode.

Drawing in OrCAD Capture Lite

B.1 LEARNING OBJECTIVES

The objectives of this appendix are to

- Familiarize students with the electronic circuit simulation software OrCAD Capture
- Explain how to draw electrical circuits with OrCAD Capture.

B.2 INSTALLING THE SOFTWARE

To install OrCAD Capture Lite, complete the following steps:

1. From Windows, enter the File Manager, and click the left mouse button (**CLICKL**) on the CD drive. Double-click on the subdirectory "Orcad Capture Lite 9.2."
2. Left click on *setup.exe*, and then select **File**, **Run**, **OK**. OK the virus protection message and click Next.
3. Select "Capture" and "PSpice" as shown in Fig. B.1(a). Click Next.
4. Click Next to select the default location C:\Program Files\OrcadLite.
5. Select **Yes** to choose *Program Folder: OrCAD Family Release 9.2 Lite Edition* as shown in Fig. B.1(b). Click Next to accept the current settings and install Capture and PSpice.
6. Click OK after reading the Acrobat Reader Installation Instructions.
7. Click Finish to "Launch the internet browser to view the Release Notes."
8. Create CaptureLite and PSpice A/D icons from the Start, Programs, OrCAD Family, and CaptureLite menus. These icons are shown in Fig. B.2.
9. Click the left mouse button once on the CaptureLite icon; the CaptureLite window will open.

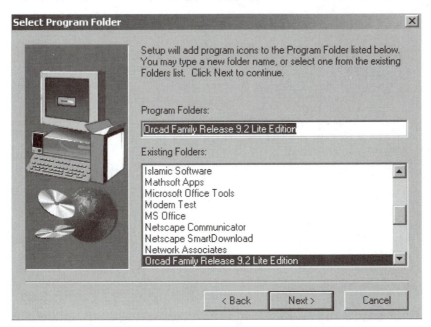

(a) Selecting products

(b) Selecting program folder

FIGURE B.1

Installation setup for OrCAD Capture.

FIGURE B.2

Icons for OrCAD Capture and PSpice A/D.   (a) OrCAD Capture Lite   (b) PSpice A/D Student Edition

B.3   OVERVIEW

The OrCAD Capture software package has three major interactive programs: *Capture*, *PSpice*, and *Probe*. *Capture* is a powerful program that lets us build circuits by drawing them within a window on the monitor. *PSpice A/D* allows us to analyze the circuit created by *Capture* and generate voltage and current solutions. *Probe* is a graphic postprocessor that lets us display plots of voltages, currents, impedance, and power.

The general layout of the Capture program is shown in Fig. B.3. The top menu shows ten main choices. The right-side menu shows the schematic "Drawing" menu for

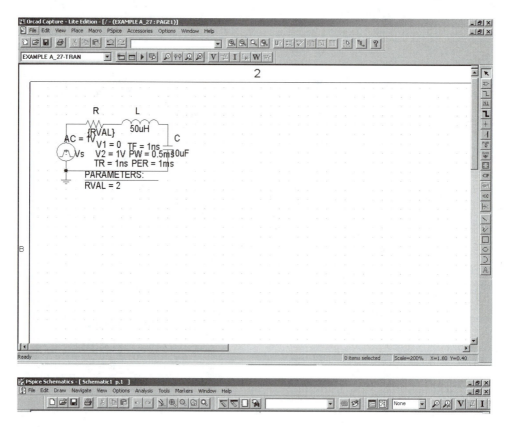

FIGURE B.3

General layout of OrCAD Capture.

selecting and placing parts. **File**, **Edit**, **View**, **Place**, and ***PSpice*** are most frequently used. A Help menu is provided for additional assistance.

B.3.1   Mouse

The mouse follows an *object–action* sequence. First you select an object, and then you perform an action:

- Click the mouse on a menu title so it stays open. Then, click on the command you want.
- A single click on the left mouse button (**CLICKL**) selects an item.
- A double click on the left mouse button (**DCLICKL**) *performs an action* such as ending a mode or editing a selection.
- To drag a selected item, *left click and hold down and move the mouse* (**CLICKLH**). Release the left button of the desired location.
- To end a mode or edit a selection, *right click once* (**CLICKR**).
- To repeat an action, *double click the right* mouse button (**DCLICKR**).

The following table lists the actions and functions for the mouse in OrCAD Capture:

Button	Action	Function
Left	Single-click	Select an item
	Double-click	End a mode
	Double-click on a selected object	Edit a selection
	Shift + single-click	Extend a selection
Right	Single-click	Abort the mode
	Double-click	Repeat an action

B.4   THE CIRCUIT ANALYSIS PROCESS

As an example, we will draw and analyze the pulse response of an *RLC* circuit as shown in Fig. B.4. The following steps are used to draw and analyze a circuit:

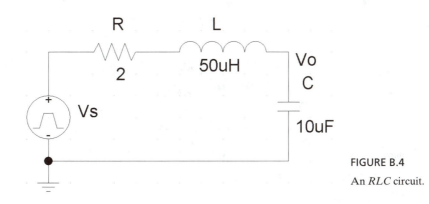

FIGURE B.4

An *RLC* circuit.

1. Draw the circuit under *Capture*.
2. Select the mode of analysis under *PSpice*.
3. Simulate the circuit under *PSpice A/D*.
4. Display the results under *Probe*.

B.5 DRAWING THE CIRCUIT

The following steps are used to draw a circuit:

- Begin a New Project.
- Get components using the Get New Parts command.
- Place them on the drawing board.
- Wire the components together.
- Level components.
- Set component values or models.
- View the schematic.
- Save the schematic.

B.5.1 Beginning a New Project

Let us begin a new project to draw and analyze, call it **Example B.3**. From the **File** menu, choose **New** and then select **Project** as shown in Fig. B.5.

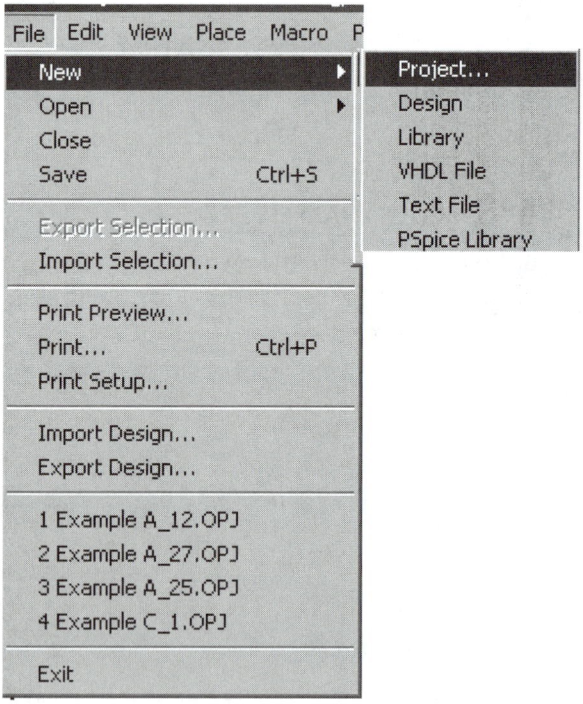

FIGURE B.5

File menu for a new project.

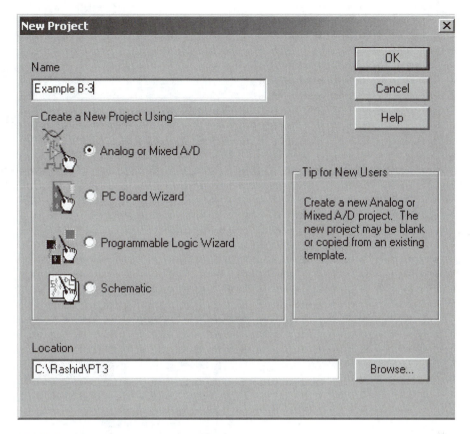

FIGURE B.6

New Project dialog box.

The **New Project** dialog box opens as shown in Fig. B.6. Choose a file name for the new project (e.g., **Example B.3**), select **Analog or Mixed A/D**, and choose a location for this new file (e.g., **C:/Rashid/PT3**). Next, select **Create a blank project** as shown in Fig. B.7.

B.5.2   Getting and Placing Components

Let us start by placing a pulse source, a resistor, an inductor, a capacitor, and a ground on the drawing board. From the capture editor, you can place parts from the component libraries onto your schematic.

1. Use the Part command from the **Place** menu as shown in Fig. B.8 or from the Place Part button on the right side.

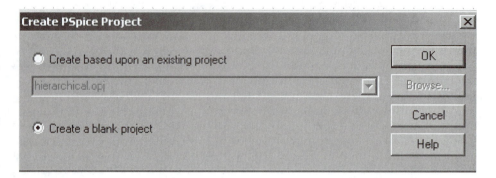

FIGURE B.7

Creating a new blank project.

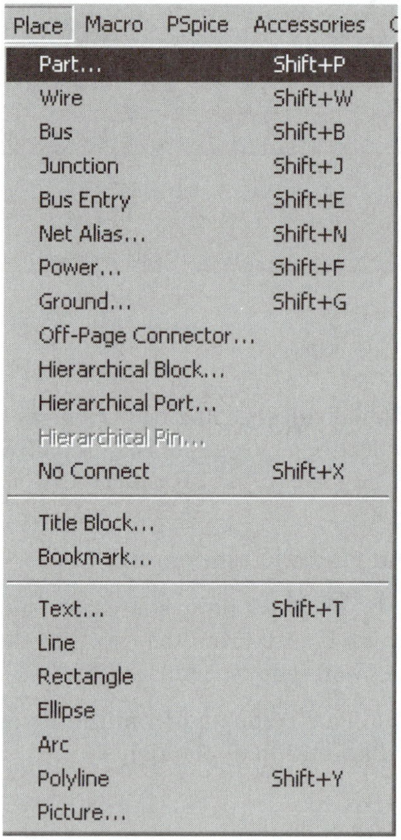

FIGURE B.8

Place menu.

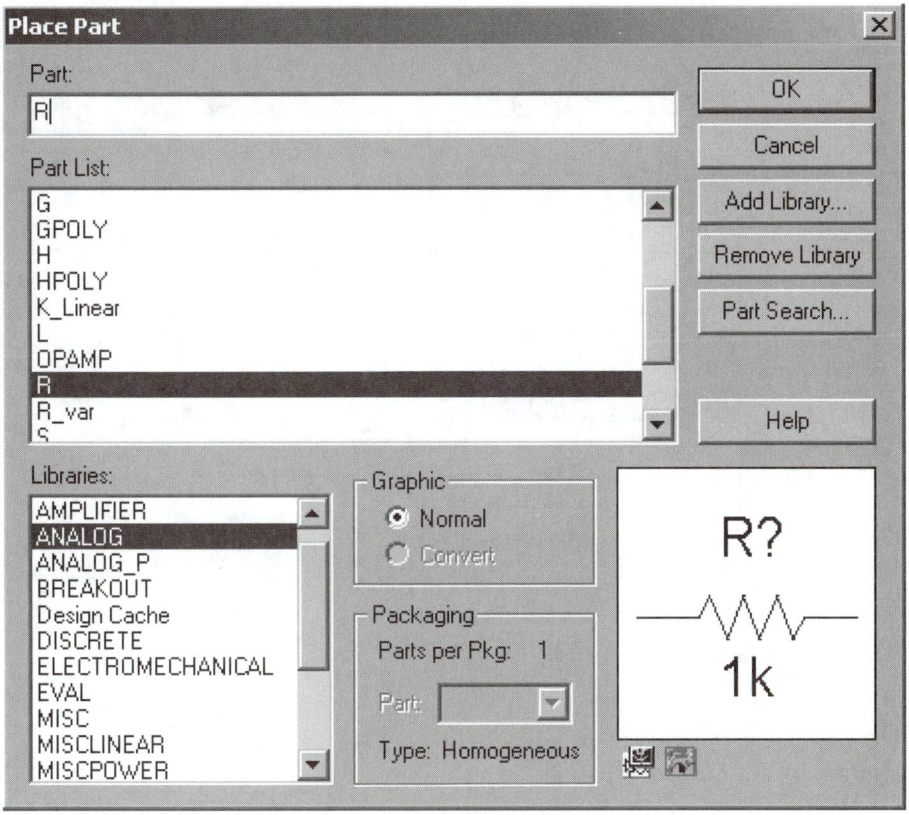

FIGURE B.9

Place part.

2. Choose Browse to browse the list of libraries or use this command to enter the name of a known part (i.e., resistor R), as shown in Fig. B.9.

3. Choose OK, or double click after selecting the part. The chosen part becomes the "current part" and is ready to be placed on your schematic.

4. The cursor is replaced by the shape of the part.

5. Left click to place an instance of the part; double click to place the part and end the mode; or right click to end the mode without placing the part. To drag, point to an object, press and hold the left mouse button, and move the mouse. When the object is where you want it, release the mouse button.

6. To move a component, point to it, press and hold the left mouse button, and drag it to a new location.

7. To remove a component, select it, and choose Delete from the **Edit** menu.

Place a pulse source (VPULSE) from source.slb library; a resistor (R), an inductor (L), and a capacitor (C) from analog.slb library. Add a ground symbol (GND) from **ground** using the **Place** menu or by choosing **Place Ground** (zero 0) from the right-side menu as shown in Fig. B.10. Arrange them as shown in Fig. B.11.

FIGURE B.10

Place ground.

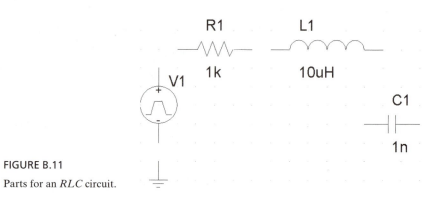

FIGURE B.11

Parts for an *RLC* circuit.

B.5.3 Rotating Components

Now rotate the capacitor so it can be wired neatly into the circuit. Each time you rotate a component, it turns clockwise 90 degrees. The **Flip** menu is used to flip a selected object to produce a mirror image.

- To rotate the capacitor (or other component), select it and choose Rotate from the **Edit** menu, as shown in Fig. B.12. If you select an area of your schematic, the area is rotated around the center of the selection box. This is done as shown in Fig. B.13.
- To flip a component, select it and choose Flip from the **Edit** menu.

Edit	Draw	Navigate	View	Option

Undo   Ctrl+Z
Redo   Ctrl+Y

Cut   Ctrl+X
Copy   Ctrl+C
Paste   Ctrl+V
Copy to Clipboard

Delete   DEL
Select All

Attributes...
Label...   Ctrl+E
Model...
Stimulus
Symbol
Graphics Properties...

Views...
Convert Block...

Rotate   Ctrl+R
Flip   Ctrl+F
Align Horizontal
Align Vertical

Replace...
Find...

FIGURE B.12

Edit menu.

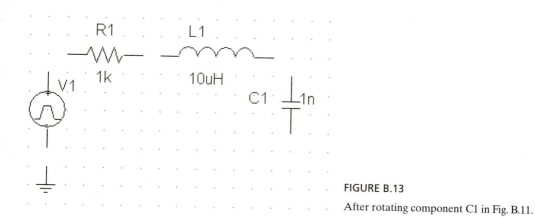

FIGURE B.13

After rotating component C1 in Fig. B.11.

Hint: If the Rotate command is dimmed, the capacitor isn't selected. Try again by pointing to it so that the pointer becomes a hand, and then click the left mouse button.

- To deselect the selected capacitor (or other selected component), click it with the *right* mouse button or click an empty spot with the left mouse button.
- To drag or rotate two or more components at once, first select them by drawing a rectangle around the components, and then drag or rotate the rectangle.

Hint: To draw a rectangle around components, point above and beside one of the components you want to select. Press and hold the left mouse button, and drag diagonally until the rest of the components are in the rectangle that appears.

- To deselect one of the selected components, click it with the *right* mouse button. To deselect everything on the marked rectangle, click an empty spot with the left mouse button.

B.5.4   Wiring Components

Once components are placed on the drawing board, you need to connect them. You can use the schematic editor to draw wires on your schematic or make vertices by completing the following steps:

- To draw a wire, choose Wire from the Place menu or wire menu on the right side.
- The cursor changes to a "wire" mode and is displayed as crosshairs. Left click to start drawing.

- Move the mouse in any direction to extend the wire.
- Left click again to end wiring the part. Left click once to start and left click again to stop. Continue until all parts are wired.
- Press the **ESC** key or double-click the left mouse or right click to end wiring mode. The cursor will change from crosshairs to normal mode.

Wire the circuit as shown in Fig. B.14.

B.5.5   Labeling a Component

You can also places a label on selected wires, bus segments, or ports. Wire and bus segments or ports may display multiple labels; however, all labels for a segment will contain the same text. Each component in a circuit can be labeled. We will assign levels R, L, C, V_s, and V_o.

1. To edit a label, select a wire, bus segment, or port for which to edit the label.
2. Double-click on the label to bring up the dialog box for Display Properties.

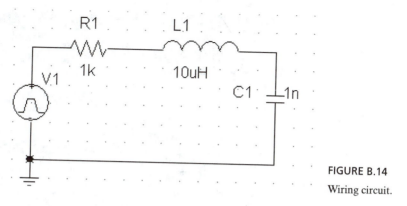

FIGURE B.14

Wiring circuit.

3. Enter the text for the label, as shown in Fig. B.15(a).
4. Choose **OK**.
5. Change all levels R, L, C, V_s, and V_o as shown in Fig. B.15(b).
   Hint: To move the level, CLICKLH on it and move it to a desired location.
6. You get text from **Text** on the **Place** menu.

B.5.6   Adding Text

You can place text anywhere on your schematic and size it to suit your needs with the following steps:

1. To add text to your schematic, choose Text from the Place Menu or Place text on the right side menu.

2. In the dialog box, type in the desired text, as shown in Fig. B.16(a).
3. To change the font size, modify the Font Size shown in the dialog box to suit your needs.
4. Choose **OK**.
5. Move the text to the desired location on the schematic, and click left to place the text. Click right to end the mode.

Place the output voltage, V_o, as shown in Fig. B.16(b).

B.5.7   Editing Attributes (Properties)

An attribute of a schematic item consists of a name/value pair.

1. To edit the properties of a selected object(s), select the object(s) to edit.
2. Choose Properties from the **Edit** menu or double click on the attribute text to bring up the Display Properties dialog box directly as shown in Fig. B.17(a).

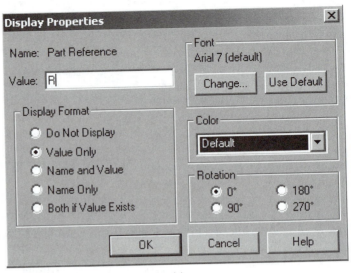

(a)

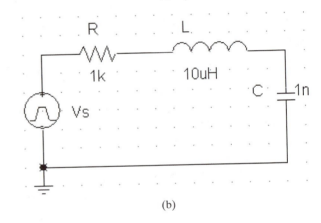

FIGURE B.15

Labeling components.

(b)

3. Select an individual attribute: A dialog box appears in which you can enter a new value for the attribute.
4. Select an entire part: A dialog box appears showing all of the attributes that may be edited for that part (e.g., *Value only*).
5. Set $R = 2\ \Omega$, $L = 50\ \mu H$, and $C = 10\ \mu F$, as shown in Fig. B.17(b).

Hint: A quick way to set a component's value is to double click it.
Another Hint: Changes made to attributes in the schematic editor occur on the particular part instance only. Changes do not affect the underlying symbol in the library.

6. To change the value of an attribute, select it. The name and value should appear in the properties of the **Edit** menu.
7. Change the value in the value edit field and press **OK**.

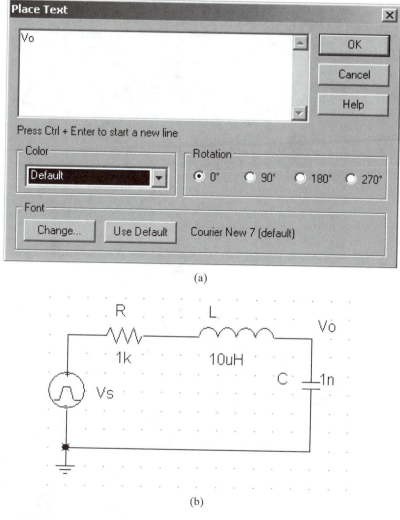

(a)

(b)

FIGURE B.16

Adding text.

8. To delete an attribute, select it from the list and press the Delete button.
9. To change whether the attribute name or value is shown on the schematic, select it and press the **OK** button.
10. To add a new attribute, type the name and value in the edit fields, and press **OK**. Set the pulse source V_s as shown in Fig. B.18. Click on the attributes, and then type v1 = 0V, v2 = 1V, TD (delay time) = 0, TR (rise time) = 1 ns, TF (fall time) = 1 ns, PW (pulse width) = 0.5 ms, PER (period) = 1 ms.
11. Choose **OK** to incorporate the change.

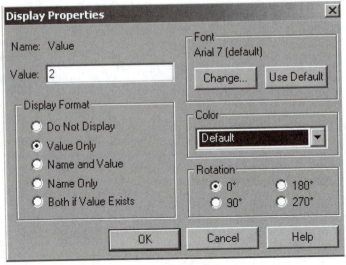

(a)

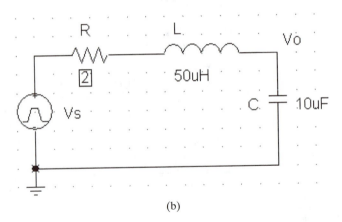

(b)

FIGURE B.17

Setting attributes.

B.5.8   Viewing and Zooming the Schematic

You can change the viewing scale on your schematic from the **View** menu as shown in Fig. B.19. These are the available options:

Menu Item	Function
In	Allows you to view an area on the schematic at closer range (i.e., to magnify). After selecting this item, a crosshair appears. Move the crosshair to the area on which you want to zoom in.
Out	Changes the viewing scale so that you can view the schematic from a greater distance (i.e., to view more of

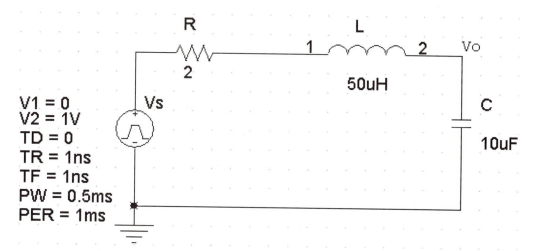

FIGURE B.18

Changing source attributes.

FIGURE B.19

View menu.

the schematic on the screen). After selecting this item, a crosshair appears. Move the crosshair to define the center of the viewing area.

Area

Allows you to select a rectangular area on your schematic to be expanded to fill the screen. If you already have a selection box on your screen when you choose Area, the contents of the selection box will be expanded. Or you can drag the mouse to form a selection box around the portion of the schematic you want to have expanded. The items contained within the selection box will be expanded to fill the screen.

All (Entire Page)

Allows you to view the entire schematic page at once.

File	Edit	View	Place	Macro	PSpice	Acce

New ▶
Open ▶
Close
Save   Ctrl+S

Export Selection...
Import Selection...

Print Preview...
Print...   Ctrl+P
Print Setup...

Import Design...
Export Design...

1 Example A_12.OPJ
2 C:\Rashid\PT3\Example B-3.opj
3 Example A_27.OPJ
4 Example A_25.OPJ

Exit

FIGURE B.20

File menu.

B.5.9   Saving the Circuit File

Your schematic files are automatically saved into the current directory, unless otherwise specified. To save a file manually, choose *Save* from the **File** menu to save changes to the current file. (See Fig. B.20.) You do not need to type a file name extension. Remember that a circuit name must be a valid DOS file name and that all schematic files are automatically given a file name extension OPJ. For example, the file name FIGB_3 will be saved as FIG B_3.OPJ.

B.6   COPYING AND CAPTURING SCHEMATICS

B.6.1   Copy to Clipboard

You can use this command from the **Edit** menu to copy an item from your schematic for use in another Windows program.

1. Use the mouse to select a rectangular area on the schematic to copy.
2. Choose **Copy to Clipboard** from the **Tools** menu.

3. Open the **Windows program** into which you want to paste the item(s).

4. Use the **Paste** command from the other Windows application to paste the item into the new file.

B.6.2   Cut, Copy, or Paste

The **Cut** command deletes items from your schematic and places them in the **Paste** buffer. The **Copy** command copies items from your schematic to the **Paste** buffer. The **Paste** command places the contents of the **Paste** buffer on the schematic.

To cut or copy an item(s) on your schematic, complete the following steps:

1. Select the item(s) on your schematic to be cut or copied.

2. Choose **Cut** or **Copy** from the **Edit** menu.

The following steps allow you to paste a cut or copied item(s) onto your schematic:

1. After cutting or copying an item(s) with the **Cut** or **Copy** commands, choose **Paste** from the **Edit** menu.

2. Place the cursor at the desired location for the item that was cut or copied and click left to place the item(s). **Click** right to end the mode.

APPENDIX C

Creating Input Circuit File

The PSpice program has a built-in-editor with shell, as shown in Fig. C.1. The Netlist or the input circuit file can be can be created by choosing Create Netlist from the Analysis menu of PSpice Schematics, as shown in Fig. C.2. The input files can also be created by text editors. The text editor that is always available is EDLIN. It comes with DOS and is described in the DOS user's guide. Notepad or WordPad are located in the

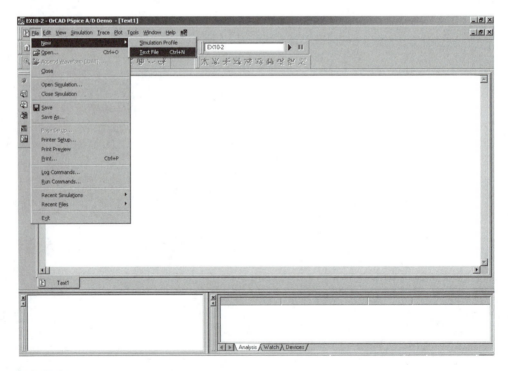

FIGURE C.1

OrCAD PSpice A/D platform.

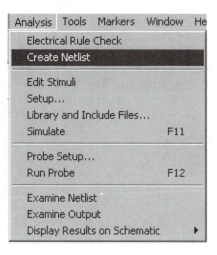

FIGURE C.2

Creating Netlist from PSpice Schematics.

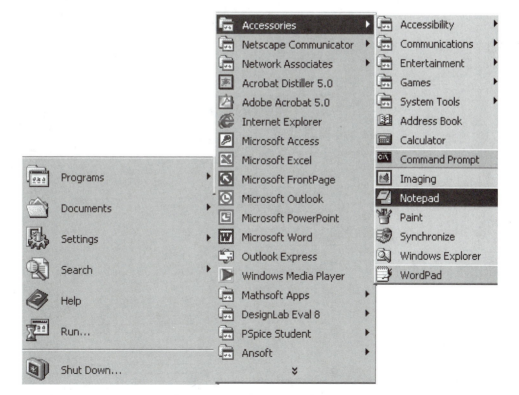

FIGURE C.3

Selecting the text editor from Windows programs.

Accessories folder within the Programs folder on the Start menu, as shown in Fig. C.3. There are other editors, such as Program Editor (from WordPerfect Corporation). Word processing programs such as WordStar 2000, WordPerfect, and MSWord may also be used to create the input file. The word processor normally creates a file that is not a text file. It contains embedded characters to determine margins, paragraph boundaries, pages, and the like. However, most word processors have a command or mode to create a text file without these control characters. For example, WordStar 2000 creates text files with UNIFORM format.

APPENDIX D

DOS Commands

The DOS commands in the Windows environment are enabled from the Command Prompt, which is selected from the Windows Start menu and the Accessories submenu as shown in Fig. D.1. Selecting Command Prompt opens an empty window for DOS commands as shown in Fig. D.2.

The DOS commands that are frequently used are as follows.

To format a new diskette in drive A, type

```
FORMAT A:
```

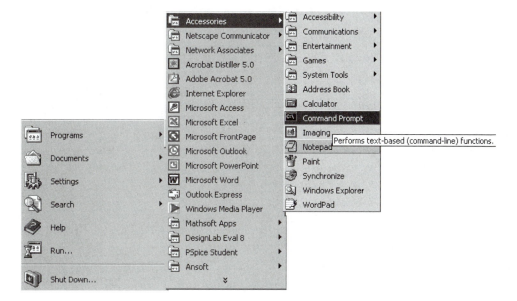

FIGURE D.1

Selecting Command Prompt from the Start menu.

FIGURE D.2

Command prompt.

To list the directory of a diskette in drive A, type

    DIR A:

To delete the file EX2-1.CIR in drive A, type

    Delete A: EX2-1.CIR (or Erase A: EX2-1.CIR)

To copy the file EX2-1.CIR in drive A to the file EX2-2.CIR in drive B, type

    COPY A:EX2-1.CIR B:EX2-2.CIR

To copy all the files on a diskette in drive A to a diskette in drive B, type

    COPY A:*.* B:

To type the contents of the file EX2-1.OUT on the diskette in drive A, type

    TYPE A:EX2-1.OUT

To print the contents of the file EX2-1.CIR in drive A to the printer, first activate the printer by pressing the Ctrl (Control) and Prtsc (Print Screen) keys together, and then type

    TYPE A: EX2-1.CIR

The printer can be deactivated by pressing the Ctrl (Control) and Prtsc (Print Screen) keys again.

APPENDIX E

Noise Analysis

Noise is generated in electronic circuits. That is, an electronic circuit will have output even without any input signal. Noise can be classified as one of five types:

Thermal noise
Shot noise
Flicker noise
Burst noise
Avalanche noise.

E.1 THERMAL NOISE

Thermal noise is generated in resistors due to random motion produced by the thermal agitation of electrons. This noise is dependent on temperature. The equivalent circuit for thermal noise in resistors is shown in Fig. E.1. The mean square value of the noise generator is expressed as

$$V_t^2 = 4kTR\Delta f \qquad \text{V}^2$$
$$I_t^2 = 4kT\Delta f/R \qquad \text{A}^2$$

where k = Boltzmann's constant (1.38×10^{-23} J/K)
 T = absolute temperature, kelvins
 R = resistance, ohms
 Δf = noise bandwidth, hertz

E.2 SHOT NOISE

Shot noise is generated by random fluctuations in the number of charged carriers that are emitted from a surface or diffused from a junction. This noise is always associated with a direct current flow and is present in bipolar transistors. The mean square value

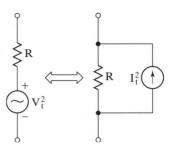

FIGURE E.1

Equivalent circuit for noise in resistors.

of the noise current is expressed by

$$I_s^2 = 2qI_D \, \Delta f \qquad \text{A}^2$$

where Δf = noise bandwidth, hertz
q = electron charge $(1.6 \times 10^{-19}$ C)
I_D = dc current, amps.

E.3 FLICKER NOISE

Flicker noise is generated due to surface imperfections resulting from the emission. This noise is associated with all active devices and some discrete passive elements such as carbon resistors. The mean square value of the noise current is expressed by

$$I_f^2 = K_f \frac{I_D^a}{f} \Delta f \qquad \text{A}^2$$

where Δf = noise bandwidth, hertz
I_D = direct current
K_f = flicker constant for a particular device
a = flicker-exponent constant in the range of 0.5 to 2

E.4 BURST NOISE

Burst noise is generated due to the presence of heavy metal ion contamination and is found in some integrated circuits and discrete transistors. The repetition rate of noise pulses is in the audio frequency range (a few kilohertz or less) and produces a "popping" sound when played through a speaker. (This noise is also known as *popcorn noise*.) The mean square value of the noise current is expressed as

$$I_b^2 = K_b \frac{I_D^c}{1 + (f/f_c)} \Delta f \qquad \text{A}^2$$

where Δf = noise bandwidth, hertz
I_D = direct current

K_b = burst constant for a particular device
c = burst-exponent constant
f_c = a particular frequency for a given noise

E.5   AVALANCHE NOISE

Avalanche noise is produced by Zener or avalanche breakdown in p–n junctions. The holes and electrons in the depletion region of a reverse-biased p–n junction acquire sufficient energy to create hole–electron pairs by collision. This process is cumulative, resulting in the production of a random series of large noise spikes. This noise is associated with direct current and is much greater than shot noise for the same current. Zener diodes are normally avoided in circuits requiring low noise.

E.6   NOISE IN DIODES

The equivalent circuit for noise in diodes is shown in Fig. E.2. There are two generators. The voltage generator is due to thermal noise in the resistance of the silicon. The current source is due to the shot noise and flicker noise. The noise voltage is given by

$$V_s^2 = 4kTr_s\Delta f \qquad V^2$$

$$I_d^2 = 2qI_D\Delta f + K_f\frac{I_D^a}{f}\Delta f \qquad A^2$$

where I_D = forward diode current, amps
r_s = resistance of the silicon, ohms
K_f = flicker constant for a particular device
a = flicker-exponent constant in the range of 0.5 to 2

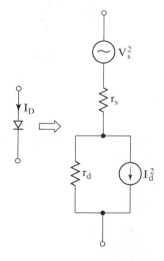

FIGURE E.2

Equivalent circuit for noise in diodes.

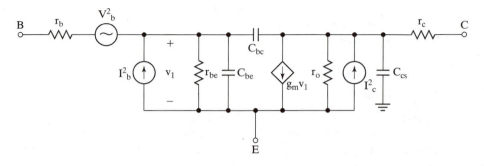

FIGURE E.3

Equivalent circuit for noise in bipolar transistors.

E.7   NOISE IN BIPOLAR TRANSISTORS

The equivalent circuit for noise in bipolar transistors is shown in Fig. E.3. The current generator in the collector is due to shot noise. The noise voltage generator in the base circuit is due to thermal noise in the base resistance. The current generator in the base circuit consists of shot noise, flicker noise, and burst noise. The noise is expressed by

$$V_b^2 = 4kTr_b\Delta f \quad V^2$$

$$I_c^2 = 2qI_C\Delta f \quad A^2$$

$$I_b^2 = 2qI_B\Delta f + K_f\frac{I_B^a}{f}\Delta f + K_b\frac{I_B^c}{1 + (f/f_c)}\Delta f \quad A^2$$

where I_B = base bias current, amps
   I_C = collector bias current, amps
   r_b = resistance at the transistor base, ohms
   K_b = burst constant for a particular device
   c = burst-exponent constant in the range of 0.5 to 2
   f_c = a particular frequency for a given noise
   K_f = flicker constant for a particular device
   a = flicker-exponent constant in the range of 0.5 to 2

E.8   NOISE IN FIELD-EFFECT TRANSISTORS

The equivalent circuit for noise in field-effect transistors (FETs) is shown in Fig. E.4. The current generator in the gate is due to shot noise, which is very small. The current generator in the drain circuit consists of thermal and flicker noise. The noise current is expressed by

$$I_g^2 = 2qI_G\Delta f \quad A^2$$

$$I_d^2 = 4kT\left(\frac{2}{3}g_m\right)\Delta f + K_f\frac{I_D^a}{f}\Delta f \quad A^2$$

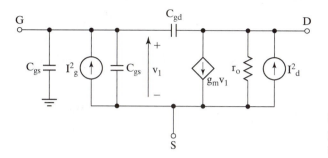

FIGURE E.4

Equivalent circuit for noise in field-effect transistors.

where I_D = drain bias current, amps

$\quad\quad I_G$ = gate leakage current, amps

$\quad\quad g_m$ = transconductance at bias point, amps per volt

$\quad\quad r_b$ = resistance of the transistor base, ohms

$\quad\quad K_f$ = flicker constant for a particular device

$\quad\quad a$ = flicker-exponent constant in the range of 0.5 to 2

E.9 EQUIVALENT INPUT NOISE

Each noise generator contributes to the output of a circuit. The effect of all the noise generators can be found by summing the mean square value of individual noise contributions. Once the total mean square noise output voltage is found, all the noises can be represented by an equivalent input noise at a desired source, as shown in Fig. E.5. This input noise is found by dividing the output voltage by the gain. The gain is the output noise voltage with respect to a defined input. PSpice calculates the output noise and the equivalent input noise by the .NOISE command, which is covered in Section 6.13. It may be noted that PSpice calculates the noise in $V/\sqrt{Hz}$ or $A/\sqrt{Hz}$. Dividing the mean square value of the noise output voltage $V_{out(noise)}$ by the noise bandwidth gives the output noise spectrum V_{out}. That is,

$$V_{out} = \frac{V_{out(noise)}}{\Delta f} \quad V/\sqrt{Hz}$$

Dividing the output noise spectrum by the gain yields the equivalent input noise spectrum

$$V_{in} = \frac{V_{out}}{G_v} \quad V/\sqrt{Hz}$$

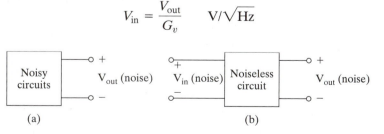

(a)　　　　　　　　　　　　　　　　(b)

FIGURE E.5

Equivalent noise input.

where G_v is the voltage gain with respect to the equivalent input source.

If the equivalent input is a current source, the equivalent input current noise spectrum becomes

$$I_{in} = \frac{V_{out}}{R_t} \quad V/\sqrt{Hz}$$

where R_t is the transresistance with respect to the equivalent input source.

Example E.1:   Noise analysis of a TTL inverter

For the TTL inverter circuit in Fig. 8.39, calculate and print the equivalent input and output noise. The frequency is varied from 1 Hz to 100 kHz with a decade increment and 1 point per decade. The input voltage is 1 V for ac analysis and 3.5 V for dc analysis.

Solution   The PSpice schematic is shown in Fig. E.6(a). The setup for finding the noise analysis is shown in Fig. E.6(b). The listing of the circuit file follows.

Example B.1   TTL inverter

```
▲   * Input voltage of 3.5 V for dc analysis and 1 V for ac analysis
    VIN  1  0  DC  3.5V  AC  1V
    VCC  13  0  5V
▲▲   RS  1  2  50
     RB1 13  3  4K
     RC2 13  5  1.4K
     RE2  6  0  1K
     RC3 13  7  100
     RB5 13  10  4K
     * BJTs with model QNP and substrate connected to ground by default
     Q1  4  3  2  QNP
     Q2  5  4  6  QNP
     Q3  7  5  8  QNP
     Q4  9  6  0  QNP
     Q5  11 10 9  QNP
     * Diodes with model DIODE
     D1  8  9  DIODE
     D2  11 12 DIODE
     D3  12  0 DIODE
     * Model of NPN transistors with model QNP
     .MODEL QNP NPN (BF=50 RB=70 RC=40 TF=0.1NS TR=10NS VJC=0.85 VAF=50
     +KF=6.5E−16 AF=1.2)
     * Diodes with model DIODE
     .MODEL DIODE D (RS=40 TT=0.1NS)
▲▲▲  * Ac sweep from 1Hz and 100 kHz with a decade increment
      * and 1 point per decade
      .AC  DEC  1  1HZ  100KHZ
        * Noise analysis between output voltage, V(9) and input voltage,
      VIN
      .NOISE  V(9)  VIN
      * Printing the results of noise analysis
      .PRINT NOISE ONOISE INOISE
    .END
```

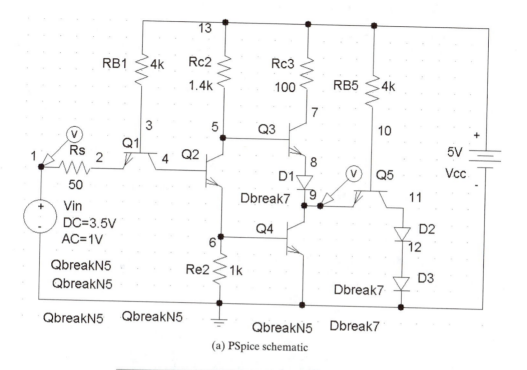

(a) PSpice schematic

(b) Setting for ac sweep and noise analysis

FIGURE E.6

PSpice schematic for Example E.1.

The results of the noise analysis are as follows:

```
****    SMALL-SIGNAL BIAS SOLUTION           TEMPERATURE =   27.000 DEG C
   NODE      VOLTAGE     NODE     VOLTAGE     NODE     VOLTAGE    NODE      VOLTAGE
(    1)     3.5000    (    2)    3.4724    (    3)    2.7564    (    4)    1.9134
(    5)     1.1702    (    6)    1.0205    (    7)    4.9998    (    8)     .5585
(    9)      .0635    (   10)     .9265    (   11)     .0816    (   12)     .0408
(   13)     5.0000
VOLTAGE SOURCE CURRENTS
NAME          CURRENT
VIN        -5.524E-04
VCC        -4.317E-03
TOTAL POWER DISSIPATION      2.35E-02   WATTS
```

```
   ****     AC  ANALYSIS                    TEMPERATURE =  27.000 DEG C
FREQ                ONOISE         INOISE
1.000E+00         8.146E-10       7.205E+02
1.000E+01         8.146E-10       7.205E+02
1.000E+02         8.146E-10       7.205E+02
1.000E+03         8.146E-10       7.205E+02
1.000E+04         8.146E-10       7.205E+02
1.000E+05         8.146E-10       7.204E+02
         JOB CONCLUDED
         TOTAL JOB TIME            22.58
```

A P P E N D I X   F

Nonlinear Magnetic Model

The nonlinear magnetic model uses MKS (metric) units. However, the results for Probe are converted to Gauss and Oersted and may be displayed using $B(Kxx)$ and $H(Kxx)$. The B–H curve can be drawn by a transient run with a slowly rising current through a test inductor and then by displaying $B(Kxx)$ against $H(Kxx)$.

Characterizing core materials may be completed by trial using PSpice and Probe. The procedures for setting parameters to obtain a particular characteristic are the following:

1. Set domain wall pinning constant, $K = 0$. The curve should be centered in the B–H loop, like a spine. The slope of the curve at $H = 0$ should be approximately equal to that when it crosses the x-axis at $B = 0$.
2. Set the magnetic saturation, MS = Bmax/0.01257.
3. Set the slope, ALPHA. Start with the mean field parameter, ALPHA = 0, and vary its values to get the desired slope of the curve. It may be necessary to change MS slightly to get the desired saturation value.
4. Change K to a nonzero value to create hysteresis; K affects the opening of the hysteresis loop.
5. Set C to obtain the initial permeability. Probe displays the permeability, which is $\Delta B/\Delta H$. Probe calculates differences, not derivatives, so the curves will not be smooth. The initial value of $\Delta B/\Delta H$ is the initial permeability.

Example F.1:   Plotting the B–H characteristic

The coupled inductors in Fig. 5.13(a) are nonlinear. (See Fig. F.1.) The parameters of the inductors are $L_1 = L_2 = 500$ turns, and $k = 0.9999$. Plot the B–H characteristic of the core from the results of transient analysis if the input current is varied very slowly from 0 A to −15 A, −15 A to 15 A, and 15 A to −15 A. The load resistance of $R_L = 1$ KΩ is connected to the secondary of the transformer. The model parameters of the core are AREA = 2.0, PATH = 62.73, GAP = 0.1, MS = 1.6E + 6, A = 1E + 3, C = 0.5, and K = 1500.

Example F.1   A typical B-H characteristic

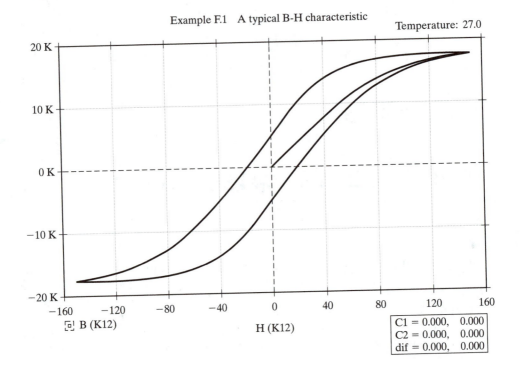

Temperature: 27.0

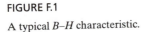

H (K12)

C1 = 0.000, 0.000
C2 = 0.000, 0.000
dif = 0.000, 0.000

FIGURE F.1

A typical *B–H* characteristic.

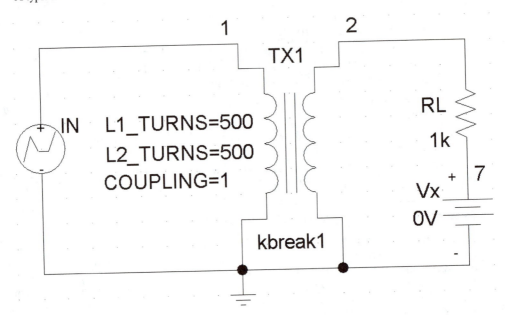

FIGURE F.2

PSpice schematic for Example F.1.

Solution   The PSpice schematic is shown in Fig. F.2. The circuit life for the coupled inductors in Fig. 5.8(a) would be as follows.

Example F.1 A typical *B–H* characteristic

```
▲    *    PWL waveform for transient analysis
     IN  1  0  PWL (0  0  1  −15  2  15  3  −15)
▲▲   *    Inductors represent the number of turns.
     L1  1  0  500
     L2  2  0  500
     R2  2  0  1000
     *    Coupled inductors with k = 0.9999 and model CMOD
     K12 L1 L2 0.9999 CMOD
     *    Model parameters for CMOD
     .MODEL CMOD CORE (AREA=2.0 PATH=62.73 GAP=0.1 MS=1.6E+6 A=1E+3
     +   C=0.5 K=1500)
▲▲▲  * Transient analysis from 0 to 3 s in steps of 0.03 s
         .TRAN 0.05 3
         .PROBE
     .END
```

APPENDIX G

PSpice Default Symbol Libraries

The student version of the PSpice allows a maximum of 10 configured (addressable) symbol libraries at a given time. It has nine default libraries (with the extension **.slb**) as shown in Fig. G-1 and as described below:

1. **abm.slb**: Analog behavior models for voltage and current sources whose outputs can be controlled by mathematical expressions or tabular data and a Laplace transform block.
2. **analog.slb**: Analog components for passive components (R, L, C), coupled inductors, and dependent voltage and current sources.
3. **analog_p.slb**: Passive components (R, L, C) with parameterized values.
4. breakout.slb: Passive and active components in a simplified generic or default configuration. It is useful for circuits when detailed models, especially for active devices, not available.
5. **eval.slb**: Detailed models for a few specific devices such as Q2N2222 for BJT, J2N3819 for JFET, uA741 op amp, and IRF150 power MOSFET.
6. **port.slb**: Different ports and pins for specifying node connections without having to have a directly drawn wire connection. It also includes ground connection or symbol.
7. **source.slb**: Independent current and voltage source models. Different models of a voltage (or a current) source are used depending on the type of analysis, such as ac sweep, dc sweep, and transient analysis.
8. **sourcstm.slb**: Specialized models of voltage and current source for use with the Edit Stimulus command on the Analysis menu.
9. **special.slb**: Specialized parts such as PARAM (short for Parameters) for making parameterized simulations.

The user can add or remove libraries as desired within the maximum limit of 10. Thus, the user can add at least one library without removing any of the PSpice default libraries. The **sourcstm.slb** and **analog_p.slb** libraries, which are not often needed, can be removed to make room for other libraries.

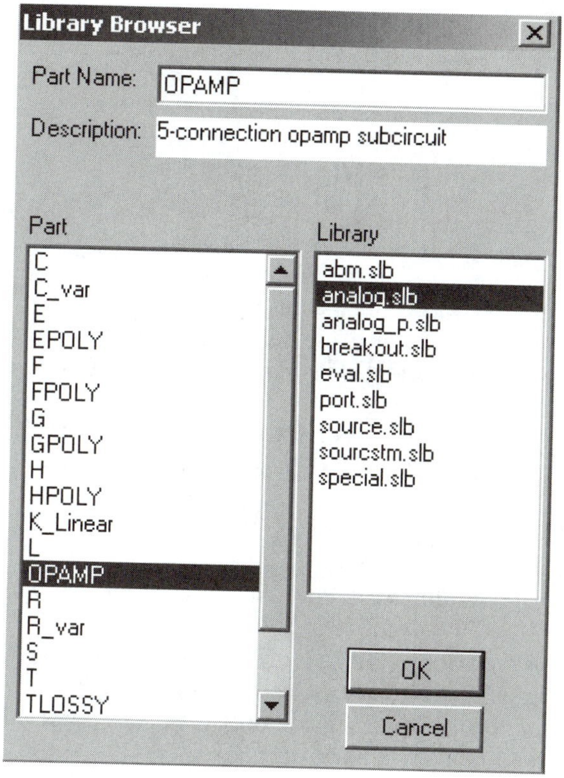

FIGURE G.1

PSpice default symbol libraries

Bibliography

[1] PSpice Books and Articles: http://www.orcadpcb.com/pspice/

[2] Allen, Phillip E., *CMOS Analog Circuit Design*. New York: Holt, Rinehart and Winston, 1987.

[3] Antognetti, Paolo, and Guiseppe Massobrio, *Semiconductor Device Modeling with SPICE*. New York: McGraw-Hill, 1988.

[4] Baliga, B. Jayant, *Power Semiconductor Devices*. Boston: PWS Publishing, 1996.

[5] Banzhaf, Walter, *Computer-Aided Circuit Analysis Using SPICE*. Englewood Cliffs, New Jersey: Prentice Hall, 1989.

[6] Benda, Vítezslav, John Gowar, and Duncan A. Grant, *Power Semiconductor Devices: Theory and Applications*. Chichester, New York: Wiley, 1999.

[7] Bugnolo, Dimitri S., *Computer Programs for Electronic Analysis and Design*. Reston, Virginia: Reston Publishing, 1983.

[8] Chattergy, Rahul, *SPICEY Circuits*. Boca Raton, Florida: CRC Press, 1992.

[9] Chua, Leon O., and Pen-Min Lin, *Computer-Aided Analysis of Electronic Circuits—Algorithms and Computational Techniques*. Englewood Cliffs, New Jersey: Prentice Hall, 1975.

[10] Connelly, J. A., J. Alvin Connelly, and Pyung Choi, *Macromodeling with SPICE*. Englewood Cliffs, New Jersey: Prentice Hall, 1992.

[11] Ferris, Clifford D. and Jerry C. Hamann, *SPICE for Electronics*. Minneapolis - St. Paul, Minnesota: West Publishing, 1995.

[12] Foty, D., *MOSFET Modeling with SPICE: Principles and Practice*. Upper Saddle River, New Jersey: Prentice Hall, 1997.

[13] Ghandi, S. K., *Semiconductor Power Devices*. New York: Wiley, 1977.

[14] Gray, Paul R., and Robert G. Meyer, *Analysis and Design of Analog Integrated Circuits*. New York: Wiley, 1984.

[15] Grove, A. S., *Physics and Technology of Semiconductor Devices*. New York: Wiley, 1967.

[16] Hodges, D. A., and H. G. Jackson, *Analysis and Design of Digital Integrated Circuits*. New York: McGraw-Hill, 1988.

[17] Jaecklin, André A., *Power Semiconductor Devices and Circuits*. New York: Plenum Press, 1992.

[18] Keown, John, *PSpice and Circuit Analysis*. New York: Macmillan, 1991.

[19] Kielkowski, Ron M., *SPICE: Practical Device Modeling*. New York: McGraw-Hill, 1995.

[20] Kielkowski, Ron M., *Inside SPICE*. New York: McGraw-Hill, 1998.

[21] Liu, William, *MOSFETs Models for SPICE Simulation, Including BSIM3v3 and BSIM4*. New York: Wiley-Interscience, 2001.

[22] McCana, William J., *Fundamentals of Computer-Aided Circuit Simulation*. Norwell, Massachusetts: Kluwer Academic, 1988.

[23] MicroSim Corporation, *PSpice Manual*. Irvine, California: MicroSim, 1992.

[24] Nagel, Laurence, W., *SPICE2—A Computer Program to Simulate Semiconductor Circuits*. Memorandum no. ERL-M520, May 1975, Electronics Research Laboratory, University of California, Berkeley.

[25] *OrCAD PSpice A/D Reference Guide: Version 9.1*. Beaverton, Oregon: OrCAD, Inc., November, 1999.

[26] *OrCAD PSpice A/D: User's Guide*. Beaverton, Oregon: OrCAD, Inc., November 15, 1999.

[27] Pillage, Lawrence T., Ronald A. Rohrer, and Chandramouli Visweswariah, *Electronic Circuit and System Simulation Methods*. New York: McGraw-Hill, 1995.

[28] Rashid, M. H., *Power Electronics—Circuits, Devices and Applications*. Upper Saddle River, New Jersey: Prentice Hall, 2003.

[29] Rashid, M. H., *SPICE for Power Electronics and Electric Power*. Upper Saddle River, New Jersey: Prentice Hall, 2004.

[30] Roberts, Gordon and Adel S. Sedra, *Spice*. New York: Oxford University Press, 1997.

[31] Sandler, Steven M., *SMPS Simulation with SPICE 3*. New York: McGraw-Hill, 1997.

[32] Spence, Robert, and John P. Burgess, *Circuit Analysis by Computer—From Algorithms to Package*. London: Prentice Hall International (UK), 1986.

[33] Spencer, Richard and Mohammed Ghausi, *Introduction to Electronic Circuit Design*. Upper Saddle River, New Jersey: Prentice Hall, 2003.

[34] Tront, Joseph G., *Elementary Circuit Analysis Using SPICE*. New York: Wiley, 1989.

[35] Tuinenga, Paul W., *SPICE: A Guide to Circuit Simulation and Analysis Using PSpice*. Upper Saddle River, New Jersey: Prentice Hall, 1995.

[36] Vladimirescu, Andrei, *The Spice Book*. New York: Wiley, 1994.

[37] Ziel, Aldert van der, *Noise in Solid State Devices*. New York: Wiley, 1986.

Index